Strategies in Sustainable Tourism, Economic Growth and Clean Energy

Daniel Balsalobre-Lorente · Oana M. Driha · Muhammad Shahbaz
Editors

Strategies in Sustainable Tourism, Economic Growth and Clean Energy

Springer

Editors
Daniel Balsalobre-Lorente
University of Castile-La Mancha
Cuenca, Cuenca, Spain

Oana M. Driha
University of Alicante
Alicante, Spain

Muhammad Shahbaz
Beijing Institute of Technology
Beijing, China

ISBN 978-3-030-59674-3 ISBN 978-3-030-59675-0 (eBook)
https://doi.org/10.1007/978-3-030-59675-0

This Springer imprint is published by the registered company Springer Nature Switzerland AG
The registered company address is: Gewerbestrasse 11, 6330 Cham, Switzerland

Contents

1 **The Impact of Tourism and Renewable Energy Use Over Economic Growth in Top 10 Tourism Destinations** 1
Daniel Balsalobre-Lorente, Nuno Carlos Leitão, Oana M. Driha, and José María Cantos-Cantos

2 **The Possible Influence of the Tourism Sector on Climate Change in the US** 15
Faik Bilgili, Yacouba Kassouri, Aweng Peter Majok Garang, and H. Hilal Bağlıtaş

3 **Tourism Sector and Environmental Quality: Evidence from Top 20 Tourist Destinations** 39
Burcu Ozcan, Seref Bozoklu, and Danish Khan

4 **The Effects of Tourism, Economic Growth and Renewable Energy on Carbon Dioxide Emissions** 67
Nuno Carlos Leitão and Daniel Balsalobre-Lorente

5 **Clean India Mission and Its Impact on Cities of Tourist Importance in India** 89
Perfecto G. Aquino Jr., Mercia Selva Malar Justin, and Revenio C. Jalagat Jr.

6 **The Effects of Globalization and Terrorism on Tourist Arrivals to Turkey** 109
Zübeyde Şentürk Ulucak and Ali Gökhan Yücel

7 **Testing the Dynamic Relationship Among CO_2 Emissions, Economic Growth, Energy Consumption and Tourism Development. Evidence for Uruguay** 125
Juan Gabriel Brida, Bibiana Lanzilotta, and Fiorella Pizzolon

8 Analyzing the Tourism Development and Ecological Footprint Nexus: Evidence From the Countries With Fastest-Growing Rate of Tourism GDP 141
Ilyas Okumus and Sinan Erdogan

9 Investigating the Tourism Originating CO_2 Emissions in Top 10 Tourism-Induced Countries: Evidence from Tourism Index 155
Asli Ozpolat, Ferda Nakipoglu Ozsoy, and Mehmet Akif Destek

10 Sustainable Tourism Production and Consumption as Constituents of Sustainable Tourism GDP: Lessons from a Typical Index of Sustainable Economic Welfare (ISEW) 177
Angeliki N. Menegaki

11 Developments and Challenges in the Greek Hospitality Sector for Economic Tourism Growth: The Case of Boutique Hotels 197
Aimilia Vlami

12 Airbnb and Overtourism: An Approach to a Social Sustainable Model Using Big Data 211
María Jesús Such-Devesa, Ana Ramón-Rodríguez, Patricia Aranda-Cuéllar, and Adrián Cabrera

13 Determination of Standard of Living for People Involved with Tourism in Digha by Ordinal Regression Analysis 235
Subhankar Parbat, Payel Chatterjee, Sourav Sen, and Adwitiraj Banerjee

14 The Validation of the Tourism-Led Growth Hypothesis in the Next Leading Economies: Accounting for the Relevant Role of Education on Carbon Emissions Reduction? 249
Festus Victor Bekun, Festus Fatai Adedoyin, Daniel Balsalobre-Lorente, and Oana M. Driha

Correction to: Developments and Challenges in the Greek Hospitality Sector for Economic Tourism Growth: The Case of Boutique Hotels C1
Aimilia Vlami

About the Editors

Dr. Daniel Balsalobre-Lorente holds a Ph.D. in Economics from the University of Castilla–La Mancha, where he is currently an Associate Professor. He has more than ten years of experience as a Professor of Economic Growth, Public Economics and Regional Sciences. His main research activities are focused on the energy economy, energy innovation, economic growth and development economics. He has co-authored several articles in various journals, including Energy Policy, Cleaner Production Magazine and Environmental Science and Pollution Research, as well as several book chapters. He regularly reviews articles for journals such as Economic Modelling and the Journal of Cleaner Production.

Dr. Oana M. Driha holds an International Ph.D. in Economics from the University of Alicante where she is currently an Assistant Professor of Applied Economics. She has nine years of experience as a Professor of International Economics and EU Economics. She has been involved as an expert in numerous EU funded projects in the field of sustainable development (green energy, climate change, sustainable tourism, etc.). Her main research activities are focused on energy economics, energy innovation, economic growth and sustainable tourism. She has co-authored several articles in various journals, including Resources Policy, Environmental Science and Pollution Research, Current Issues in Tourism or International Journal of Contemporary Hospitality Management, as well as several book chapters. She regularly reviews articles for journals such as Journal of Cleaner Production or Technological Forecasting & Social Change.

Dr. Muhammad Shahbaz is a Full Professor at the School of Management and Economics, Beijing Institute of Technology, China. He is also an Affiliated Visiting Scholar at the Department of Land Economy, University of Cambridge, UK, and an Adjunct Professor at COMSATS Institute of Information Technology, Lahore, Pakistan. He previously served as a Chair Professor of Energy and Sustainable Development at Montpellier Business School, France, and Principal Research Officer at COMSATS. He received his Ph.D. in Economics from the National College of Business Administration and Economics, Lahore, Pakistan. His research focuses on financial economics, energy finance, energy economics, environmental economics, development economics and tourism economics. He has published more than 300 research papers in peer-reviewed international journals, is among the world's top 15 economics authors as ranked by IDEAS, and was selected as one of the top 5 authors on economics in developing countries by David McKenzie, Chief Economist of the World Bank. Dr. Shahbaz has published papers in various journals, including Applied Economics, Social Indicators Research, Renewable Energy and the Journal of Cleaner Production.

Chapter 1
The Impact of Tourism and Renewable Energy Use Over Economic Growth in Top 10 Tourism Destinations

Daniel Balsalobre-Lorente, Nuno Carlos Leitão, Oana M. Driha, and José María Cantos-Cantos

Abstract During the last six decades, economic growth has been closely influenced by tourism, energy use and environmental degradation. This connection has involved several effects over energy mix, like, for example, a rising share of renewable energy sources or more efficient management in the tourism industry, which has enhanced a sustainable economic growth with lower carbon emissions. To explore these effects over economic growth for a panel of Top 10 between 1995 and 2015, we explore the role of international tourism, renewable energy use and carbon emissions. The aim of this study is to validate the Tourism-Led Growth Hypothesis (TLGH) for selected Top 10 tourism destinations. Furthermore, how structural changes impact the energy mix and their effect over income levels is also tested via the driving mentioned above forces (i.e. renewable energy use, international tourism and CO_2 emissions). Through FMOLS and DOLS econometric estimations, the TLGH is confirmed. The same methodology endorses the existence of a dampening effect which raise the moderation effect between renewable energy sources and carbon emissions over economic growth. Thus, a moderating effect of the promotion of renewable sources over economic growth, via scale effect, is also endorsed.

D. Balsalobre-Lorente (✉) · J. M. Cantos-Cantos
Department of Political Economy and Public Finance, Economic and Business Statistics and Economic Policy, University of Castilla-La Mancha, Ciudad Real, Spain
e-mail: Daniel.Balsalobre@uclm.es

J. M. Cantos-Cantos
e-mail: JoseMaria.Cantos@uclm.es

N. C. Leitão
Polytechnic Institute of Santarém, Center for Advanced Studies in Management and Economics, Évora University, Évora, Portugal
e-mail: nunocarlosleitao@gmail.com

Center for African and Development Studies, Lisbon University, Lisbon, Portugal

O. M. Driha
Department of Applied Economics, International Economy Institute, Institute of Tourism Research, University of Alicante, Alicante, Spain
e-mail: oana.driha@ua.es

D. Balsalobre-Lorente et al. (eds.), *Strategies in Sustainable Tourism, Economic Growth and Clean Energy*, https://doi.org/10.1007/978-3-030-59675-0_1

Keywords Tourism-led growth hypothesis · Renewable energy use · Carbon emissions · Sustainable tourism

JEL Z32 · Q40 · Q20 · Q01 · C33 · Q53

1.1 Introduction

Since the middle of the last century, the tourism industry has emerged as an essential driving force in enhancing income levels for both developed and developing economies. The tourism industry presents a pivotal role in the economic development of countries with a tourism-related infrastructure. Hence, the present analysis has Top 10 tourism destinations in the spotlight. In 2017, the World Travel and Tourism Council quantified around 10.4% of the tourism sector's overall contribution to the global economy gross domestic product and 9.9% of total employment (WTTC 2018). This expansion of international tourism has boosted revenues advanced in household spending, with encouraging long-run effects over economic growth (Chou 2013).

Moreover, environmental regulations brought a restructured tourism sector, improving sustainable practices (Govdeli and Direkci 2017). Under such a context, the analysis of the main driving forces that have jammed the connection between economic growth and international tourism seems relevant. For this purpose, confirming the tourism-led growth hypothesis (hereafter TLGH) is the main objective. Additionally, to omit biased effects, other variables are considered for the ten main touristic destinations between 1995 and 2015. In this line, additional explanatory variables are renewable energy use and environmental degradation. It is also tested the dampening effect among the additional variables, under a TLGH scenario.

Traditionally, it has been assumed that in the early stages of economic development, it has appeared to overexploitation of energy sources with low environmental restrictions (Zuo and Huang 2017, 2018). Some studies have evidenced the direct impact that energy use and ecological damage exert over economic growth (Aitken et al. 1997; Turner and Witt 2001; Shahbaz et al. 2016; Balsalobre et al. 2020a, b). Furthermore, additional empirical evidence has demonstrated that in the early stages of economic growth, environmental damage has contributed positively to increase income levels, due to industrialisation, modernisation, or urbanisation process (Azam et al. 2016).

In our attempt to validate the TLGH, environmental damage and renewable energy use over the economic growth process are also considered. In line with Zuo and Huang (2017), we assume that the led growth process implies a long-run specialisation process in the tourism sector, where the stimulation of this industry would contribute reducing poverty as well as environmental damage, but also to increase more potent effects over local economies (Lee and Chang 2008; Li et al. 2018).

The way tourism reacts to environmental challenges and energy advances, where technical advances and environmental regulations foster a more efficient energy process, boosting a sustainable tourism sector (Scott 2011; Weaver 2011; Li et al. 2018; Balsalobre et al. 2020a) allows a better understanding of how sustainability and competitiveness impact tourism. Tourism is related to local infrastructures and services that distress the environment (Gössling 2002; Gössling et al. 2002, 2015; Lee et al. 2018).

By contrast, some literature has revealed that tourism infrastructures can also generate adverse effects over local economies as a consequence of inefficient, traditional tourism (Shan and Wilson 2001; Blake et al. 2003; Smorfitt et al. 2005; Zhang and Lee 2007; Dwyer et al. 2006; Li et al. 2018, Balsalobre et al. 2020a, b). The absence of progress in tourism can also generate harmful effects over local businesses and the environment (Long et al. 1990). They are analysing the environmental results and how the energy sector impacts on economic growth under a TLGH scenario might bring some more light not just for academics, but also for practitioners.

Traditionally, empirical literature has assumed that the use of fossil fuels boosts both economic growth and tourism. Still, recent studies assert that clean energy sources can be considered as a necessary alternative to attract tourism (Balsalobre et al. 2020a, b). When assuming that environmental degradation contributes to expanding economic growth (though scale effect), it is also considering that dirty energy sources appear in the first stage of economic growth. By contrast, energy efficiency and renewable sources promotion in tourism support new services attraction as well as sustainable economic growth, where the coherent utilisation of capital and new capital investment should accompany energy-saving technology and is essential for sustainable tourism (Becken and Cavanagh 2003; He et al. 2020).

The chapter is organised as follows. The second section is dedicated to the previous empirical literature, and the third one describes the empirical methodology. The estimation results are given in the fourth section, while their discussion is included in the fifth section. The final section covers the conclusions and some recommendations.

1.2 Literature Review

Although the linkage between tourism and economic growth is not new in the economic literature (Balaguer and Cantavella-Jordá 2002; Chen and Chiou-Wei 2009; Chang et al. 2009; Zhao and Mao 2013; Balsalobre et al. 2020a, b), our study tries to shed some light by exploring how energy use, environmental degradation and the interaction between them influences economic growth. Even if the main objective of our study is to explore the connection between economic growth and tourism sector (though TLGH) for Top 10 tourism destinations, we also consider the effects that environmental degradation exerts over economic growth, trained by inefficient energy use (Lee and Brahmasrene 2013; Turner and Witt 2001). This detrimental impact indirectly confirms the need to implement renewable energy strategies and apply more efficient energy technologies (Álvarez et al. 2017).

The TLGH assumes that tourism sector is an essential economic engineering strategy (Chen and Chiou-Wei 2009; Chang et al. 2012; Zhao and Mao 2013; Zuo and Huang 2017), where its dynamics contribute generating numerous macroeconomic effects, drafting valuable policy recommendations (Dogru and Bulut 2018; Brida and Pereyra 2009; Brida et al. 2016). Some studies have predicted the existence of the TLGH, through the presence of energy shocks or environmental factors, which have inferred over economic growth (Dunn and Dunn 2002; Smorfitt et al. 2005; Zhang and Lee 2007; Pham et al. 2010; Agarwal 2012; Groizard and Santana-Gallego 2018). Additional literature argued that implementing energy strategies is required for sustainable tourism. It helps to correct the pernicious effects that the expansion of a traditional and inefficient tourism sector can exert over economic growth (Sequeira and Campos 2007; Balsalobre et al. 2020a, b). When tourism industry generates diminishing returns (e.g. reduction in income levels for hosting countries, or dirty overexploitation of natural resources), the linkage between tourism and economic growth becomes negative (Essletzbichler and Rigby 2007; Po and Huang 2008), causing a crowding-out effect, which reflects the damaging impact of external corporations over local economies (Zuo and Huang 2017). Governments should urge regulations related to energy innovation strategies and clean energy source in the host tourism industry (Zuo and Huang 2017), avoiding or at least mitigating damaging effects of the tourism industry over economic growth. Katircioglu (2014) showed that tourism development increases energy capability and pollution levels, given the expansion of tourism-related activities. This study confirms the existence of an interaction between tourism and the energy sector, environment, or economic growth. Liu et al. (2011) demonstrated that energy use impacts directly over economic growth. More recent studies have shown that international tourism boosts economic growth and increases energy consumption and carbon emissions (Scott et al. 2016; Lee et al. 2018). Therefore, the promotion of a cleaner energy mix and putdowns of fossil sources will, at first, reduce income levels via scale effect. This extra cost would be due to modifications in the energy mix and the promotion of energy innovation processes (Álvarez et al. 2017).

1.3 Empirical Methodology

As already mentioned previously, the main objective is to test the connection between international tourism and economic growth, validating the tourism-led growth hypothesis (TLGH) for Top 10 tourism destinations between 1995 and 2015. As a complementary effect, we also assume the existence of a direct connection between environmental degradation and economic growth and renewable energy use and economic growth. By considering environmental regulatory, we measure the presence of a dampening effect between environmental degradation and renewable energy use, contributing this way to empirical literature and EKC methodology. The result aims at confirming the impact of the promotion of renewable sources on environmental degradation and the effect over economic growth. To do so, Fully Modified Least Squares (FMOLS) and Dynamic Ordinary Least Squares (DOLS) econometric methods are used. This way, endogeneity and serial correlation problem are tackled (Narayan and Narayan 2005).

We propose (Eq. 1.1), as follows (Table 1.1):

$$\begin{aligned} LGDP_{it} = & \alpha_0 + \alpha_1 \, LIT_{it} + \alpha_3 \, LRNW_{it} \\ & + \alpha_4 \, LCO_{2it} + \alpha_5 \, LRNW_{it} * LCO_{2it} + \varepsilon_{it} \end{aligned} \quad (1.1)$$

Equation 1.1 considers $LGDP_{it}$ (logarithm of per capita gross domestic product) and its relationship with LIT_{it} (logarithm of the international tourism) for testing the TLGH for selected top 10 tourism destinations, during the period 1995–2015. Some additional explanatory variables are also included: the share of renewable energy consumption $LRNW_{it}$, per capita carbon emissions LCO_{2it}, as a proxy of environmental damage. Aiming to test also environmental energy regulations and their effect on the interaction between renewable energy and carbon emissions $LCO_{2it} * LRNW_{it}$

Table 1.1 Expected relationships between independent and dependent variables

Dependent variable $LGDP_{it}$ (gross domestic product, per capita current USD)			
Independent variables	Measure	Notation	Expected relationship
LIT	The logarithm of International tourism, passengers	LIT_{it}	Positive: confirming TLGH
LRNW	The logarithm of renewable energy consumption (% total final energy consumption)	$LRNW_{it}$	Positive
LCO_2	The logarithm of carbon emissions per capita, as a proxy of environmental damage	LCO_{2it}	Positive
LCO_2 * LRNW	Logarithm interaction between renewable energy and environmental damage (a proxy of energy regulations)	$LCO_{2it} * LRNW_{it}$	Negative

Correlation matrix

	LGDP	LIT	LCO_2	LRNW
LGDP	1.000000			
LIT	0.716638	1.000000		
LCO_2	0.799409	0.761618	1.000000	
LRNW	−0.648637	−0.377188	−0.627902	1.000000

Sources WDI (2020)

is also included (Abrell and Weigt 2008). This variable will allow exploring the dampening effect that the promotion of renewable energy sources exerts over environmental damage and its impact over economic growth. A negative connection is expected, i.e. a reduction in income levels, due to energy transition efforts in encouraging renewable energy use as it would mitigate accumulative environmental degradation, via scale effect.

First, a traditional LLC (Levin et al. 2002), ADF-Fisher and PP-Fisher (Choi 2001) panel unit root tests are employed for checking if the variables ($LGDP_{it}$, $LGDP_{it}$, LIT_{it}, $LRNW_{it}$, LCO_{2it}) are cointegrated I(1) based on the presence of unitary roots I(1) in the panel variables (Apergis and Payne 2009a, b).

While LLC (2002) assumes that ρ is constant across the panel, individual time series regressions are carried out via ADF and PP tests through each cross section and the p-value for each series from their unit root tests is combined, instead of averaging individual test statistics (Im et al. 2003). If these tests confirm that the variables are cointegrated I(1), all the series are non-stationary at levels and null hypothesis would be accepted. We reject the null hypothesis a priori at the first difference between them, I(1).

The Pedroni (1999), Kao (1999) and Johansen (1991) cointegration tests the existence of a long-run relationship among proposed variables. While Pedroni (1999) tests assume heterogeneous intercepts and trend coefficients across cross sections, Kao (1999) proposes cross-sectional intercepts and homogeneous coefficients on the first-stage regressors. Fisher-Johansen's cointegration test (Johansen 1991) combines individual tests and connects tests from individual cross sections.

Finally, FMOLS and DOLS methodologies are necessary to check our main hypotheses.

1.4 Empirical Results

Preliminary tests establish that all variables are cointegrated I(1) as depicted in Table 1.2.

A long-run relationship between the variables is also confirmed (see Table 1.3).

The FMOLS (Phillips and Hansen 1990) and DOLS (Saikkonen 1991; Stock and Watson 1993) methodologies (Table 1.4) offer an adjustment for serial correlation and endogeneity due to the presence of cointegrating relationships (Phillips 1995).

The empirical results confirm the TLGH ($\alpha_1 > 0$), where international tourism (LIT_{it}) promotes economic growth ($LGDP_{it}$,), in selected Top 10 tourism destinations during the period 1995 and 2015. A positive connection between renewable energy use ($LRNW_{it}$) and economic growth ($\alpha_2 > 0$), and environmental damage (LCO_{2it}) and economic growth ($\alpha_3 > 0$) are also validated. Finally, a dampening effect between renewable energy use and environmental damage ($LCO_{2it}*LRNW_{it}$), as a proxy of environmental regulation (Álvarez et al. 2017), is confirmed by the negative connection with economic growth ($\alpha_4 > 0$).

Table 1.2 Panel unit root test

	(A) Null: unit root (assumes common unit root process)		(B) Null: unit root (assumes individual unit root process)			
	Levin, Lin and Chu t		ADF—Fisher Chi-square		PP—Fisher Chi-square	
	t-Statistic	Prob.	t-Statistic	Prob.	t-Statistic	Prob.
At level						
LGDP	3.16968	(0.9992)	2.10509	(1.0000)	1.55307	(1.0000)
LIT	4.03332	(1.0000)	1.91924	(1.0000)	1.87205	(1.0000)
LRNW	0.16448	(0.5653)	15.8417	(0.7264)	119.094	(0.0000)
LCO_2	−3.07870*	(0.0010)	61.6284*	(0.0000)	32.6317***	(0.0370)
At first difference						
ΔLGDP	−4.97406*	(0.0000)	71.4071*	(0.0000)	88.2298*	(0.0000)
ΔLIT	−5.34429*	(0.0000)	61.1305*	(0.0000)	75.9990*	(0.0000)
ΔRNW	−5.30588*	(0.0000)	68.5245*	(0.0000)	119.094*	(0.0000)
ΔCO2	−4.83034*	(0.0000)	61.6284*	(0.0000)	130.894*	(0.0000)

Notes (*) Significant at the 10%; (**) Significant at the 5%; (***) Significant at the 1%.*MacKinnon (1996) one-sided *p*-values. **Probabilities for Fisher tests are computed using an asymptotic Chi-square distribution. All other tests assume asymptotic normality. *Note* *, **, and *** significance at 10%, 5%, and 1%

1.5 Discussion of Empirical Results

Based on the econometric results obtained from both FMOLS and DOLS regressions (Fig. 1.1; Table 1.4), TLGH is confirmed for selected Top 10 tourism destinations during the period 1995–2015. Consequently, international tourism leads to economic growth in these Top 10 tourism destinations, in line with previous empirical literature (Gössling and Hall 2006; Scott 2006; Peeters 2007; WTTC 2011; OECD 2018). Additionally, a positive connection between environmental degradation and economic growth is related to scale effect. This scale effect reflects that, in initial stages of economic development, ascending income levels are obtained through fossil sources' overexploitation. The positive impact that renewable energy use exerts over economic growth is confirmed, suggesting the existence of mixed composition and technical effects as a consequence of more efficient energy uses and reduced dependence of fossil sources (Balsalobre and Álvarez 2016).

Finally, the interaction between renewable energy use and environmental damage moderates economic growth. Thus, the promotion of renewable energy sources aimed to correct environmental degradation might reduce the rhythm of economic growth for these Top 10 tourism destinations. Renewable energy use has a positive and negative impact on economic growth and carbon emissions (Bhattacharya et al. 2017), depending on the stage of investment and promotion of renewables. Governments need to promote the use of renewable energy across economic activities to ensure

Table 1.3 Cointegration tests

Pedroni Residual Cointegration Test				
Alternative hypothesis: common AR coefficients (within-dimension)				
			Weighted	
	Statistic	Prob.	Statistic	Prob.
Panel v-Statistic	0.254844	(0.3994)	−1.325253	(0.9075)
Panel rho-Statistic	1.029602	(0.8484)	0.442286	(0.6709)
Panel PP-Statistic	−0.366854**	(0.3569)	−1.935090**	(0.0265)
Panel ADF-Statistic	−0.451629***	(0.3258)	−1.462103**	(0.0719)
Alternative hypothesis: individual AR coefficients (between-dimension)				
	Statistic	Prob.		
Group rho-Statistic	2.478507	(0.9934)		
Group PP-Statistic	−2.234653*	(0.0127)		
Group ADF-Statistic	−0.627356	(0.2652)		

Kao Residual Cointegration Test		
	t-Statistic	Prob.
ADF	−1.347980***	(0.0888)
Residual variance	0.022498	
HAC variance	0.029135	

(continued)

Table 1.3 (continued)

Johansen Fisher Panel Cointegration Test **Unrestricted Cointegration Rank Test (Trace and Maximum Eigenvalue)**				
Hypothesised	Fisher Stat.*		Fisher Stat.*	
No. of CE(s)	(from trace test)	Prob.	(from max-eigen test)	Prob.
None	113.4*	(0.0000)	90.55*	(0.0000)
At most 1	39.47*	(0.0001)	35.19*	(0.0004)
At most 2	16.54	(0.1677)	10.20	(0.5984)
At most 3	25.12**	(0.0143)	25.12**	(0.0143)

*Probabilities are computed using asymptotic Chi-square distribution. *Notes* (*) Significant at the 1%; (**) Significant at the 5%; (***) Significant at the 10%. Individual cross-sectional results. **MacKinnon-Haug-Michelis (1999) *p*-values

Table 1.4 Panel Fully Modified Least Squares (FMOLS) and Dynamic Ordinary Least Square (DOLS) econometric results

Dependent variable: LGDP	FMOLS	DOLS
LIT	0.163975*	0.193936*
	[2.862738]	[3.058170]
	(0.0050)	(0.0043)
LCO_2	3.094308*	3.174471*
	[3.867859]	[3.124726]
	(0.0002)	(0.0036)
LRNW	1.496738*	1.917434*
	[2.650230]	[2.823301]
	(0.0092)	(0.0078)
LRNW * LCO_2	−0.606851**	−0.782374*
	[−2.562336]	[−2.817419]
	(0.0117)	(0.0079)
R-squared	0.994064	0.998638
Adjusted R-squared	0.992887	0.995798
S.E. of regression	0.092559	0.063255
Log likelihood	0.011840	0.000989
Mean dependent var	9.688190	9.899146
S.D. dependent var	1.097439	0.975867
Sum squared resid	0.993795	0.140042

Notes (*) Significant at the 1%; (**) Significant at the 5%; (***) Significant at the 10%, and (no) Not Significant

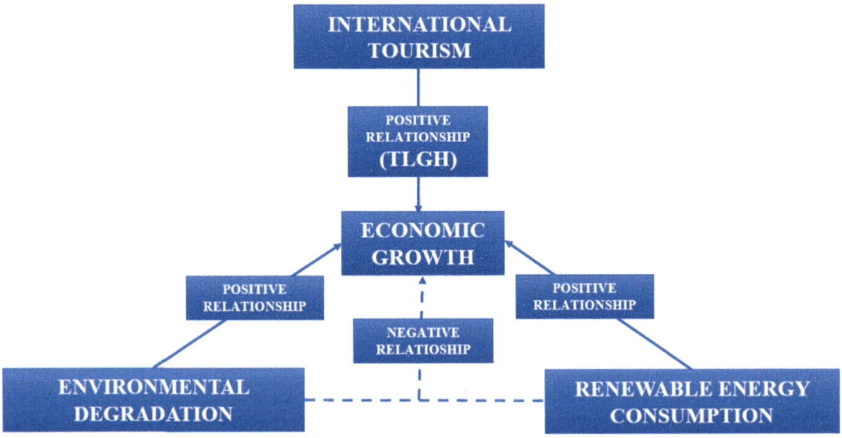

Fig. 1.1 Empirical scheme

sustainable economic development. As such, in the first stage of renewables promotion, economic growth might be reduced due to budgetary/investment efforts, but it is expected to recover the rhythm of economic growth in the next stage.

1.6 Final Conclusions

Since the decade of the '60s, the tourism sector has emerged as a fundamental driving force of economic growth in both developed and developing countries. Having in the spotlight the Top 10 tourism destination countries, tourism-led growth hypothesis was checked. The empirical results underline the relevance of tourism and its impact on economic growth in these ten countries, confirming the TLGH. As economic growth seems to intensify not only by international tourism, additional explanatory variables were considered: the impact of renewable energy use, carbon emissions, and the interaction between renewable energy use and carbon emissions.

To test the TLGH, we use FMOLS and DOLS econometric techniques. The selection of these methods is based on their robustness when an adjustment for serial correlation and endogeneity is requested due to cointegrating relationships, configured to be an asymptotically efficient estimator and to eliminate feedback in the cointegrating system.

The econometric results validate a direct connection between the tourism sector, renewable energy uses, environmental degradation, and economic growth. Thus, the results confirm the TLGH for Top 10 tourism destinations. By contrast, a negative relationship between the interaction of renewable energy use and environmental degradation as a proxy of environmental regulation is obtained. This result reveals the existence of a dampening effect of renewable energy over carbon emissions, showing how the transition to a more renewable energy mix impacts, via scale effect, by dropping the direct effect that emissions exert over economic growth. Therefore, the promotion of renewable energy sources is to be considered by policymakers aiming to diminish the impact of fossil fuels and carbon emissions on the environment and the way towards more sustainable tourism. It is essential to design and implement energy efficiency, more renewable sources and increased awareness of society as a whole to reach sustainable tourism. Strong public support is compulsory in promoting more efficient and green energy and the transition process from fossil to renewable energy sources. The positive effects of renewable energy use on economic growth and the possibility of long-run planning should motivate and facilitate the design and implementation of sustainable development policies. Reaching optimum levels requires a certain period, but the benefits are soon to come after the investment in sustainable actions and infrastructures. This impact is even more relevant for industries like tourism where mass tourism is no longer an option (at least not in the medium or long run) with a very high connection with other sectors and a significant impact on the economy.

Future studies should focus on the effects of globalisation and energy innovation on the tourism-economic growth relationship. The non-linear relationship should be tested between economic growth and (1) international tourism and (2) the interaction effect renewable energy use—carbon emissions. This connection would allow a more in-depth analysis able to guide policymakers in designing environmental and sustainable regulations.

References

Abrell, J., & Weigt, H. (2008). The interaction of emissions trading and renewable energy promotion. Dresden University of Technology Working Paper no. WP-EGW-05.

Agarwal, S. (2012). Resort economy and direct economic linkages. *Annals of Tourism Research, 39*(3), 1470–1494.

Aitken, B., Hanson, G. H., & Harrison, A. E. (1997). Spillovers, foreign investment, and export behaviour. *Journal of International Economics, 43,* 103–132.

Álvarez, A., Balsalobre-Lorente, D., Shahbaz, M., & Cantos, J. M. (2017). Energy innovation and renewable energy consumption in the correction of air pollution levels. *Energy Policy, 105,* 386–397.

Apergis, N., & Payne, J. E. (2009a). Energy consumption and economic growth in Central America: Evidence from a panel cointegration and error correction model. *Energy Economics, 31,* 211–216.

Apergis, N., & Payne, J. E. (2009b). Energy consumption and economic growth: Evidence from the Commonwealth of Independent States. *Energy Economics, 31,* 641–647.

Azam, M., Khan, A. Q., Abdullah, Abdullah, H. B., & Qureshi, M. E. (2016). The impact of CO_2 emissions on economic growth: Evidence from selected higher CO_2 emissions economies. *Environmental Science and Pollu tion Research, 23*(7), 6376–6389.

Balaguer, J., & Cantavella-Jordá, M. (2002). Tourism as a long-run economic growth factor: The Spanish case. *Applied Economics, 34*(7), 877–884.

Balsalobre, D., & Álvarez, A. (2016). Economic growth and energy regulation in the environmental Kuznets curve. *Environment Science and Pollution Research, 23*(16), 16478–16494.

Balsalobre-Lorente, D., Driha, O.M., Shahbaz, M., & Sinha, A. (2020a). The effects of tourism and globalisation over environmental degradation in developed countries. *Environmental Science and Pollution Research, 27*(7), 7130–7144.

Balsalobre-Lorente, D., Driha, O. M., Bekun, F. V., & Adedoyin, F. F. (2020b). The asymmetric impact of air transport on economic growth in Spain: Fresh evidence from the tourism-led growth hypothesis. *Current Issues in Tourism,* 1–17.

Becken, S., & Cavanagh, J. A. (2003). Energy efficiency trend analysis of the tourism sector. Landcare Research Contract Report LC0203/180 prepared for the Energy Efficiency and Conservation Authority. Retrieved February 12, 2004, from https://www.landcareresearch.co.nz/research/sustain_business/tourism.

Bhattacharya, M., Churchill, S. A., & Paramati, S. R. (2017). The dynamic impact of renewable energy and institutions on economic output and CO_2 emissions across regions. *Renewable Energy, 111,* 157–167.

Blake, A., Sinclair, M. T., & Sugiyarto, G. (2003). Quantifying the impact of foot and mouth disease on tourism and the UK economy. *Tourism Economics, 9*(4), 449–465.

Brida, J. G., & Pereyra, J. S. (2009). Tourism taxation and environmental quality: A model with vertical differentiation of the accommodation industry. *Tourismos: An International Multidisciplinary Journal of Tourism, 4*(1), 45–62.

Brida, J. G., Cortés-Jiménez, I., & Pulina, M. (2016). Has the tourism-led growth hypothesis been validated? A literature review. *Current Issues in Tourism, 19*(5), 394–430.

Chang, C. L., Khamkaew, T., & McAleer M. J. (2009). A Panel Threshold Model of Tourism Specialization and Economic Development (No. EI 2009–40). Erasmus School of Economics (ESE). Retrieved July 13, 2015, from https://hdl.handle.net/1765/17310.

Chen, C. F., & Chiou-Wei, S. Z. (2009). Tourism expansion, tourism uncertainty and economic growth: New evidence from Taiwan and Korea. *Tourism Management, 30,* 812–818.
Choi, I. (2001). Unit root tests for panel data. *Journal of International Money and Finances, 20,* 249–272.
Chou, M. (2013). Does tourism development promote economic growth in transition countries? A panel data analysis. *Economic Modelling, 33,* 226–232.
Dogru, T., & Bulut, U. (2018). Is tourism an engine for economic recovery? Theory and empirical evidence. *Tourism Management, 67,* 425–434.
Dunn, H. S., & Dunn, L. (2002). *People and tourism: Issues and attitudes in the Jamaican hospitality*
Dwyer, L., Forsyth, P., Spurr, R., & VanHo, T. (2006). Economic effects of the world tourism crisis on Australia. *Tourism Economics, 12*(2), 171–186.
Essletzbichler, J., & Rigby, D. L. (2007). Exploring Evolutionary Economic Geographies. *Journal of Economic Geography, 7*(5), 549–571.
Gössling, S., & Hall, C. M. (2006). An introduction to tourism and global environmental change. In S. Gossling & C. M. Hall (Eds.), *Tourism and global environmental change* (pp. 1–34). London: Routledge.
Gössling, S. (2002). Global environmental consequences of tourism. *Global Environmental Change, 12*(4), 283–302.
Gössling, S., Hansson, C. B., Horstmeier, O., & Saggel, S. (2002). Ecological footprint analysis as a tool to assess tourism sustainability. *Ecological Economics, 43,* 199–211.
Govdeli, T., & Direkci, T. B. (2017). The Relationship between Tourism and Economic Growth: OECD Countries. *International Journal of Academic Research in Economics and Management Sciences, 6,* 104–113.
Groizard, J. L., & Santana-Gallego, M. (2018). The destruction of cultural heritage and international tourism: The case of the Arab countries. *Journal of Cultural Heritage, 33,* 285–292.
He, L., Zha, J., & Loo, H. A. (2020). How to improve tourism energy efficiency to achieve sustainable tourism: Evidence from China. *Current Issues in Tourism, 23*(1), 1–16.
Im, K. S., Pesaran, M. H., & Shin, Y. (2003). Testing for unit roots in heterogeneous panels. *Journal of Econometrics, 115,* 53–74.
Johansen, S. (1991). Estimation and hypothesis testing of cointegration vectors in Gaussian vector autoregressive models. *Econometrica: Journal of the Econometric Society,* 1551–1580.
Kao, C. (1999). Spurious regression and residual-based tests for cointegration in panel data. *Journal of Econometrics, 90*(1), 1–44.
Katircioglu, S. (2014). Testing the tourism-induced EKC hypothesis: The case of Singapore. *Economic Modelling, 41,* 383–391.
Lee, C. C., & Chang, C. P. (2008). Tourism development and economic growth: A closer look at panels. *Tourism Management, 29*(1), 180–192.
Lee, J. W., & Brahmasrene, T. (2013). Investigating the influence of tourism on economic growth and carbon emissions: Evidence from panel analysis of the European Union. *Tourism Management, 38,* 69–76.
Lee, S. H., Wu, S. C., & Li, A. (2018). Low-carbon tourism of small islands responding to climate change. *World Leisure Journal, 60*(3), 235–245.
Levin, A., Lin, C. F., & Chu, C. S. J. (2002). Unit root tests in panel data: Asymptotic and finite-sample properties. *Journal of Econometrics, 108*(1), 1–24.
Li, K. X., Jin, M., & Shi, W. (2018). Tourism as an important impetus to promoting economic growth: A critical review. *Tourism Management Perspectives, 26,* 135–142.
Liu, J., Feng, T., & Yang, X. (2011). The energy requirements and carbon dioxide emissions of tourism industry of Western China: A case of Chengdu city. *Renewable and Sustainable Energy Reviews, 15,* 2887–28944.
Long, P. T., Perdue, R. R., & Allen, L. (1990). Rural resident tourism perceptions and attitudes by community level of tourism. *Journal of Travel Research, 28,* 3–9.
Narayan, P. K., & Narayan, S. (2005). Estimating income and price elasticities of imports for Fiji in a cointegration framework. *Economic Modelling, 22*(3), 423–438.

OECD. (2018). *Tourism trends and policies 2018*. Publishing, Paris. https://doi.org/10.1787/tour-2018-en
Pedroni, P. (1999). Critical values for cointegration tests in heterogeneous panels with multiple regressors. *Oxford Bulletin of Economic Statistics, 61,* 653–670.
Peeters, P. (2007). Tourism and climate change mitigation methods, greenhouse gas reductions and policies. In *NHTV Academics Studies*, No. 6, NHTV. Breda, The Netherlands: Breda University.
Pham, T. D., Simmons, D. G., & Spurr, R. (2010). Climate change-induced economic impacts on tourism destinations: The case of Australia. *Journal of Sustainable Tourism, 18*(3), 449–473.
Phillips, P. C. B. (1995). Fully modified least squares and vector autoregression. *Econometrica, 63,* 1023–1078.
Phillips, P. C., & Hansen, B. E. (1990). Statistical inference in instrumental variables regression with I (1) processes. *The Review of Economic Studies, 57*(1), 99–125.
Po, W., & Huang, B. (2008). Tourism development and economic growth–a non-linear approach. *Physica A: Statistical Mechanics and Its Applications, 387,* 5535–5542.
Saikkonen, P. (1991). Asymptotically efficient estimation of cointegration regressions. *Econometric Theory, 7*(1), 1–21.
Scott, D. (2006). Climate change and sustainable tourism in the 21st century. In J. Cukier (Ed.), *Tourism research: Policy, planning, and prospects* (pp. 175–248). Waterloo: Department of Geography Publication Series, University of Waterloo.
Scott, D. (2011). Why sustainable tourism must address climate change. *Journal of Sustainable Tourism, 19*(1), 17–34.
Scott, D., Gössling, S., Hall, C. M., & Peeters, P. (2016). Can tourism be part of the decarbonised global economy? The costs and risks of alternate carbon reduction policy pathways. *Journal of Sustainable Tourism, 24*(1), 52–72.
Sequeira, T. N., & Campos, C. (2007). International tourism and economic growth: A panel data approach. In A. Matias, P. Nijkamp, & P. Neto (Eds.), *Advances in modern tourism research* (pp. 153–63). New York: Physica-Verlag Heidelberg.
Shahbaz, M., Loganathan, K., Muzaffar, A. T., Ahmd, K., & Jabran, M. A. (2016). How urbanisation affects CO_2 emissions in Malaysia? The application of STIRPAT model. *Renewable and Sustainable Energy Reviews, 57,* 83–93.
Shan, J., & Wilson, K. (2001). Causality between Trade and Tourism: Empirical Evidence from China. *Applied Economics Letters, 8,* 279–283.
Smorfitt, D. B., Harrison, S. R., & Herbohn, J. L. (2005). Potential economic implications for regional tourism of a foot and mouth disease outbreak in North Queensland. *Tourism Economics, 11*(3), 411–430.
Stock, J. H., & Watson, M. W. (1993). A simple estimator of cointegrating vectors in higher order integrated systems. *Econometrica: Journal of the Econometric Society*, 783–820.
Turner, L. W., & Witt, S. F. (2001). Factors influencing demand for international tourism: Tourism demand analysis using structural equation modelling, revisited. *Tourism Economics, 7*(1), 21–38.
WDI. (2020). *World development indicators. World bank database.* https://databank.worldbank.org/source/world-development-indicators (Accessed May 2020).
Weaver, D. (2011). Can sustainable tourism survive climate change? *Journal of Sustainable Tourism, 19*(1), 5–15.
WTTC. (2011). *World travel and tourism council. Travel and tourism economic impact 2011: European Union.* London, UK: World Travel and Tourism Council.
WTTC. (2018). *World travel and tourism council. Travel & tourism-economic impact 2018-World.*
Zhang, J., & Lee, D. J. (2007). The effect of wildlife recreational activity on Florida's economy. *Tourism Economics, 13*(1), 87–110.
Zhao, L., & Mao, R. (2013). Tourism development, threshold effects and economic growth-evidence from China. *Chinese Journal of Shanxi University of Finance and Economics, 12,* 60–67.
Zuo, B., & Huang, S. (2018). Revisiting the tourism-led economic growth hypothesis: The case of China. *Journal of Travel Research, 57*(2), 151–163.

Chapter 2
The Possible Influence of the Tourism Sector on Climate Change in the US

Faik Bilgili, Yacouba Kassouri, Aweng Peter Majok Garang, and H. Hilal Bağlıtaş

Abstract The effect of tourism development on GHG has been a controversial research topic, and the existing literature fails to provide satisfactory evidence about the impact of tourism on climate change. To the best of our knowledge, this work is the first to study the dynamics of tourism development with several climate-changing substances through time- and regime (state)-varying analysis. Therefore, this article aims at contributing towards a novel analysis of the behaviour of carbon emissions and tourism development in the US following Markov regime-switching VAR (MS-VAR) models. This book chapter will observe the estimates to understand the effect of tourism on air pollution (CO_2 emissions) at different regimes/states. The stochastic process generating the unobservable regimes is an ergodic Markov chain with a finite number of states ($s_t = 1......N$) which is defined by the transition probabilities. Most of the current studies provide mixed evidence on the relationship between tourism and climate change through time- and regime-invariant parameter estimations. In contrast, MS-VAR model predictions reveal the constant term and other parameter coefficients, which are also subject to change from one regime to another regime, to explore the effects of explanatory variables on CO_2 in the US. The explanatory variables of this work are the Number of Tourist Visiting the US, Energy Consumption of Transportation Sector, and Industrial Production. MS-VAR models also monitored seasonality effects. In the estimations, we aim at observing accurately the impact of tourism on CO_2 emissions, as well as the effects of industrial production and transportation sector's energy usage on emissions, in the US.

Keywords Tourism · CO_2 emissions · Transportation sector · MS-VAR models; the US

F. Bilgili (✉) · H. H. Bağlıtaş
Faculty of Economics and Administrative Sciences, Erciyes University, 38039, Melikgazi Kayseri, Turkey
e-mail: fbilgili@erciyes.edu.tr

Y. Kassouri · A. P. M. Garang
SSI, Ph.D. Program in Economics, Erciyes University, 38039, Melikgazi Kayseri, Turkey

D. Balsalobre-Lorente et al. (eds.), *Strategies in Sustainable Tourism, Economic Growth and Clean Energy*, https://doi.org/10.1007/978-3-030-59675-0_2

2.1 Introduction

With the increasing population of the World, the increasing level of production/consumption of commodities and services contributed much to climate change and global warming. Especially for the last three decades, administrators and policy-makers have been implementing some regulations and energy policy acts to mitigate environmental degradation, e.g. through (a) renewable energy usage and (b) new production, heating, lighting, air conditioning, and transportation technologies with low carbon dioxide (CO_2) emissions. The relevant literature of climatic change has focused intensively to observe if renewables could diminish CO_2 emissions as investigated in Zafar et al. (2020), Sharif et al. (2020), and Bilgili et al. (2016a, b, 2017, 2019a, b).

Throughout the discussions on energy policies for environmental quality at country, region, and/or continental level, the sector-specific discussions and potential new regulations and policies to mitigate the carbon emissions have become almost a priority. Among other sectors, the nexus between the growth-environment and tourism sector has begun to attract much attention and importance as explored in Balsalobre-Lorente et al. (2020a, b).

As one of the largest destinations in international tourist arrivals during the last few decades, the United States travel and tourism industry is projected to welcome 95.5 million international visitors annually by 2030 (UNWTO 2019). According to the World Tourism Organization (2013), the U.S. tourism industry accounts for $740 billion in direct travel expenditures by both domestic and international travellers. This dynamism of the U.S. tourism industry is expected to continue providing significant benefits in terms of socioeconomic development, employment, and tax revenue (Aratuo and Etienne, 2019; Lim and Won 2020). Despite the salient benefits and substantial importance of the U.S. tourism industry to the economy, the tourism-led production and consumption activities may also have several negative impacts (Alexandrakis et al. 2015; Gil-Alana et al. 2019; Melián-González and Bulchand-Gidumal 2020). Arguably, one of the most severe negative impacts of tourism is its environmental impact, and one of the main challenges facing the tourism sector today is to decouple its projected dynamics from anthropogenic greenhouse gases (GHG) emissions to ensure a sustainable tourism development in the United States.

In particular, a global assessment of the emissions from tourism activities indicates that emissions from the three subsectors of tourism including aviation, transportation, and luxury accommodation are projected to grow by 135% over the next three decades from 2005 to 2035 (UNWTO 2019), which could substantially increase global temperature and accelerate global warming. In this context, the adoption of effective tourism policies to ensure the long-term sustainability of the sector is of high relevance for the tourism industry as this sector is expected to experience significant growth in the next decades. To this end, tourism stakeholders should understand the dynamics of tourism and the behaviour of GHG triggers, namely, air pollution, gasoline consumption, and fossil fuel consumption, which are the different discharges this study focuses on.

There have been several studies about the influence of tourism activities on climate change from the carbon emanation perspective. However, much of these studies do not provide satisfactory evidence about the impact of tourism on climate change and leave several problems unsolved. First, previous studies have examined the environmental impact of tourism from the single effect on carbon dioxide emissions, which may be too restrictive as it ignores other major pollutants which too contribute to climate change (Galli et al. 2014; Ulucak and Bilgili 2018). Thus, emphasis should also be focused on other major climate-changing pollutants which have been proved to be sensitive to tourism development such as gasoline consumption, fossil fuel consumption, and air pollution (Saenz-de-Miera and Rosselló 2014; Sajjad et al. 2014). Second, the literature reports inconclusive results on the linkage between tourism and emissions. For instance, tourism-driven emissions have been extensively explored by several studies (Filimonau et al. 2014; Jin et al. 2018; Katircioglu et al. 2014; Munday et al. 2013; Tao and Huang 2014; Tsai et al. 2014; Verbeek and Mommaas 2008). However, other studies failed to find a positive relationship between tourism and emissions (Cerutti et al. 2016; Jamal et al. 2011; Katircioğlu 2014; Lee and Brahmasrene 2013). Recently, another strand of the literature found evidence for the EKC nexus between tourism and emissions (Ozturk et al. 2016; Paramati et al. 2017; Zaman et al. 2016). The empirical controversy can partly be explained by the narrow ground of previous empirical studies assuming that the impact of tourism is stable over time, ignoring thus the sensitivity of tourism indicators to shocks related to political violence or terrorist attack. Third, empirical investigations on tourism provide a temporal pattern of the linkage between tourism indicators and climate change. Such a model specification is not without criticism since the dynamics of tourism indicators and is more aptly described by structural change and nonlinear dynamics (Gu et al. 2018). The assumption of temporal pattern in the relationship between tourism and climate change is too simplistic and does not capture the full nature of the stochastic behaviour of tourism- and climate-related hazards.

The current study aims to address the aforementioned gaps in the existing literature by examining the influence of tourism on air pollution in the United States following Markov regime-switching VAR (MS-VAR) models. The MS-VAR approach is particularly useful for situations where the stochastic behaviour of the series is allowed to vary between a discrete number of regimes where regime switch is driven by an observable state variable. The authors argue that this type of modelling is particularly relevant given the structural uncertainty in climate data records and tourism indicators. The choice of the United States is quite natural given its importance in the global tourism industry and climate change policy. Naturally, the evidence from this inquiry allows us to track the significant capacity to change the pattern of global climate change stemming from the tourism sector, which is unfortunately seen as an essential energy intense, difficult to decarbonize sector.

Concerning the existing literature, this study has two important innovations: (i) while the previous studies have mainly focused on the temporal pattern of the tourism and climate change nexus, this study is motivated by the dynamic dependence within the regimes, which is a particular characteristic of anthropogenic gas variables (Kuşkaya and Bilgili 2020; Şahin 2019). Thus, this study provides a realistic modelling framework by focusing on to what extent linkages in tourism emissions are affected by structural shifts caused, for instance, by shocks affecting the dynamics of tourism activities such as political violence and terrorist attacks and other climate-related shocks such as the U.S. withdrawal from the Paris agreement. (ii) To the best of our knowledge, this study is the first to study the dynamics of tourism development with several climate-changing substances through a regime-dependent analysis. In this context, this study presents different scenarios of the dynamics of climate-changing metrics and tourism development in the U.S. Our empirical analysis allows us to identify where the bulk of tourism-driven emissions comes from across different time horizons, which is highly relevant to design effective climate policy objectives.

2.2 Literature Review

The theoretical framework holds that tourism affects climate change through energy consumption and greenhouse gas emissions (Becken 2002; Becken et al. 2003). As recently estimated by (Russo et al. 2020), the total contribution of tourism activities to global emissions was 67.6% (for both NOx and PM10 for aviation), followed by 15.1% (for PM10 in the transport sector). These figures show that tourism has a significant impact on atmospheric emissions, which raises concerns about tourism sustainability. Apart from carbon emissions as a result of combusting fossil fuels and energy use/demand in the transport and accommodation sectors, changes in land-use management due to tourism activities increase pressures on natural conditions (such as climate and water resources, carbon sequestration, and cropland use), resulting in changes in climatic conditions (Bai et al. 2011; Kindu et al. 2016; Li et al. 2020). In light of theoretical explanations, it can be seen that tourism activities tend to increase climate change vulnerability. Empirically, much has been discussed about the relationship between tourism and climate change through the effects of tourism on CO_2 emissions (Al-Mulali et al. 2015; Balli et al. 2019; Gössling et al. 2015; León et al. 2014a, b; Nepal et al. 2019; Shakouri et al. 2017; Sharif et al. 2017; Solarin 2014). Recently, many scholars have studied carbon footprints associated with tourism consumption (Dwyer et al. 2010; Lenzen et al. 2018; Paiano et al. 2020; Sharp et al. 2016).

Several papers have analysed the nexus between tourism and carbon emissions using various empirical tools and have provided conflicting results. Although many papers provide evidence for the negative effect of the tourism industry on carbon emissions, other scholars front cases to the contrary. For instance, using a host of variables such as CO_2 emissions per capita (henceforth CO_2 emissions pc), population, tourist arrivals, and GDP pc from a sample of 45 countries, León et al. 2014a, b; show that tourism industry contributes significantly to carbon emissions by employing Panel GMM model. With the same method, similar results were reported by (Qureshi et al. 2017) from a sample of 37 countries and by Shakouri et al. (2017) in 12 Asia–Pacific countries using variables on Health expenditures, GDP pc, FDI inflows, trade, and CO_2 emissions; and CO_2 emissions pc, real GDP pc, energy use, tourist arrivals, respectively. An application of the Panel Granger-causality by Al-Mulali et al. (2015) revealed a significant impact of tourist arrivals on CO_2 emissions from the transportation sector during 1995–2009 in 48 top international tourism destinations. In 34 developed and developing countries, tourism-induced emissions hypothesis proofed to be valid in a study conducted by Zaman et al. (2016) between 2005 and 2013 using panel two-stage least squares. Zaman et al. (2017) investigated the same issue in 11 transition countries using Panel fixed effect and reported that international tourism receipts and tourism expenditures increase CO_2 emissions. Panel DOLS also shows that tourist arrivals contribute to emissions in OECD countries according to Dogan et al. (2017). Evidence from ARDL co-integration reported by Işik et al. (2017), Sharif et al. (2017), Nepal et al. (2019), and Akadiri et al. (2019) in various countries over different periods reveal positive interaction between tourism arrivals and CO_2 emissions. Other co-integration methods like panel Granger causality show that in 10 major tourism countries, tourism investments improve environmental quality by curbing carbon emanations (Alam and Paramati 2017). In a sample of the top 10 most visited countries, Koçak et al. (2020) applied CUP-FM and CUP-BC to show feedback effects between tourism and CO_2 emissions, thus implying that tourism arrivals positively affect carbon emissions. Much as these distinguished scholarly contributions show pieces of evidence of the negative influence of tourism on carbon emissions, other studies indicate that tourism indeed reduces carbon emission. For example, Lee and Brahmasrene (2013) use the panel fixed effects method to show that tourism reduces CO_2 emissions in countries of the European union during the period 1988–2009. His findings are backed by Dogan and Aslan (2017) who employ Panel DOLS to demonstrate that tourist arrivals contribute to emissions in 25 EU countries between 1995 and 2010. Various empirical methods examining this question have revealed contrasting results. However, results reported by employing Panel DOLS and FMOLS mostly confirm that tourism sector improves environmental quality by reducing CO_2 emissions (Zhang and Gao 2016; Danish and Wang 2018; Ben Jebli and Hadhri 2018). Therefore, consensus remains to be built on whether indeed the tourism sector adversely affects climate change through carbon emissions using sophisticated empirical tools on new globally influential samples.

In Table 2.1, we provided readers with clear information about how literature has evolved.

Based on the survey on the existing literature, one observation is that several empirical papers examine the effects of tourism on CO_2 emissions, with little attention paid to other greenhouse gases. Using only CO_2 emissions to assess the impact of tourism on climate change may not yield accurate outcomes owing to the use of inappropriate indicators. Despite some exceptions, in terms of the econometric approach, panel data estimation techniques have been extensively used. Therefore, this specification may lead to misguiding results since the estimation approaches assumed that tourism and CO_2 emissions have stable properties, which is hard to validate practically. This study is one of the few attempts to use advanced time-series techniques to inform sustainable tourism policy decisions concerning the relationship between tourism and greenhouse gases responsible for climate change. This research gathers disparate sources of GHG.

Table 2.1 Literature review

Author(s)	Country	Data	Variables used	Method(s)	Main finding(s)
Lee and Brahmasrene (2013)	EU countries	1988–2009	FDI, CO_2 emissions, real GDP pc, tourism receipts	Panel fixed effects	Tourism reduces CO_2 emissions
León et al. (2014a, b)	45 countries	1998–2006	CO_2 emissions pc, population, tourist arrivals, GDP pc	Panel GMM	The tourism industry contributes significantly to carbon emissions
Al-Mulali et al. (2015)	48 top international tourism destinations	1995–2009	CO_2 emissions from the transportation sector, GDP, urban population, tourist arrivals	Panel Granger-causality	A significant impact of tourist arrivals on CO_2 emissions from the transportation sector
Zaman et al. (2016)	34 developed and developing countries	2005–2013	Tourist development index, per capita GDP, GFCF, health expenditure, energy use	Panel two-stage least squares	Tourism-induced emission hypothesis is valid
Qureshi et al. (2017)	37 countries	1995–2015	Health expenditures, GDP pc, FDI inflows, trade, and CO_2 emissions	Panel GMM	The positive relationship between inbound tourism and environmental degradation
Zhang and Gao (2016)	30 provinces in China	1995–2011	GDP pc, tourism receipts, CO_2 emissions, energy consumption and trade	Panel FMOLS	Tourism sector improves environmental quality by reducing CO_2 emissions
Zaman et al. (2017)	11 transition countries	1995–2013	CO_2 emissions from transport, tourism expenditures, FDI, energy consumption, trade, urban population, tourism receipts	Panel fixed effect	International tourism receipts and tourism expenditures increase CO_2 emissions

(continued)

Table 2.1 (continued)

Author(s)	Country	Data	Variables used	Method(s)	Main finding(s)
Dogan and Aslan (2017)	25 EU countries	1995–2011	CO_2 emissions, GDP pc, energy consumption, and tourist arrivals	Panel FMOLS and DOLS; Panel fixed effects	Tourism arrivals negatively influence carbon emissions
Danish and Wang (2018)	BRICS	1995–2014	Carbon dioxide emissions, real GDP pc, trade openness, globalization, tourism receipts, and investment in tourism	Panel DOLS and FMOLS	Feedback effect between tourism receipts and CO_2 emissions. Investment in tourism reduces the level of CO_2 emissions
Işik et al. (2017)	Greece	1970–2014	Carbon dioxide emissions, real GDP pc, financial development, and trade	ARDL	Tourism expenditure positively affects CO_2 emissions
Dogan et al. (2017)	OECD countries	1995–2010	CO_2 emissions, GDP pc, tourist arrivals, energy consumption and trade	Panel DOLS	Tourist arrivals contribute to emissions
Shakouri et al. (2017)	12 Asia–Pacific countries	1995–2013	CO_2 emissions pc, real GDP pc, energy use, tourist arrivals	Panel GMM approach	Tourist arrivals positively influence CO_2 emissions in the long run
Sharif et al. (2017)	Pakistan	1972–2013	CO_2 emissions, tourist arrivals, real GDP pc,	ARDL	Positive interaction between tourism arrivals and CO_2 emissions
Alam and Paramati (2017)	10 Major tourism countries	1995–2013	Tourism investment, GDP pc, CO_2 emissions, trade, and total population	Panel Granger-causality	Tourism investments improve environmental quality by curbing carbon emanations
Ben Jebli and Hadhri (2018)	Top 10 international tourism destinations	1995–2013	CO_2 emissions from transport, real GDP pc, energy use	Panel FMOLS and DOLS	International tourism contributes to the reduction of CO_2 from transport

(continued)

Table 2.1 (continued)

Author(s)	Country	Data	Variables used	Method(s)	Main finding(s)
Nepal et al. (2019)	Nepal	1975–2014	Tourist arrivals, CO_2 emissions, Gross fixed capital formation, real GDP pc, energy consumption	ARDL	Tourist arrivals positively influence CO_2 emissions
Akadiri et al. (2019)	Turkey	1970–2014	CO_2 emissions, globalization, real GDP pc, tourists inbound	ARDL	One-way causation ranging from tourism to CO_2 emissions
Koçak et al. (2020)	Top 10 most visited countries	1995–2014	CO_2 emissions, urban population, GDP pc, tourism developments, energy intensity	CUP-FM and CUP-BC	Feedback effects between tourism and CO_2 emissions. Tourism arrivals positively affect carbon emissions

Notes ARDL: Autoregressive Distributed Lag, CUP-BC: Continuously Updated Bias Corrected; CUP-FM: Continuously Updated Fully Modified, DOLS: Dynamic Ordinary Least squares; GDP: Gross Domestic Product, FDI: Foreign Direct Investments; GFCF: Gross Fixed Capital Formation; GMM: Generalized Method of Moments

2.3 Estimation Methodology: MS-VAR

Markov switching (MS) or Markov regime-switching models measure the behaviours of the nonlinear time series in different regimes/states with multiple equations (Bilgili et al. 2019a, b; Doornik 2019; Doğan and Bilgili 2014; Bilgili et al. 2012). Besides, Vector Auto-Regressive (VAR) models can analyse all variables symmetrically and endogenously for multivariate time series (Enders 2009; Camacho 2013). Using together MS and VAR models makes it possible to analyse the regime changes with equation systems (Krolzig 1998).

Markov switching VAR (MS-VAR) models might be more suitable if there is a time-varying component (Krolzig 1997). In MS-VAR models, complex dynamic relationships could be analysed throughout switching characters between the equations. Unobservable state variable governs switching and follows a first-order Markov process (Sinica 2002).

Original MS models are related to the mean behaviours of variables (Sinica 2002). Timmermann (2000) classifies three models reflecting the mean, variance, and dynamics properties of the series with discrete regimes. The first model has a simple pattern including two variables which are ergodic and k-state unobservable regime indicators with identically and independently distributed (iid) errors.

$$y_t = \mu_{S_t} + \sigma_{S_t} u_t \qquad s_t \in \{1, \ldots, M\} \tag{2.1}$$

MS-VAR can be depicted through the VAR (*p*, *q*) model where ρ and *k* represent the degree of order and dimension of time series, respectively (Timmermann 2000; Krolzig 1997).

$$y_t = \mu_{S_t} + \sum_{k=1}^{p} A_{S_t}^{(k)} y_{t-k} + u_t \qquad s_t = 1, \ldots, T \tag{2.2}$$

In Eq. (2.2), $A_{S_t}^{(k)}$ shows state-independent dynamic and when this coefficient is replaced by $A_{S_{t-k}}^{(k)}$, it represents the state-dependent condition (Timmermann 2000). Irreducible ergodic s_t follows the Markov chain, and each regime (m) has its VAR(p) representation with parameters (Krolzig 1998).

While μ_m parameter demonstrates permanent regime shift in the mean, υ_{S_t} is the intercept term and similar to a shock in u_t series. If a regime shift occurs, there will be a new level after the transition.

Finally, MS-VAR models employing regime shifts in constant (mean) term can be exhibited by Eqs. (2.3) and (2.4) as depicted by Krolzig (1997, 1998).

$$E|y_t|Y_{t-1}, s_t| = m(s_t) + \sum_{k=1}^{n} A_k y_{t-k} \tag{2.3}$$

$$y_t = m(s_t) + \sum_{k=1}^{n} A_k y_{t-k} + u_t \tag{2.4}$$

where E,y_t, m, s_t, and $y_{t\text{-}k}$ denote expected term, time-series vector, mean, regime, and lagged variables of time-series vector, respectively.

The model given in (2.4) might be expanded when all parameter coefficients are subject to change from one regime to another regime.

$$E|y_t|Y_{t-1}, s_t| = m(s_t) + \sum_{k=1}^{n} A_{k,st}, y_{t-k,st} \tag{2.5}$$

Therefore, in MS-VAR models, it might be modelled that some parameters are regime-sensitive, while others are not through either fixed or scaling or switching variances (Krolzig 1998; Doornik 2019).

Besides regime-dependent coefficients, one might also need to observe the transition probabilities of MS-VAR. In the case of two unobservable regimes, the first-order Markov process manages the transition between the regimes (Hamilton 1989). If there exist more than two regimes and all probabilities are taken into consideration, transition probabilities of unobservable regime variable s_t are as follows (Krolzig 1997):

$$p_{ij} = \Pr(s_{t+1} = j | s_t = i), \quad \sum_{j=1}^{M} p_{ij} = 1 \quad \forall_{i,j} \in \{1, \ldots, M\} \tag{2.6}$$

In Eq. (2.6), the probability of transition from state i to state j (p_{ij}) and the sum of each row of the probability matrix (P_{M*M}) is equal to 1 (Droumaguet et al. 2015). Regime history affects transition probabilities. If there exist M states, the probability density of y_t can be given in Eq. (2.7) as presented in Krolzig (1998).

$$p(y_t | Y_{t-1}, s_t) = \begin{cases} f(y_t | Y_{t-1}, \mu_1) & if\ s_t = 1 \\ \vdots \\ (y_t | Y_{t-1}, \mu_M) & if\ s_t = M \end{cases} \tag{2.7}$$

where error terms are normally and independently distributed; $u_t \sim nid(0, \sum(s_t))$, $m = 1, \ldots, M$ and Y_{t-1} are time-series observations.

The main focus of this article is the effects of tourism on emissions which might change across time and state. Modelling these time series might need to assume the discontinued changes in uncertain points in time. Kim (1994) defines regime shifts from one regime to another like model instability. This uncertainty and instability reflect the dynamic property of the time series (Shumway and Stoffer 1991). Holst et al. (1999) propose switching auto-regression with Markov regime for this kind of dynamical system analysis (p. 489). Under these assumptions, theorems, and/or postulates, the next section will exhibit the estimation outputs of Markov regime-switching VAR models.

2.4 MS-VAR Estimation Output

This paper employs Markov regime-Switching VAR (MS-VAR) models to yield the effects of explanatory variables on carbon dioxide emissions (CO_2) in the US. The paper followed the logarithmic forms of the variables: CO_2 (LNCO_2), the Number of Tourists Visiting the US (LNTOURISTS), Energy Consumption of Transportation Sector (LNTRANS_EN), and Industrial Production (LNIP). MS-VAR models also considered seasonality effects in estimations in which we aim at observing accurately the impact of tourism on CO_2 emissions, besides the effects of industrial production and transportation sectors energy usage on emissions, in the US. All data span from January 2000 to February 2020.

The data for CO_2 emissions and transportation energy usage has been extracted from the U.S. Energy Information Administration (2020). The data for the number of tourist arrivals has been obtained from the National Travel and Tourism Office (2020). The industrial production data has been obtained from the Federal Reserve Economic Data (2020).

To avoid the potential unit root issue, the models employed the differenced logarithms (DL) of the variables. In Table 2.2, all MS-VAR model estimations, the vector (y_t) includes the variables of $DLNCO_2$, DLNTOURISM, DLNTRANS_EN, and DLNIP, respectively.

Table 2.2 MS-VAR estimations in which constant is regime-dependent

$DLNCO_2$ eq. from MS-VAR	MS-VAR1 Variance type: fixed variance		MS-VAR2 Variance type: switching scale		MS-VAR3 Variance Type: Switching Variance	
	Coefficient	T-Prob	Coefficient	T-Prob.	Coefficient	T-Prob.
$DLNCO_2$_1@$DLNCO_2$	−0.463099	0.000	−0.459830	0.000	−0.472501	0.000
$DLNCO_2$_2@$DLNCO_2$	−0.410908	0.000	−0.448625	0.000	−0.451605	0.000
$DLNCO_2$_3@$DLNCO_2$	−0.235394	0.003	−0.248349	0.002	−0.266669	0.001
$DLNCO_2$_4@$DLNCO_2$	−0.060903	0.403	−0.035086	0.584	−0.063957	0.384
DLNTOURISTS_1@$DLNCO_2$	0.013678	0.645	0.035354	0.243	0.003423	0.896
DLNTOURISTS_2@$DLNCO_2$	0.057181	0.069	0.080296	0.007	0.061725	0.021
DLNTOURISTS_3@$DLNCO_2$	0.084267	0.008	0.088856	0.003	0.068305	0.016
DLNTOURISTS_4@$DLNCO_2$	−0.016963	0.587	0.024176	0.387	0.018578	0.488
DLNIP_1@$DLNCO_2$	0.178485	0.424	0.248455	0.120	0.071311	0.694
DLNIP_2@$DLNCO_2$	0.487346	0.026	0.363099	0.020	0.411605	0.025
DLNIP_3@$DLNCO_2$	0.070898	0.741	−0.040418	0.806	−0.017983	0.921
DLNIP_4@$DLNCO_2$	−0.195656	0.346	−0.179359	0.254	−0.154771	0.384
DLNTRANS_EN_1@$DLNCO_2$	−0.194404	0.134	0.006749	0.958	−0.097481	0.439
DLNTRANS_EN_2@$DLNCO_2$	0.294953	0.045	0.422150	0.004	0.419363	0.004
DLNTRANS_EN_3@$DLNCO_2$	0.470675	0.002	0.539503	0.000	0.490922	0.001
DLNTRANS_EN_4@$DLNCO_2$	0.331581	0.010	0.289497	0.005	0.236076	0.065
Seasonal@$DLNCO_2$	−0.036469	0.106	−0.050422	0.014	−0.021177	0.326
Seasonal_1@$DLNCO_2$	−0.203933	0.000	−0.214221	0.000	−0.184897	0.000
Seasonal_2@$DLNCO_2$	−0.151137	0.000	−0.141778	0.000	−0.124826	0.000
Seasonal_3@$DLNCO_2$	−0.214313	0.000	−0.258787	0.000	−0.206037	0.000
Seasonal_4@$DLNCO_2$	−0.197743	0.000	−0.224688	0.000	−0.207688	0.000
Seasonal_5@$DLNCO_2$	−0.191164	0.000	−0.226591	0.000	−0.192993	0.000
Seasonal_6@$DLNCO_2$	−0.127065	0.000	−0.157575	0.000	−0.124721	0.000
Seasonal_7@$DLNCO_2$	−0.111014	0.000	−0.133179	0.000	−0.105002	0.000
Seasonal_8@$DLNCO_2$	−0.228561	0.000	−0.253149	0.000	−0.215379	0.000
Seasonal_9@$DLNCO_2$	−0.215518	0.000	−0.213129	0.000	−0.198378	0.000
Seasonal_10@$DLNCO_2$	−0.125224	0.000	−0.148680	0.000	−0.120519	0.000
Constant(0)@$DLNCO_2$	0.147225	0.000	0.162879	0.000	0.137978	0.000
Constant(1)@$DLNCO_2$	0.153412	0.000	0.168597	0.000	0.142089	0.000

The selection of VAR lag length is based on the Schwarz information criterion (SC) lag length criteria. MS-VAR lag length is determined as four due to SC among other criteria of LR Likelihood ratio (LR), Final prediction error (PE), AIC Akaike information criterion (AIC), and Hannan-Quinn information (HQ) criterion. Lag length (L) $= 4$ was preferred by following SC rather than other criteria indicating $L = 8$ or $L = 10$ to avoid the over-parameterization problem. Some evidence from Monte Carlo studies yields that SC is superior to all other criteria in the VAR process as is noted in Köse and Uçar (2003). All MS-VAR estimations are conducted by following two different constant (mean) terms. Constant 0 and constant 1 correspond to 'constant at state 0' and 'constant at state 1'. State 0 and state 1 are also called 'regime 0' and 'regime 1', respectively.

Table 2.2 reveals the results from three separate MS-VAR model estimations. As the first column yields the explanatory variables, the 2nd, 4th, and 6th columns depict the coefficients from MS-VAR models with fixed variance, scaling variance, and switching variance, respectively. The 3rd, 5th, and 7th columns exhibit the probability values of t statistics.

In Table 2.3, in terms of AIC, SC, and LR, one might consider that MS-VAR3 is the best among other MS-VAR models given in the table. The convergence of all MS-VAR models is strong by SQPF. For MS-VAR1, the probability of staying at regime 0 (as the current regime is 0) is 0.948776. For MS-VAR2, the probability of staying at regime 1 (as the current regime is 1) is 0.816296. For MS-VAR3, the probability of switching from regime 1 to regime 0 is quite low ($= 0.108810$). The linearity tests reveal that null of linearity of three MS-VAR models is rejected at 1% level.

In Table 2.2, the constant terms at regimes 0 and 1 and all seasonal terms, except seasonality at time t, are significant at 1% levels. The outputs of 6th and 7th columns, under switching variance structure, are as follows:

Table 2.3 Linearity tests, convergence, transition probabilities, and goodness–of-fit test statistics

	MS-VAR1	MS-VAR2	MS-VAR3
Linearity test (Chi^2)	(6) $= 5894.5$ [0.0000]	(10) $= 5932.8$ [0.0000]	(16) $= 6004.1$ [0.0000]
Convergence	Strong by SQPF	Strong by SQPF	Strong by SQPF
Log-Likelihood	2485.02768	2504.15475	2539.81288
P(0|0)	0.948776	0.802746	0.947177
P(1|0)	0.051224	0.19725	0.052823
P(0|1)	0.57691	0.18370	0.108810
P(1|1)	0.423091	0.816296	0.891185
AIC	−19.8905289	−20.0181836	−20.2684631
SC	−18.0174838	−18.0866058	−18.2490863
Number of observations	237	237	237

(a) The current $DLNCO_2$ in the US is negatively affected by the 1st, 2nd, and 3rd lagged values of $DLNCO_2$.
(b) 2nd and 3rd lagged values of DLNTOURISTS have positive impacts on the current value of $DLNCO_2$ in the US.
(c) The 2nd lagged value of DLNIP has also a positive influence on the current value of $DLNCO_2$ in the US.
(d) Similarly, the 2nd, 3rd, and 4th lagged values of DLNTRANS_EN have positive effects on the current value of $DLNCO_2$ in the US.

In Fig. 2.1, regime 1 corresponds to high volatile periods as regime 0 corresponds to less volatile periods of the $DLNCO_2$ equation of MS-VAR3. Regime 1 is denoted by the grey area in the figure. Other areas, except the grey areas, exhibit regime 0 periods. Regime 0 consists of 156 months (65.82%) with an average duration of 22.29 months (Table 2.4a). Regime 1 includes 81 months (34.18%) with an average duration of 11.57 months (Table 2.4b).

Table 2.5 reveals the estimation outputs from MS-VAR4, which is a revised form of MS-VAR3 model, in which constant term, as well as parameter estimations of explanatory variables, is regime-dependent. The grey rows indicate regime 1 outcomes as white rows depict regime 0 outputs. Table 2.6a, b yield the regime 0 classification and regime 1 classification, respectively.

Table 2.7 explores linearity tests, convergence, transition probabilities, and goodness-of-fit test statistics. The convergence of MS-VAR4 models is also found

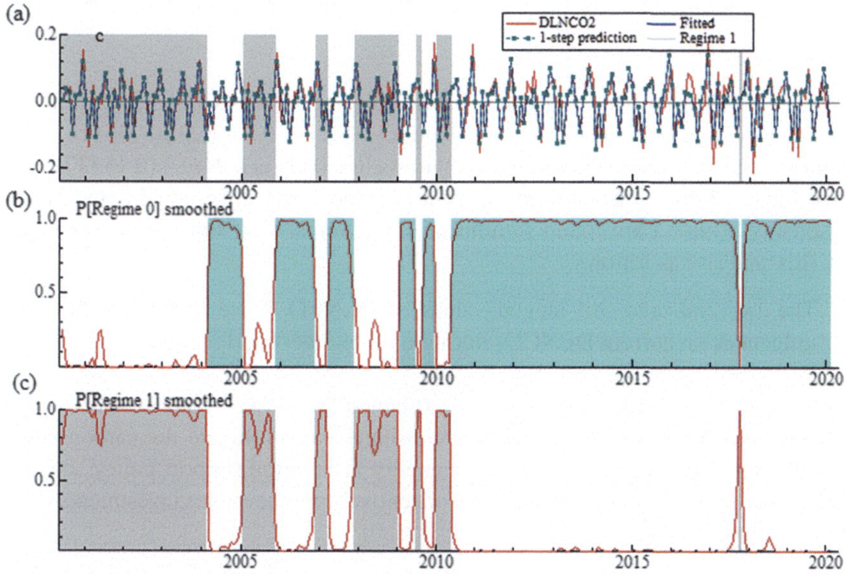

Fig. 2.1 The P [Regime 0] smoothed and the P [Regime 1] smoothed of MS-VAR3. (**a**) Actual, Fitted, Prediction, and Regime 1 area of $DLNCO_2$, (**b**) Probability of smoothed regime 0, and (**c**) Probability of smoothed regime 1

Table 2.4a MS-VAR3 Regime (0) classification based on smoothed probabilities

Regime 0	Months	Avg. prob
2004(3)–2005(1)	11	0.945
2005(12)–2006(11)	12	0.961
2007(4)–2007(11)	8	0.898
2009(2)–2009(6)	5	0.943
2009(9)–2009(12)	4	0.894
2010(6)–2017(9)	88	0.988
2017(11)–2020(2)	28	0.981

Total: 156 months (65.82%) with an average duration of 22.29 months

Table 2.4b MS-VAR3 Regime (1) classification based on smoothed probabilities

Regime 1	Months	Avg. prob
2000(6)–2004(2)	45	0.977
2005(2)–2005(11)	10	0.859
2006(12)–2007(3)	4	0.940
2007(12)–2009(1)	14	0.910
2009(7)–2009(8)	2	0.996
2010(1)–2010(5)	5	0.987
2017(10)–2017(10)	1	0.999

Total: 81 months (34.18%) with an average duration of 11.57 months

strong by SQPF following analytical derivatives. For MS-VAR4, the probability of staying at regime 0 (1) as the current regime is 0 (1) is 0.846629 (0.753114). The probability of jumping from regime 1 (0) to regime 0 (1) is 0.24689 (0.15337).

In Table 2.5, the constant terms at regimes 0 and 1 and all seasonal terms except seasonality at time t are found significant. Some remarkable outputs appear in (i), (ii), (iii), and (iv) as follows.

(i) The 1st, 2nd, and 3rd lagged values of $DLNCO_2$ have significant negative influences on current $DLNCO_2$ during both regime 1 and regime 0.

The outputs from (a) and (i) might be explained by EPACT 2005 in the US that promotes renewable energy production/consumption to mitigate the carbon emissions through tax incentives, renewable energy installation support policy, energy-efficient technology policy, subsidies and incentives for renewables investments since the 1990s and 2000s.

(ii) The 1st, 2nd, and 3rd lagged values of the number of tourists visiting the US (DLNTORISTS) have significant positive impacts on $DLNCO_2$ during regime 0.

Table 2.5 MS-VAR4 estimations in which constant and all coefficients are regime-dependent

$DLNCO_2$ eq. from MS-VAR	Coefficient	T-Prob
$DLNCO_2$_1(0)@$DLNCO_2$	−0.346561	0.000
$DLNCO_2$_1(1)@$DLNCO_2$	−0.512036	0.000
$DLNCO_2$_2(0)@$DLNCO_2$	−0.183352	0.010
$DLNCO_2$_2(1)@$DLNCO_2$	−0.568417	0.000
$DLNCO_2$_3(0)@$DLNCO_2$	−0.137578	0.028
$DLNCO_2$_3(1)@$DLNCO_2$	−0.318763	0.010
$DLNCO_2$_4(0)@$DLNCO_2$	−0.072059	0.206
$DLNCO_2$_4(1)@$DLNCO_2$	0.042442	0.771
DLNTOURISTS_1(0)@$DLNCO_2$	0.063673	0.024
DLNTOURISTS_1(1)@$DLNCO_2$	0.010570	0.828
DLNTOURISTS_2(0)@$DLNCO_2$	0.069934	0.009
DLNTOURISTS_2(1)@$DLNCO_2$	0.064377	0.239
DLNTOURISTS_3(0)@$DLNCO_2$	0.086838	0.004
DLNTOURISTS_3(1)@$DLNCO_2$	0.048860	0.327
DLNTOURISTS_4(0)@$DLNCO_2$	0.000437	0.987
DLNTOURISTS_4(1)@$DLNCO_2$	−0.070461	0.189
DLNIP_1(0)@$DLNCO_2$	0.404334	0.044
DLNIP_1(1)@$DLNCO_2$	0.267422	0.450
DLNIP_2(0)@$DLNCO_2$	0.485197	0.013
DLNIP_2(1)@$DLNCO_2$	0.311298	0.323
DLNIP_3(0)@$DLNCO_2$	−0.127840	0.484
DLNIP_3(1)@$DLNCO_2$	0.298718	0.406
DLNIP_4(0)@$DLNCO_2$	−0.481492	0.010
DLNIP_4(1)@$DLNCO_2$	−0.171810	0.624
DLNTRANS_EN_1(0)@$DLNCO_2$	−0.027427	0.827
DLNTRANS_EN_1(1)@$DLNCO_2$	−0.529028	0.007
DLNTRANS_EN_2(0)@$DLNCO_2$	0.347266	0.009
DLNTRANS_EN_2(1)@$DLNCO_2$	0.073808	0.798
DLNTRANS_EN_3(0)@$DLNCO_2$	0.361074	0.005
DLNTRANS_EN_3(1)@$DLNCO_2$	0.402468	0.123
DLNTRANS_EN_4(0)@$DLNCO_2$	0.318095	0.002
DLNTRANS_EN_4(1)@$DLNCO_2$	0.331771	0.218
Seasonal@$DLNCO_2$	−0.013600	0.494
Seasonal_1@$DLNCO_2$	−0.185051	0.000
Seasonal_2@$DLNCO_2$	−0.160239	0.000
Seasonal_3@$DLNCO_2$	−0.201501	0.000
Seasonal_4@$DLNCO_2$	−0.173317	0.000

(continued)

Table 2.5 (continued)

$DLNCO_2$ eq. from MS-VAR	Coefficient	T-Prob
Seasonal_5@$DLNCO_2$	−0.147214	0.000
Seasonal_6@$DLNCO_2$	−0.116979	0.000
Seasonal_7@$DLNCO_2$	−0.122128	0.000
Seasonal_8@$DLNCO_2$	−0.236241	0.000
Seasonal_9@$DLNCO_2$	−0.169538	0.000
Seasonal_10@$DLNCO_2$	−0.100037	0.000
Constant(0)@$DLNCO_2$	0.130547	0.000
Constant(1)@$DLNCO_2$	0.131794	0.000

Table 2.6a MS-VAR4 Regime (0) classification based on smoothed probabilities

Regime 0	Months	Avg. prob
2000(6)–2000(10)	5	0.975
2001(4)–2001(8)	5	1.000
2001(11)–2001(11)	1	0.999
2002(5)–2002(11)	7	0.991
2003(1)–2003(1)	1	0.998
2003(3)–2003(11)	9	0.996
2004(2)–2004(11)	10	0.993
2005(5)–2005(11)	7	0.999
2006(4)–2006(12)	9	0.986
2007(5)–2007(11)	7	0.963
2008(5)–2008(7)	3	1.000
2009(4)–2009(11)	8	0.988
2010(5)–2010(11)	7	0.933
2011(4)–2011(12)	9	0.978
2012(2)–2012(4)	3	0.967
2012(6)–2012(10)	5	0.967
2013(5)–2013(10)	6	0.965
2014(5)–2015(1)	9	0.966
2015(5)–2016(11)	19	0.991
2017(4)–2017(10)	7	0.996
2018(5)–2018(9)	5	0.975
2019(5)–2019(10)	6	0.936

Total: 148 months (62.45%) with an average duration of 6.73 months

Table 2.6b MS-VAR4 Regime (1) classification based on smoothed probabilities

Regime 1	Months	Avg. prob
2000(11)–2001(3)	5	0.914
2001(9)–2001(10)	2	1.000
2001(12)–2002(4)	5	0.998
2002(12)–2002(12)	1	0.994
2003(2)–2003(2)	1	1.000
2003(12)–2004(1)	2	0.971
2004(12)–2005(4)	5	1.000
2005(12)–2006(3)	4	1.000
2007(1)–2007(4)	4	0.907
2007(12)–2008(4)	5	0.981
2008(8)–2009(3)	8	0.999
2009(12)–2010(4)	5	1.000
2010(12)–2011(3)	4	0.996
2012(1)–2012(1)	1	0.990
2012(5)–2012(5)	1	0.994
2012(11)–2013(4)	6	0.991
2013(11)–2014(4)	6	0.912
2015(2)–2015(4)	3	0.996
2016(12)–2017(3)	4	1.000
2017(11)–2018(4)	6	0.926
2018(10)–2019(4)	7	0.971
2019(11)–2020(2)	4	0.996

Total: 89 months (37.55%) with an average duration of 4.05 months

Table 2.7 Linearity tests, convergence, transition probabilities, and goodness-of-fit test statistics

Linearity test (Chi^2)	(80) = 6064.8 [0.0000]
Convergence	Strong by SQPF
Log-Likelihood	2570.15384
P(0|0)	0.846629
P(1|0)	0.15337
P(0|1)	0.24689
P(1|1)	0.753114
Number of observations	237

The outputs from (b) and (ii) might be underpinned by the fact that the higher the number of tourists visiting the US, the higher the demand for accommodation (hence, the higher the demand, e.g. for electricity and chemical products through the usages of electrical appliances, lighting, heating, air conditioning, room maintenance, and cleaning services, dry-cleaning, and laundry services, etc.) will be.

(iii) The 1st and 2nd lagged values of industrial production (DLNIP) lead the $DLNCO_2$ to increase at regime 0, while its 4th lagged value influences the $DLNCO_2$ negatively. The cumulative effect (from lag 1 to lag 4) of DLNIP on $DLNCO_2$ appears to be positive.

The results from (c) and (iii) might be interpreted as follows: The industrial production implies eventually the demand of business people for investment goods, intermediate goods, land, electricity and chemical products, and, demand for labour which in turn result in an increasing demand for food, transportation, accommodation, lighting, heating, air conditioning, and health-medical services, etc. All these factors are considered the elements to boost the Greenhouse gas (GHG), or shortly, carbon emissions.

(iv) The transportation sector's energy usage (DLNTRANS_EN) has a negative impact (with lag 1 at regime 1) and positive impacts (with lags 2, 3, and 4 at regime 0) on $DLNCO_2$. The cumulative impact (from lag 1 to lag 4) of DLNTRANS_EN on $DLNCO_2$ is positive.

The results from (d) and (iv) might underpin the fact that the transportation sector was expected often to increase the GHG because of its demand for fossil fuel energy. Today, there is a slight shift from fossil fuels to non-fossil fuels (renewables) for the engines of transportation sector vehicles. However, still, there exist the expectations that the transportations sector, primarily through the usage of aviation gasoline, will contribute to the carbon emissions significantly today and in the next decade(s).

In Fig. 2.2, regime 1 refers to high volatile periods as regime 0 indicates less volatile periods of the $DLNCO_2$ equation of MS-VAR4. Regime 1 is depicted by the grey area in the figure. Other areas, except the grey areas, show regime 0 periods. Regime 0 comprises 148 months (62.45%) with an average duration of 6.73 months as given in Table 2.6a. Regime 1 involves 89 months (37.55%) with an average duration of 4.05 months as depicted in Table 2.6b.

2.5 Summary and Conclusion

This book chapter aimed at analysing the movements of carbon emissions (CO_2) in the US, and hence, to understand the paths and dynamics of CO_2, estimated several nonlinear models of CO_2 equations in which the impacts of potential explanatory variables on carbon emissions are determined through Markov regime-switching VAR (MS-VAR) models. The explanatory variables are lagged values of CO_2, the number of tourists visiting the US, energy usage of the transportation sector, and industrial production in the US. All variables were translated into differenced logarithmic forms to avoid possible biased estimations and/or spurious regression problem. The seasonal effects were also employed in MS-VAR models.

Initially, three MS-VAR models were estimated to observe the results under regime-dependent constant terms with fixed variance, scaling variance, and switching

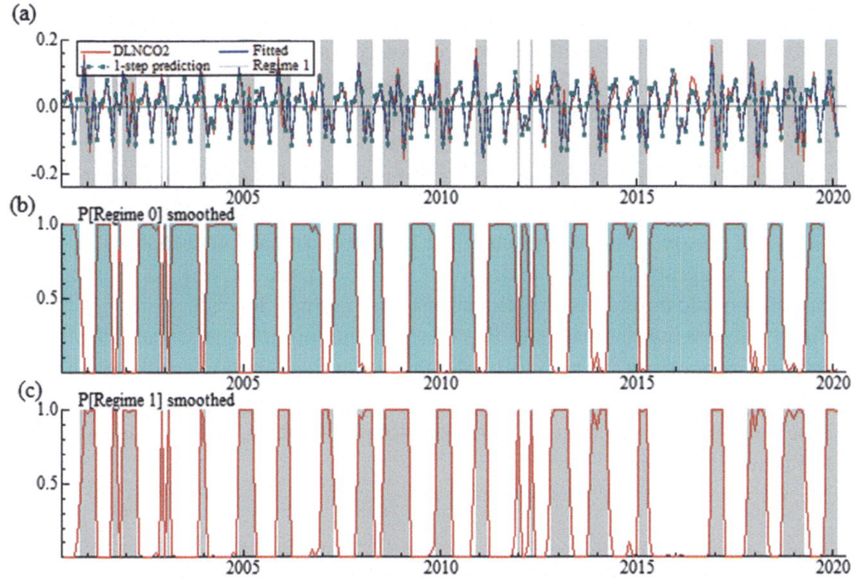

Fig. 2.2 The P [Regime 0] smoothed and the P [Regime 1] smoothed of MS-VAR4. (**a**) Actual, Fitted, Prediction, and Regime 1 area of DLNCO$_2$, (**b**) Probability of smoothed regime 0, and (**c**) Probability of smoothed regime 1

variance structures. Following goodness-of-fit statistics and convergence criteria, the MS-VAR3 model was chosen best among three to understand the impact of tourism (DLNTOURISTS) and other explanatory variables' influences on DLNCO$_2$. Later this work, following the MS-VAR3 model, launched a new model (MS-VAR4) in which constant term, as well as parameter estimations of explanatory variables, is regime-dependent. The constant terms at regimes 0 and 1, and all seasonal terms except seasonality at time t are found significant. The highlights of the findings are as follows:

- The DLNCO$_2$ in the US is negatively affected by lagged values of DLNCO$_2$,
- DLNTOURISTS has a positive impact on DLNCO$_2$,
- Industrial production (DLNIP) also has a positive influence on DLNCO$_2$, and
- Transportation energy usage (DLNTRANS_EN) has positive effects on DLNCO$_2$ in the US.

These results are noteworthy since.

- It is observed that levels of carbon dioxide emissions are well explained by not only other variables' dynamics but also the dynamics of itself. The lagged values of carbon dioxide emissions are found significant during both regime 1 and regime 0.
- The tourism sector leads to an increase in the level of carbon dioxide emissions. This output was observed clearly during less volatile periods (regime 0 periods) of the MS-VAR equation for carbon dioxide emissions.

- The positive influences of transportation energy usage and industrial production on carbon dioxide emissions also appeared mostly during regime 0 periods in the US.
- If we had not observed different regimes (states), we would not have obtained the positive contributions of tourism, transportation, and industrial production to carbon dioxide emissions at regime 0.
- Upon the outcomes of this work, one might suggest that policymakers of the US observe thoroughly the regime 0 and regime 1 time periods in the US to monitor better the reason/facts underpinning the high level of carbon emissions.
- One also suggests that administrators of the US expand EPACT 2005 to mitigate carbon dioxide emissions at both national and sectoral levels.
- Finally, following the results of this chapter, one might suggest that US officials consider supporting the new technologies to diminish carbon emissions through the considerable switch from fossils to renewable energy usage in transportation/aviation, accommodation, air-conditioning, heating, lighting, laundry, and telecommunication.

References

Akadiri, S. S., Alola, A. A., & Akadiri, A. C. (2019). The role of globalization, real income, tourism in environmental sustainability target Evidence from Turkey. *Science of the Total Environment, 687,* 423–432.

Al-Mulali, U., Fereidouni, H. G., & Mohammed, A. H. (2015). The effect of tourism arrival on CO_2 emissions from transportation sector. *Anatolia, 26,* 230–243.

Alam, M. S., & Paramati, S. R. (2017). The dynamic role of tourism investment on tourism development and CO_2 emissions. *Annals of Tourism Research, 66,* 213–215.

Alexandrakis, G., Manasakis, C., & Kampanis, N. A. (2015). Valuating the effects of beach erosion to tourism revenue a management perspective. *Ocean & Coastal Management, 111,* 1–11.

Aratuo, D. N., & Etienne, X. L. (2019). Industry-level analysis of tourism-economic growth in the United States. *Tourism Management, 70,* 333–340.

Bai, Y., Zhuang, C., Ouyang, Z., Zheng, H., & Jiang, B. (2011). Spatial characteristics between biodiversity and ecosystem services in a human-dominated watershed. *Ecological Complexity, 8,* 177–183.

Balli, E., Sigeze, C., Manga, M., Birdir, S., & Birdir, K. (2019). The relationship between tourism, CO_2 emissions and economic growth: A case of Mediterranean countries. *Asia Pacific Journal of Tourism Research, 24,* 219–232.

Balsalobre-Lorente, D., Driha, O., & Sinha, A. (2020a). The dynamic effects of globalization process in analysing N-shaped tourism led growth hypothesis. *Journal of Hospitality and Tourism Management, 43,* 42–52.

Balsalobre-Lorente, D., Driha, O., Shahbaz, M., & Sinha, A. (2020b). The effects of tourism and globalization over environmental degradation in developed countriesEnvironmental Science and Pollution Research*27*(7), 7130–7144.

Becken, S. (2002). Analysing international tourist flows to estimate energy use associated with air travel. *Journal of Sustainable Tourism, 10,* 114–131.

Becken, S., Simmons, D. G., & Frampton, C. (2003). Energy use associated with different travel choices. *Tourism Management, 24,* 267–277.

Ben Jebli, M., & Hadhri, W. (2018). The dynamic causal links between CO_2 emissions from transport, real GDP, energy use and international tourism. *International Journal of Sustainable Development & World Ecology, 25,* 568–577.

Bilgili, F., Kuşkaya, S., Ünlü, F., & Gençoğlu, P. (2019a). Does waste energy usage mitigate the CO_2 emissions? A time-frequency domain analysis.*Environmental Science and Pollution Research, 27,* 5056–5073.

Bilgili, F., Koçak, E., Bulut, Ü, & Kuşkaya, S. (2017). Can biomass energy be an efficient policy tool for sustainable development? *Renewable & Sustainable Energy Reviews, 71,* 830–845.

Bilgili, F., Öztürk, İ., Koçak, E., Bulut, Ü., Pamuk, Y., Muğaloğlu, E., & Bağlıtaş, H. (2016a). The influence of biomass energy consumption on CO_2 emissions: A wavelet coherence approach. *Environmental Science and Pollution Research,23*(19), 19043–19061.

Bilgili, F., Koçak, E., & Bulut, Ü. (2016b). The dynamic impact of renewable energy consumption on CO_2 emissions: A revisited environmental Kuznets curve approach. *Renewable & Sustainable Energy Reviews, 54,* 838–845.

Bilgili, F., Ulucak, R., Koçak, E., & İlkay, Ç. (2019b). Does globalization matter for environmental sustainability? Empirical investigation for Turkey by Markov regime switching models. *Environmental Science and Pollution Research.* https://doi.org/10.1007/s11356-019-069.

Bilgili, F., Tülüce, N. S., & Doğan, İ. (2012). The determinants of FDI in Turkey: A Markov regime-switching approach. *Economic Modelling, 29*(4), 1161–1169.

Camacho, M. (2013). Mixed-frequency VAR models with Markov-switching dynamics. *Economics Letters, 121*(3), 369–373.

Cerutti, A. K., Beccaro, G. L., Bruun, S., Donno, D., Bonvegna, L., & Bounous, G. (2016). Assessment methods for sustainable tourism declarations: The case of holiday farms. *Journal of Cleaner Production, 111,* 511–519.

Danish, & Wang, Z. (2018). Dynamic relationship between tourism, economic growth, and environmental quality. *Journal of Sustainable Tourism, 26,* 1928–1943.

Dogan, E., & Aslan, A. (2017). Exploring the relationship among CO_2 emissions, real GDP, energy consumption and tourism in the EU and candidate countries: Evidence from panel models robust to heterogeneity and cross-sectional dependence. *Renewable and Sustainable Energy Reviews.* https://doi.org/10.1016/j.rser.2017.03.111

Dogan, E., Seker, F., & Bulbul, S. (2017). Investigating the impacts of energy consumption, real GDP, tourism and trade on CO_2 emissions by accounting for cross-sectional dependence: A panel study of OECD countries. *Current Issues in Tourism, 20,* 1701–1719.

Doğan, İ, & Bilgili, F. (2014). The non-linear impact of high and growing government external debt on economic growth: A Markov Regime-switching approach. *Economic Modelling, 39,* 213–220.

Doornik, J. A. (2019). Econometric Analysis with Markov-Switching Models, PcGiveTM 14 OxMetrics 7.

Droumaguet, M., Warne, A., & Woźniak, T. (2015). Granger causality and regime inference in Bayesian Markov-Switching VARs. ECB Working Paper (No. 1794).

Dwyer, L., Forsyth, P., Spurr, R., & Hoque, S. (2010). Estimating the carbon footprint of Australian tourism. *Journal of Sustainable Tourism, 18,* 355–376.

Enders, W. (2009). *Applied econometrics time series.* Wiley, United States. ISBN: 978-1-118-80856-6.

Energy Information Administration. Retrieved May 2020, from https://www.eia.gov/.

Federal Reserve Economic Data. Retrieved May 2020, from https://fred.stlouisfed.org.

Filimonau, V., Dickinson, J., & Robbins, D. (2014). The carbon impact of short-haul tourism: A case study of UK travel to Southern France using life cycle analysis. *Journal of Cleaner Production, 64,* 628–638.

Galli, A., Wackernagel, M., Iha, K., & Lazarus, E. (2014). Ecological footprint: Implications for biodiversity. *Biological Conservation, 173,* 121–132.

Gil-Alana, L. A., dos Santos Figueiredo, O. H., & Wanke, P. (2019). Structural breaks in Brazilian tourism revenues: Unveiling the impact of exchange rates and sports mega-events. *Tourism Management, 74,* 207–211.

Gössling, S., Scott, D., & Hall, C. M. (2015). Inter-market variability in CO_2 emission-intensities in tourism: Implications for destination marketing and carbon management. *Tourism Management, 46,* 203–212.
Gu, X., Wu, J., Guo, H., & Li, G. (2018). Local tourism cycle and external business cycle. *Annals of Tourism Research, 73,* 159–170.
Hamilton, J. D. (1989). A new approach to the economic analysis of nonstationary time series and the business cycle. *Econometrica: Journal of the Econometric Society*, 357–384.
Holst, U., Lindgren, G., Holst, J., & Thuvesholmen, M. (1999). Recursive estimation in switching autoregressions with a Markov regime. *Journal of Time Series Analysis, 15*(5), 489–506.
Işik, C., Kasımatı, E., & Ongan, S. (2017). Analyzing the causalities between economic growth, financial development, international trade, tourism expenditure and/on the CO 2 emissions in Greece. *Energy Sources, Part B: Economics, Planning, Policy, 12,* 665–673.
Jamal, T., Taillon, J., & Dredge, D. (2011). Sustainable tourism pedagogy and academic-community collaboration: A progressive service-learning approach. *Tourism and Hospitality Research, 11,* 133–147.
Jin, C., Cheng, J., Xu, J., & Huang, Z. (2018). Self-driving tourism induced carbon emission flows and its determinants in well-developed regions: A case study of Jiangsu Province China. *Journal of Cleaner Production, 186,* 191–202.
Katircioğlu, S. T. (2014). Testing the tourism-induced EKC hypothesis: The case of Singapore. *Economic Modelling, 41,* 383–391.
Katircioglu, S. T., Feridun, M., & Kilinc, C. (2014). Estimating tourism-induced energy consumption and CO_2 emissions: The case of Cyprus. *Renewable and Sustainable Energy Reviews.* https://doi.org/10.1016/j.rser.2013.09.004.
Kim, C. J. (1994). Dynamic linear models with Markov-switching. *Journal of Econometrics, 60*(1–2), 1–22.
Kindu, M., Schneider, T., Teketay, D., & Knoke, T. (2016). Changes of ecosystem service values in response to land use/land cover dynamics in Munessa-Shashemene landscape of the Ethiopian highlands. *Science of the Total Environment, 547,* 137–147.
Koçak, E., Ulucak, R., & Ulucak, Z. Ş. (2020). The impact of tourism developments on CO_2 emissions: An advanced panel data estimation. *Tourism Management Perspectives, 33,* 100611.
Köse, N., Uçar, N. (2003). Durağan Vektör Otoregressif Modellerde Gecikme Uzunluğunu Belirleyen Kriterlerin Karşılaştırılması. VI. Ulusal Ekonometri ve İstatistik Sempozyumu, Gazi Üniversitesi, Ankara
Krolzig, H.M. (1997). *Markov-Switching vector autoregressions, modelling, statistical inference, and application to business cycle analysis.* Springer.
Krolzig, H. M. (1998). Econometric modelling of Markov-switching vector autoregressions using MSVAR for Ox. Retrieved March 29, 2020, from https://www.researchgate.net/publication/246006071.
Kuan, C. M. (2002). Lecture on the Markov switching model. Institute of Economics Academia Sinica, 1–30. Retrieved March 24, 2020, from https://homepage.ntu.edu.tw/~ckuan/pdf/Lec-Markov_note.pdf.
Kuşkaya, S., & Bilgili, F. (2020). The wind energy-greenhouse gas nexus: The wavelet-partial wavelet coherence model approach. *Journal of Cleaner Production, 245,* 118872.
Lee, J. W., & Brahmasrene, T. (2013). Investigating the influence of tourism on economic growth and carbon emissions: Evidence from panel analysis of the European Union. *Tourism Management, 38,* 69–76.
Lenzen, M., Sun, Y. Y., Faturay, F., Ting, Y. P., Geschke, A., & Malik, A. (2018). The carbon footprint of global tourism. *Nature Climate Change, 8,* 522–528.
León, C. J., Arana, J. E., & Hernández Alemán, A., (2014a). CO_2 emissions and tourism in developed and less developed countries. *Applied Economics Letters, 21,* 1169–1173.
León, C. J., Arana, J. E., & Hernández Alemán, A. (2014b). CO_2 emissions and tourism in developed and less developed countries. *Applied Economics Letters21,* 1169–1173.

Li, J., Bai, Y., & Alatalo, J. M. (2020). Impacts of rural tourism-driven land-use change on ecosystems services provision in Erhai Lake Basin China. *Ecosystem Services, 42,* 101081.

Lim, J., & Won, D. H. (2020). How Las Vegas' tourism could survive an economic crisis? *Cities, 100,* 102643.

Melián-González, S., & Bulchand-Gidumal, J. (2020). Employment in tourism: The jaws of the snake in the hotel industry. *Tourism Management, 80,* 104123.

Munday, M., Turner, K., & Jones, C. (2013). Accounting for the carbon associated with regional tourism consumption. *Tourism Management, 36,* 35–44.

National Travel and Tourism Office. Retrieved May 2020, from https://travel.trade.gov/view/m-2017-I-001/index.asp.

Nepal, R., & Indra al Irsyad, M., & Nepal, S.K. . (2019). Tourist arrivals, energy consumption and pollutant emissions in a developing economy–implications for sustainable tourism. *Tourism Management, 72,* 145–154.

Ozturk, I., Al-Mulali, U., & Saboori, B. (2016). Investigating the environmental Kuznets curve hypothesis: The role of tourism and ecological footprint. *Environmental Science and Pollution Research, 23,* 1916–1928.

Paiano, A., Crovella, T., & Lagioia, G. (2020). Managing sustainable practices in cruise tourism: The assessment of carbon footprint and waste of water and beverage packaging. *Tourism Management, 77,* 104016.

Paramati, S. R., Alam, M. S., & Chen, C. F. (2017). The effects of tourism on economic growth and CO_2 emissions: A comparison between developed and developing economies. *Journal of Travel Research, 56,* 712–724.

Qureshi, M. I., Hassan, M. A., Hishan, S. S., Rasli, A. M., & Zaman, K. (2017). Dynamic linkages between sustainable tourism, energy, health and wealth: Evidence from top 80 international tourist destination cities in 37 countries. *Journal of Cleaner Production, 158,* 143–155.

Russo, M. A., Relvas, H., Gama, C., Lopes, M., Borrego, C., Rodrigues, V., et al. (2020). Estimating emissions from tourism activities. *Atmospheric Environment, 220,* 117048.

Saenz-de-Miera, O., & Rosselló, J. (2014). Modeling tourism impacts on air pollution: The case study of PM10 in Mallorca. *Tourism Management, 40,* 273–281.

Sajjad, F., Noreen, U., & Zaman, K. (2014). Climate change and air pollution jointly creating nightmare for tourism industry. *Environmental Science and Pollution Research, 21,* 12403–12418.

Shakouri, B., Khoshnevis Yazdi, S., & Ghorchebigi, E. (2017). Does tourism development promote CO_2 emissions? *Anatolia, 28,* 444–452.

Sharif, A., Mishra, S., Sinha, A., Jiao, Z., Shahbaz, M., & Afshan, S. (2020). The renewable energy consumption-environmental degradation nexus in Top-10 polluted countries: Fresh insights from quantile-on-quantile regression approach. *Renewable Energy, 150,* 670–690.

Sharif, A., Afshan, S., & Nisha, N. (2017). Impact of tourism on CO_2 emission: Evidence from Pakistan. *Asia Pacific Journal of Tourism Research, 22,* 408–421.

Sharp, H., Grundius, J., & Heinonen, J. (2016). Carbon footprint of inbound tourism to Iceland: A consumption-based life-cycle assessment including direct and indirect emissions. *Sustainability, 8.*

Shumway, R. H., & Stoffer, D. S. (1991). Dynamic linear models with switching. *Journal of the American Statistical Association, 86*(415), 763–769.

Solarin, S. A. (2014). Tourist arrivals and macroeconomic determinants of CO_2 emissions in Malaysia. *Anatolia, 25,* 228–241.

Şahin, U. (2019). Forecasting of Turkey's greenhouse gas emissions using linear and nonlinear rolling metabolic grey model based on optimization. *Journal of Cleaner Production, 239,* 118079.

Tao, Y. G., & Huang, Z. F. (2014). Review of accounting for carbon dioxide emissions from tourism at different spatial scales. *Acta Ecologica Sinica, 34,* 246–254.

Timmermann, A. (2000). Moments of Markov switching models. *Journal of Econometrics, 96*(1), 75–111.

Tsai, K. T., Lin, T. P., Hwang, R. L., & Huang, Y. J. (2014). Carbon dioxide emissions generated by energy consumption of hotels and homestay facilities in Taiwan. *Tourism Management, 42,* 13–21.

Ulucak, R., & Bilgili, F. (2018). A reinvestigation of EKC model by ecological footprint measurement for high, middle and low income countries. *Journal of Cleaner Production, 188,* 144–157. https://doi.org/10.1016/J.JCLEPRO.2018.03.191

UNWTO. (2019). Home | UNWTO [WWW Document]. Retrieved December 18, 2019, from https://www.unwto.org/.

Verbeek, D., & Mommaas, H. (2008). Transitions to sustainable tourism mobility: The social practices approach. *Journal of Sustainable Tourism, 16,* 629–644. https://doi.org/10.1080/09669580802159669

World Tourism Organization. (2013). World Tourism Organization.

Zaman, K., Moemen, M. A., & Islam, T. (2017). Dynamic linkages between tourism transportation expenditures, carbon dioxide emission, energy consumption and growth factors: Evidence from the transition economies. *Current Issues in Tourism, 20,* 1720–1735. https://doi.org/10.1080/13683500.2015.1135107

Zafar, M. W., Shahbaz, M., Sinha, A., Sengupta, T., & Qin Q. (2020). How renewable energy consumption contribute to environmental quality? The role of education in OECD countries. *Journal of Cleaner Production, 268,* 122149. https://doi.org/10.1016/j.jclepro.2020.122149.

Zaman, K., Shahbaz, M., Loganathan, N., & Raza, S. A. (2016). Tourism development, energy consumption and environmental Kuznets curve: Trivariate analysis in the panel of developed and developing countries. *Tourism Management, 54,* 275–283. https://doi.org/10.1016/j.tourman.2015.12.001

Zhang, L., & Gao, J. (2016). Exploring the effects of international tourism on China's economic growth, energy consumption and environmental pollution: Evidence from a regional panel analysis. *Renewable and Sustainable Energy Reviews.* https://doi.org/10.1016/j.rser.2015.08.040.

Chapter 3
Tourism Sector and Environmental Quality: Evidence from Top 20 Tourist Destinations

Burcu Ozcan, Seref Bozoklu, and Danish Khan

Abstract This chapter aims to analyze the effect of the tourism sector on environmental quality for France, Spain, the United States, China, Italy, Turkey, Mexico, Germany, Thailand, the United Kingdom, Japan, Austria, Greece, Honk-Kong, Malaysia, Russian Federation, Portugal, Canada, Poland and the Netherlands for 1995–2018. The empirical results indicate that the development of the tourism sector reduces both CO_2 emission and ecological footprint and the environmental Kuznets curve hypothesis is confirmed for CO_2 emission but not for ecological footprint. The findings also reveal that energy consumption degrades the environment in both forms of air pollution and social pressure, and financial development alleviates CO_2 emission; however, it does not affect ecological footprint. All these results suggest that the tourism sector should be supported by the collaboration of the private sector and government.

B. Ozcan (✉)
Faculty of Economics and Administrative Sciences, Department of Economics, Firat University, Elazig 23119, Turkey
e-mail: bozcan@firat.edu.tr

S. Bozoklu
Faculty of Economics, Department of Economics, Istanbul University, Istanbul 34452, Turkey
e-mail: sbozoklu@istanbul.edu.tr

D. Khan
School of Economics and Trade, Gaungdong University of Foreign Studies, Gaungzhou 510006, China
e-mail: khan.danishkhan@hotmail.com

D. Balsalobre-Lorente et al. (eds.), *Strategies in Sustainable Tourism, Economic Growth and Clean Energy*, https://doi.org/10.1007/978-3-030-59675-0_3

3.1 Introduction

In this chapter, we explore the impacts of tourism sector on the natural environment[1] and consider the growing awareness of the need to become more sustainable. The type of relationship between tourism and the environment can be defined as two-sided impact and dependency. Pintassilgo and Silva (2007) states that the reason for this dependency is that the tourism sector uses the natural environment as an input in its production function.

Globalization, increasing population movement and developments in transportation and communication have made the tourism sector[2] as one of the key and largest sector for both developing and developed countries over the last few decades (Bramwell and Lane 1993; Choi and Sirakaya 2006; Paramati et al. 2017). The tourism sector has made significant contributions to the national economies through its ability to create income and jobs and therefore called 'Smokeless Industry' (Mathieson and Wall 1982). It also has significant indirect effects by contributing to the balance of payments, improving living standards of domestics, increasing production of goods and services, and rising the tax revenue for the government. The tourism sector also contributes to sustainable economic development if it operates in natural capacities for the renewal and future efficiency of natural resources and is an essential factor for poverty reduction (Dwyer and Forsyth 2008; Kim et al. 2006; Tang and Abosedra 2014; Santana-Gallego et al. 2011; Fereidouni and Al-Mulali 2014). The economic policymakers support the tourism sector by attracting international tourists to tourism destinations, making marketing campaigns and introducing the presence of country at many international tourism exhibitions, fairs and forums. In this regard, Balaguer and Cantavella-Jorda (2002) proposes the Tourism-Led Growth Hypothesis (TLGH hereafter) that argues there may be a causality relationship between tourism development and economic growth and tourism may stimulate economic activities.

Although the expansion of the tourism sector with increase in the number of tourists has positive impacts on the economy, it may create negative externalities on environmental quality. The economic activities have long been associated with environmental degradation due to energy production and consumption; however, the tourism sector is recently considered as a potential cause of environmental problems, and scientific researches have emerged recent decades (Neto 2003; Casler and Blair 1997; Peeters et al. 2007; Holden 2009; Perch-Nielsen et al. 2010; Dubois et al. 2011; Gössling 2013; Tsai et al. 2014). The environmental quality is, therefore, one of the

[1]Natural environment has 'common pool resource' characteristics which means the exploitation by one user reduces resource availability for others (subtractability) and exclusion of additional users is especially difficult and costly (difficulty of exclusion) (Ostrom and Field 1999). Hardin's (1968) seminal paper argues that the users of these resources are caught in a process that leads to the destruction of resources upon which they depend. The author entitled this as "the tragedy of the commons".

[2]The origins of tourism officially date back to the late sixteenth century with the beginning of the "Grand Tour" which was a trip taken by young affluent males in order to broaden their knowledge of the arts and sciences in cities of culture and history such as London, Paris, Venice, and Rome (Sorabella 2003; Towner 1985).

critical topics concerning tourism nowadays, and there is an increasing awareness of the economic and environmental significance of tourism.

The relationship between the tourism sector and environmental quality is a complex issue and involves many activities that have both detrimental and beneficial impacts on it. The detrimental impacts of tourism on the environment are related to the construction of infrastructures, such as roads and airports, and tourism facilities, such as including resorts, hotels, restaurants, shops, golf courses, and marinas. The activities essential for tourism such as accommodation, catering, entertainment, transportation cause energy consumption and greenhouse gas emission to rise, and therefore harm the natural environment (Becken, 2002; Başarir and Çakir, 2015). As a result, the tourism sector is partly responsible for global warming. The negative externalities of tourism on the environment have led to the creation of the concept of sustainable tourism[3] (and its other forms like eco-tourism and green tourism[4]) and the development of some criteria for the measurement of it. On the other hand, tourism may provide beneficial impacts on the environment by contributing to environmental protection and conservation through raising environmental awareness and providing finance for the protection of natural areas. In addition, a growing number of tourists may have worries about the environment at tourist destinations (Hsieh et al. 2017). For those reasons, understanding the relationship between economic growth, tourism, and environmental quality becomes essential for an efficient and sustainable tourism sector.

The adverse effects of tourism arise since resource usage by tourists is higher than the capability of the environment to balance it at an adequate level. The unrestrained tourism has potential threats on the environment such as soil erosion, increased pollution, discharges into the sea, natural habitat loss, increased pressure on endangered species and heightened vulnerability to forest fires. The expansion of the tourism sector causes pollution such as air emissions, noise, solid waste, or even architectural/visual pollution. It is argued in recent decades that the tourism sector is one of the important contributors to CO_2 emission and therefore climate change, which attracts scholars and policymakers to evaluate and examine their results (Rosselló-Batle et al. 2010; Saenz-de-Miera and Rosselló 2014; Zaman et al. 2016). As a result, global warming and greenhouse gas emissions have become a dominant debate in the literature and the regions experiencing a lot of tourism activities make this debate an interesting case study for the link between tourism and air pollution.

Following the pioneering work of Kuznets (1955), the relationship between economic growth and pollution has been extensively investigated in the context of the environmental Kuznets curve (EKC hereafter) hypothesis literature (Grossman and Krueger 1991; Dinda 2004). The EKC presents an inverse U-shaped relationship and argues that per capita income causes environmental damages during the early stages

[3]Sustainable Tourism can be defined as "tourism that takes full account of its current and future economic, social and environmental impacts, addressing the needs of visitors, the industry, the environment and host communities" (UNEP and UNWTO 2005).

[4]Unlike sustainable tourism, eco-tourism or green tourism focus specifically on the environmental sustainability of tourism within a destination landscape rather than incorporating many aspects of tourism such as social and economic sustainability (Nelson 2013).

of economic growth. As the level of economic development reaches into a threshold point, it is expected that environmental degradation decreases as the per capita income increases. Acknowledging the EKC hypothesis, Magnani (2001) further argues that not only per capita income but also sectoral economic effects may be the driver behind the EKC. In this context, the tourism sector has attracted considerable attention, and it has been accepted as the engine of 'the tourism-induced EKC hypothesis' through the TLGH. The TLGH supports concerns about the impacts of the tourism sector on the environment since it is mainly based on infrastructure facilities such as roads, harbors, airports, and hotels, and inevitably requires exploitation of the natural resources. Furthermore, tourism activities are associated with higher energy demand and therefore increase environmental degradation. As a result, deterioration of the natural resources due to the tourism sector may create negative impacts on environment such as pollution, soil erosion, and increased emissions of CO_2 (Katircioğlu 2014; Becken et al. 2001; Gössling 2002; Berrittella et al. 2006). Consequently, similar to other industries, the tourism sector is a part of environmental degradation and causes problems for environmental sustainability.

3.2 Literature Review

The section analyzes studies in the related literature and divides them into two kinds of classification as time series and panel data studies. Among panel data studies, Lee and Brahmasrene (2013) examined the effect of tourism on economic growth and CO_2 emissions in the European Union (EU). The results found evidence that tourism contributes to CO_2 emission reduction. Dogan and Aslan (2017) explored the linkage between energy consumption, real GDP, tourism, and CO_2 emissions in EU countries. The findings indicated that tourism helps in CO_2 emission mitigation. In another study for EU countries, Paramati et al. (2018) explored the influence of tourism investment on tourism development and CO_2 emissions. The long-run elasticity result confirms that tourism investment has a positive and significant impact on tourism development and helps in CO_2 emission reduction in EU countries.

Further, in a comparative study of Eastern and Western European Union, Paramati et al. (2017) explored the effect of tourism and economic growth on environmental quality. The long-run elasticity results recommend that development in tourism promotes economic growth in both Eastern and Western European Union countries. However, the tourism increases CO_2 emissions in the Eastern EU but decreases in the Western EU. Zaman et al. (2016) investigated the role of tourism development and energy consumption in the EKC for developing and developed countries substantiated tourism-induced CO_2 emissions. Exploring the role of international tourism on economic growth, energy consumption, and CO_2 emissions in China et al. (2016) used a multivariate framework for the data spanning from 1995 to 2011. The findings of the study suggested that the tourism-induced EKC hypothesis does not hold in central China and is weakly supported for eastern and western parts of China. Danish and Wang (2018) considered the role of tourism in CO_2 emission during the role of globalization for BRICS countries. The panel data estimator found that tourism promotes economic growth but degrades environmental quality. However, globalization has an insignificant impact on CO_2 emissions. Akadiri et al. (2020) examined the causal relationship between economic growth, globalization, tourism, and CO_2 emission in the tourism island territories. The findings of the study confirm the demand flowing and supply leading hypotheses, and internal factors mainly contribute to pollution. In a study for the top 10 tourist destinations, Katircioglu et al. (2018) examined the impact of tourism development on ecological footprint. With the help of panel data estimation tools, it is concluded that tourism development reduces environmental degradation. In another study for the top 10 tourism induced countries, Shaheen et al. (2019) investigated the dynamic linkages between energy, economic growth, tourism, and environment. The empirical results confirmed not only the EKC and pollution heaven hypotheses but established a feedback hypothesis between tourism and environmental pollution. Koçak et al. (2020) studied the impact of tourism on CO_2 emissions in most visited countries using panel data techniques. The findings reveal that a number of tourist arrivals have a detrimental effect on the environment through increases in CO_2 emissions. Recently, Khan et al. (2020), conducted a study on the role of natural resource and tourism development in the energy-growth-CO_2 emission nexus in Belt and Road Initiative countries and the results validated the tourism push emission hypothesis.

Besides panel data studies, several studies conducted a single country analysis. For instance, Saenz-de-Miera and Rosselló (2014) modeled the impact of tourism on air pollution for Mallorca. The results show that the daily stock of tourists is a significant predictor of air pollution. Katircioglu et al. (2014) estimated the tourism-induced energy consumption and CO_2 emissions hypotheses for Cyprus. Results confirm that tourism harmed the environment. In another study for Turkey, Katircioglu (2014) analyzed the nexus between international tourism, energy consumption and CO_2 emissions for Cyprus. Results found that tourism development in Turkey not only increases energy consumption but considerably increases environmental pollution as well.

Azam et al. (2018) studied the impact of tourism on CO_2 emissions for Malaysia, Singapore, and Thailand. Results reveal a positive and significant impact of tourism on environmental pollution in Malaysia, whereas a reverse relationship is found for Singapore and Thailand. Bella and Bella (2018) tested for the tourism induced EKC in France utilizing the error correction model. The outcomes corroborate the existence of a long-run diminishing relationship between CO_2 emissions and economic growth driven by tourism. Conducting trivariate analysis for Tunisia, Egypt, and Morocco, Sghaier et al. (2018) investigated the relationship between energy consumption, tourism growth, and CO_2 emissions. The results reveal that tourism development has a negative impact on CO_2 emission in Egypt, although a positive influence in Tunisia and a neutral effect in Morocco. Further, Indra and Kumar (2019) explored the relationship between tourist arrivals, energy consumption, and CO_2 emissions in Nepal. Empirical results provide evidence that both energy consumption and tourism contribute to environmental pollution. Saint et al. (2019) considered the role of real income and tourism in environmental pollution, regarding globalization for Turkey.

With the help of the autoregressive distributive lag method, the empirical results indicate that increases in the number of international tourist arrivals lead to increases in CO_2 emissions in Turkey. A brief summary of related literature is presented in Table 3.1.

Table 3.1 A Brief Summary of Literature

Reference	Country	Time	Method used	Tourism-induced pollution
Lee and Brahmasrene (2013)	EU countries	1988–2009	Fixed effect model	No
Dogan and Aslan (2017)	EU countries	1995–2011	FMOLS and DOLS	No
Paramati et al. (2018)	EU countries	1990–2013	Panel ARDL	No
Paramati et al. (2017)	Western and eastern EU countries	1991–2013	FMOLS	No
Danish and Wang (2018)	BRICS countries	1995–2014	DSUR	No
Akadiri et al. (2020)	Tourism island territories	1995–2014	SUR	Confirm demand-flowing and supply-leading hypotheses
Katircioglu et al. (2018)	Top 10 tourist destinations	1995–2014	Panel random effect	No
Shaheen et al. (2019)	Top 10 tourist destinations	1995–2016	FMOLS	Feedback hypothesis
Koçak et al. (2020)	Most visited countries	1995–2014	CUP-FM & CUP-BC models	Yes
Khan et al. (2020)	BRI countries	1990–2016	GMM method	Yes
Saenz-de-Miera and Rosselló (2014)	Mallorca	2003–2007	Semi-parametric GAM estimations	Yes
Katircioglu et al. (2014)	Cyprus	1990–2009	ARDL	Insignificant relationship
Katircioglu (2014)	Turkey	1960–2010	ARDL	Yes
Zhang and Gao (2016)	China	1995–2011	FMOLS	No in all region
Zaman et al. (2016)	Developing and developed countries	2005–2013	Panel 2SLS regression method	Yes

(continued)

Table 3.1 (continued)

Reference	Country	Time	Method used	Tourism-induced pollution
Azam et al. (2018)	Malaysia, Singapore, and Thailand	1990–2014	FMOLS	Positive relationship is found for Malaysia whereas a reverse relationship is found for Singapore and Thailand
Bella and Bella (2018)	France	1995–2014	ARDL	No
Sghaier et al. (2018)	Tunisia, Egypt and Morocco	1980–2014	ARDL	Negative impact on CO_2 emission in Egypt although a positive influence in Tunisia and a neutral effect in Morocco
Indra and Kumar (2019)	Nepal	1975–2014	ARDL	Yes
Saint et al. (2019)	Turkey	1970–2014	ARDL	Yes

Note FMOLS = Fully modified ordinary least squares; DOLS = dynamic ordinary least squares; ARDL = Auto-regressive distributive lag model; DSUR = Dynamic Seemingly Unrelated Cointegrating Regressions; CUP-FM = continuously updated fully modified; CUP-BC = continuously updated bias-corrected; GMM = Generalized method of moments model; 2SLS = Two stage least squares

3.3 Data and Model

3.3.1 Data

Our sample consists of the world's top 20 tourist attractive countries based on the 2019 ranking of the World Tourism Organization[5]. These are France, Spain, the United States, China, Italy, Turkey, Mexico, Germany, Thailand, the United Kingdom, Japan, Austria, Greece, Honk-Kong, Malaysia, Russian Federation, Portugal, Canada, Poland, and the Netherlands.[6] As proxies for environmental degradation, we utilize ecological footprint (EF hereafter) and CO_2 emissions (in terms of per capita) to assess the effect of tourism development on the environment. EF was first introduced by Rees (1992) and further developed by Wackernagel and Rees (1996). Wackernagel et al. (1999, 2002) define EF as the area of biologically productive land and water necessary to produce the resources consumed and to absorb the wastes generated by humanity, based on the predominant management and production practices in a

[5]https://www.e-unwto.org/toc/unwtotfb/current.

[6]We had to exclude Macau SAR, China from the sample because of data unavailiabity.

given year. EF is a consumption-based indicator because it includes all-natural capital directly or indirectly used for the production of the goods and services consumed by the local population, regardless of the location of the supplying area (Bagliani et al. 2008). Total EF is the sum of the footprint of the six types of productive areas, including cropland, grazing land, forest, fishing ground, built-up land, and the land area required to absorb CO_2 emissions from the use of fossil fuels (York et al. 2003). The higher values of EF a country has the environmental pressure comes from its residents. The national EF per capita data is obtained from the Global Footprint Network[7] as a global hectare. CO_2 emissions (metric tons per capita) from fossil fuels are from the Emission Database for Global Atmospheric Research[8]. Additionally, there exist two different sample periods: The first one is from 1995 to 2018 (in case of CO_2 emissions) and the second one is from 1995 to 2016 (in case of EF) based on the unavailability of some data.

Following the previous studies in the literature (Katircioglu 2009a, b, c; Katircioglu et al. 2018; Katircioğlu and Taşpinar 2017; Lee and Brahmasrene 2013), three most used indicators, the number of international tourist arrivals, the international tourism expenditures (constant 2010 US$) and the international tourism receipts (constant 2010 US$) and a composite index, are employed as proxies for tourism development. Tourism data are obtained from the Tourism Statistics Database of the United Nations World Tourism Organization[9] . The current values of tourism expenditures and receipts are converted to constant values with the US consume price index (2010 = 100). Furthermore, based on the natural logarithm values of tourist arrivals, tourism expenditures, and receipts, the principal component analysis was employed to obtain a composite tourism index. Four different models were estimated by using tourism variables, separately.

Regarding the other variables of the model, similar to previous studies, gross domestic product (GDP) per capita (constant 2010, US$ dollar) from the World Development Indicators Database[10] of World Bank and the primary energy consumption (measured Gigajoule per capita) are from the British Petroleum's Statistical Review of World Energy,[11] are included into model as essential determinants of CO_2 emissions. In doing so, the presence of the EKC, representing an inverted U-shaped relationship between income and environmental pollution, was analyzed (see Grossman and Krueger 1991). Lastly, as a control variable, financial development, represented by the domestic credit to the private sector (as a percentage of GDP) from the Global Financial Development Database[12] of World Bank, was added to the model. Financial development may have an impact on the environment because

[7]https://www.footprintnetwork.org/.

[8]https://edgar.jrc.ec.europa.eu/.

[9]https://www.e-unwto.org/toc/unwtotfb/current.

[10]https://databank.worldbank.org/source/world-development-indicators.

[11]https://www.bp.com/en/global/corporate/energy-economics/statistical-review-of-world-energy.html.

[12]https://datacatalog.worldbank.org/dataset/global-financial-development.

the listed enterprises may adopt new technologies and support technological innovations to increase energy efficiency and lower CO_2 emissions level (Tamazian et al. 2009). Additionally, financial development can ease financing at lower costs investment in the environmental project (Zhang 2011). Due to the well-developed financial systems, clean and environmentally friendly technologies might be adopted into the production processes. However, financial development may also have adverse effects on environmental quality by rising energy demand. As such, consumers can easily borrow money from well-developed financial systems, and thereby they can buy big durables such as automobiles, refrigerators, air conditioners, and washing machines that consume much energy, which in turn increases national energy demand and air pollution level (Sadorsky 2012). The net effect of financial development on the environment, therefore, is not known, a priori.

3.3.2 *Model*

Based on the studies in the related literature (see, among others, Katircioglu 2009a, b, c; Katircioglu et al. 2018; Lee and Brahmasrene 2013), the following model was defined:

$$lnEnv_{it}=\beta_{0i}+\beta_{1i}lnGDP_{it}+\beta_{2i}lnGDP_{it}^{2}+\beta lnEC_{it}+\beta_{4i}lnT_{it}+\beta_{5i}lnF_{it}+\varepsilon_{it} \tag{3.1}$$

where $i = 1, 2, \ldots\ldots\ldots, 20$ indicates individual units of panel and $t = 1995, \ldots\ldots\ldots\ldots\ldots 2018$ denotes time. Env_{it} represents CO_2 emissions per capita or EF per capita referring to environmental degradation, GPD_{it} and GPD_{it}^2 are GDP per capita and its square, respectively. EC_{it}, T_{it}, and F_{it} refer to energy consumption per capita, tourism variables, and financial development, respectively. All variables are in their natural logarithms. Thus, the coefficients of β_{1i}, β_{2i}, β_{3i}, β_{4i} and β_{5i} are the long-run elasticities for EF and CO_2 emissions for the related variables. β_{0i} is a constant term and ε_{it} is idiosyncratic errors. The signs of β_{4i} and β_{5i} could positive or negative. However, the signs of β_{1i}, β_{2i}, and β_{3i} are, respectively, expected to be positive, negative, and positive. In the framework of the EKC, an inverted U-shaped relationship is assumed between environmental degradation and economic growth. Besides, the more energy a country demands, the more air pollution it will have, e.g., β_{3i} is theoretically expected to be positive. However, as mentioned in previous lines, the environmental effects of tourism and financial development are not certain. Before proceeding to empirical analysis, some essential data statistics are tabulated in Table 3.2.

As seen in Table 3.2, the mean per capita GDP is 27,262 US dollars. Among those 20 countries, the Netherlands (55,022 US dollars) has the highest per capita GDP, while China has the lowest (1,224 US dollar). The mean of per capita CO_2 emissions is 8.32 Mt; the highest level of CO_2 emissions per capita belongs to the US (20.76 Mt in 2000) while the lowest belongs to China (2.65 Mt in 1996). Concerning

Table 3.2 Some statistical features of data

Variables Statistics	PerGDP (US $)	CO_2 (mto)	Energy (gj)	Finance (% of GDP)	T.arrivals (thousand)	T. expenditures (million)	T. receipts (million)
Mean	27,62	8.32	150.52	104.5	27,751	30,200	31,900
Median	30,100	7.87	132.4	104.19	21,600	17,100	18,800
Maximum	55,022	20.76	409.22	233.21	89,322	241,000	229,000
Minimum	1,224	2.65	30.08	8.33	3,345	1,300	4,330
Std. Dev.	16,045	3.97	84.31	51.31	19,450	36,600	35,300
Skewness	−0.1	1.13	1.37	0.04	1.27	2.55	3.26
Kurtosis	1.54	4.13	4.85	2.4	3.9	11.49	15.28
Jarque-Bera	43.6	127.29	219.26	7.46	145.31	1963.01	3870.11
Probability	0	0	0	0.02	0	0	0

Notes PerGDP, T.arrivasl, T.expenditures, T.revenues denote GDP per capita, tourist arrivals, tourism expenditures, and tourism revenues, respectively

the primary energy consumption per capita, its mean is about 150 gigajoules (gj). The highest level of energy consumption per capita was seen in Canada with 409 gj in 2000 while its lowest level was realized in China with 30 gj in 1995. The mean of financial development is about 104% of GDP; Hong-Kong has the highest percentage (233% in 2014) while Russia has the lowest (8% in 1996). Regarding tourism variables, the mean of tourist arrivals is 27,751 thousand; France was the most tourist attractive country (about 89 million in 2018) while Japan was the least attractive country (3.3 million in 1996). Tourism expenditures range between 1,300 million dollars (Turkey in 1995) and 241,000 million dollars (China in 2018), and its meaning is about 30,200 million dollars. Countries with the highest and lowest tourism receipts were the US (with 229,000 million dollars in 2015) and Malaysia (with 4,333 million US dollars in 1998), respectively. The mean of tourism receipts is about 32 billion dollars. The values of standard deviation (variation) of variables range between 36,600 million in tourism expenditures and 3.92 in CO_2 emissions per capita.

3.4 Methodology and Empirical Results

3.4.1 Methodology

3.4.1.1 Panel Unit Root Tests

Under this subsection, four-panel unit root tests developed by Levin et al. (2002) (LLC unit root test); Im et al. (2003) (IPS unit root test); Maddala and Wu (1999) (Fisher-ADF unit root test); and Choi (2001) (Fisher-PP unit root test) are explained alongside Pedroni's (1999, 2004) panel cointegration tests and the Pedroni's (2000) fully modified ordinary least squares (FMOLS) approach.

LLC Panel Unit Root Test

LLC panel unit root test assumes a common unit root process. It considers the following basic augmented Dickey and Fuller (1981) (ADF) test equation (see Levin et al. 2002):

$$\Delta y_{it}=\alpha y_{it-1}+\sum_{j=1}^{p_i}\beta_{ij}\Delta y_{it-j}+X'_{it}\delta+\varepsilon_{it} \tag{3.2}$$

where we assume a common $\alpha = p-1$, but allow the lag order for the different terms (p_i) to vary across individual units of the panel. The null and alternative hypotheses of LLC test are as follows:

H_0:$\alpha = 0$ (unit root)
H_1:$\alpha < 0$ (no unit root)

A unit root null hypothesis is tested versus the not-unit root alternative hypothesis. In the testing procedure of LLC, the estimates of α are derived from proxies for Δy_{it} and y_{it} which are standardized and free of autocorrelations and deterministic components. First, based on a given set of lag orders, two additional sets of equations are estimated regressing both Δy_{it} and y_{it-1} on the lag terms (Δy_{it-1}) (for $j - 1$, ………, p_i) and the exogenous variables (x_{it}). The estimated coefficients of regressions are represented by $(\hat{\beta}, \hat{\delta})$ and $(\dot{\beta}, \dot{\delta})$, respectively.

Then $\Delta \bar{y}_{it}$ is defined by taking Δy_{it} and removing the autocorrelations and deterministic components utilizing the first set of auxiliary estimates:

$$\Delta \bar{y}_{it} = \Delta y_{it} - \sum_{j=1}^{p_i} \hat{\beta}_{ij} \Delta y_{it-j} - X'_{it}\hat{\delta} \tag{3.3}$$

Similarly, the analogous $\bar{y}_{it-1}$ might be defined based on the second set of coefficients:

$$\bar{y}_{it-1} = y_{it-1} - \sum_{j=1}^{p_i} \dot{\beta}_{ij} \Delta y_{it-j} - X'_{it}\dot{\delta} \tag{3.4}$$

As a further step, the proxies are obtained with the standardization of both $\Delta \bar{y}_{it}$ and $\bar{y}_{it-1}$ dividing them by the regression standard errors as follows:

$$\Delta y_{it} = (\Delta \bar{y}_{it} / s_i)$$

$$y_{it-1} = (\bar{y}_{it-1} / s_i)$$

where s_i refer to estimated standard errors from each ADF regression in Eq. (3.2). Finally, an estimate of the coefficient α could be achieved from the following pooled proxy equation:

$$\Delta y_{it} = \alpha y_{it-1} + \eta_{it} \tag{3.5}$$

Under the unit root null hypothesis, a modified t-statistic for the $\hat{\alpha}$ has an asymptotic normal distribution, as defined in Eq. (3.5).

$$\mathrm{t}_\alpha^* = \frac{\mathrm{t}_\alpha - \left(N\tilde{T}\right)\mathrm{S_N}\hat{\sigma}^{-2}\mathrm{se}\left(\hat{\alpha}\right)\mu_{m\tilde{T}}*}{\sigma_{m\tilde{T}}*} \rightarrow N(0, 1) \tag{3.6}$$

where t_α denotes the standard t-statistic for $\hat{\alpha} = 0, \hat{\sigma}^2$ indicates the estimated variance of the error term (η), se($\hat{\alpha}$) is the standard error of $\hat{\alpha}$, and $\tilde{T} = T - \left(\sum_i \frac{p_i}{N}\right) - 1$.

The average standard deviation ratio (S_N) is the mean of the ratios of the long-term standard deviation to the innovation standard deviation for each individual. Applying the Kernel-based techniques, its estimate is derived. The terms $\mu_{mT}*$ and

σ_{mT}* are adjustment terms for the mean and standard deviation. The number of lags (p_i) used in each cross-section ADF regression and the exogenous variables (individual constants, individual constants, and trends, or no terms) used in the test equations must be specified.

IPS Panel Unit Root Test

Im et al. (2003) specify a separate ADF regression for each cross section as

$$\Delta y_{it} = \alpha y_{it-1} + \sum_{j-1}^{p_i} \beta_{ij} \Delta y_{it-j} + X'_{it}\delta + \varepsilon_{it} \tag{3.7}$$

The null and alternative hypotheses are defined as follows:

$$H_0 : \alpha_i = 0, for\ all\ i$$

$$H_1 : \begin{cases} \alpha_i = 0 for\ i = 1, 2, \ldots\ldots N_1 \\ \alpha_i < 0 for\ i = N+1, N+2, \ldots N \end{cases}$$

Based on the estimations of the separate ADF regressions, the average of the t-statistics for α_i from the individual ADF regressions, $t_{iTi}(p_i)$, is specified as

$$\bar{t}_{NT} = \frac{\left(\sum_{i=1}^{N} t_{iTi}(p_i)\right)}{N} \tag{3.8}$$

In the case where the lag order is always zero ($p_i = 0, for\ all\ i$). The simulated critical values for $\bar{t}_{NT}$ are provided in Im et al. (2003) for different numbers of cross section and observation as well as for test equations including either intercepts or intercepts and linear trends. A properly standardized $\bar{t}_{NT}$ statistic has an asymptotic standard normal distribution as

$$W_{\overline{t_{NT}}} = \frac{\sqrt{N}\left(\overline{t_{NT}} - N^{-1}\sum_{i=1}^{N} E(\overline{t_{iT}})(p_i)\right)}{\sqrt{N^{-1}\sum_{i=1}^{N} Var(\overline{t_{iT}})(p_i)}} \to N(0, 1) \tag{3.9}$$

The terms for the expected mean and variance of the ADF regression t-statistics $[E(\overline{t_{iT}})(p_i)]$ and $[Var(\overline{t_{iT}})(p_i)]$ are provided by Im et al. (2003) for different numbers of T and $p$and test equations including either intercepts or intercepts and linear trend. The specification of the number of lags and the deterministic component for each cross-sectional ADF equation should be defined.

Fisher-ADF and Fisher-PP

Utilizing the results obtained by Fisher (1932) to derive tests that combine the p-values from individual unit root tests, Maddala and Wu (1999) and Choi (2001) suggested two different panel unit root tests. In case that π_i is defined as the p-value from any individual unit root test for cross section i, under the unit root null hypothesis for all cross sections, the asymptotic result is specified as

$$-2\sum_{i=1}^{N} log(\pi_i) \rightarrow \chi^2_{2N} \tag{3.10}$$

Additionally, Choi (2001) denotes that

$$Z = \frac{1}{\sqrt{N}}\sum_{i=1}^{N} \phi^{-1}(\pi_i) \rightarrow N(0, 1) \tag{3.11}$$

where ϕ^{-1} is the inverse of the standard normal cumulative distribution function. The null and alternative hypotheses of both Fisher tests are the same as the IPS test. The exogenous variables (individual constants, individual constant and trend terms, or none) for the test equations must be specified. Besides, the number of lags used in each cross-sectional ADF regression for the Fisher-ADF test and a method for the estimation of f_0 (e.g., Bartlett, Parzen, and Quadratic Spektral) must be selected.

As seen in Table 3.3, all panel unit root tests except with the LLC test in some cases indicate that all variables have a unit root (nonstationary) in their levels; however, they do not have unit root (stationary) in their first-differences, i.e. they are integrated of order one (I (1)). Therefore, after confirming that all variables are integrated of order (1), the presence of a long-run relationship (cointegrating relationship) should be searched for among the variables in Eq. (3.1).

3.4.1.2 Pedroni's Residual-Based Cointegration Tests

Pedroni (1999, 2004) proposed seven residual-based cointegration tests that consider heterogeneity in both the short-run dynamics and the long-run slope coefficients across cross-sectional units of the panel. Of these seven statistics, four are based on pooling along within-dimension, and three are based on pooling along between-dimension.

First, the residuals of the assumed cointegrating regression defined in Eq. (3.12) are computed:

$$Y_{i,t} = \alpha_i + \beta_{1,i}X_{1i,t} + \beta_{2,i}X_{2i,t} + \cdots\cdots\cdots\cdots \beta_{M,i}X_{Mi,t} + \varepsilon_{i,t} \tag{3.12}$$

Table 3.3 Panel unit root test results

Variable Tests	CO_2	EF	Energy	GDP	GDP^2	T.arrivals	T.expenditures	T.receipts	T.index	Finance
Level										
LLC	−1.09 (0.13)	−0.08 (0.47)	−2.17[b] (0.01)	−2.60[a] (0.00)	−2.09[b] (0.02)	1.43 (0.92)	−2.81[a] (0.00)	−0.50 (0.31)	−0.50 (0.31)	−3.18[a] (0.00)
IPS	0.87 (0.80)	1.31 (0.91)	0.62 (0.73)	0.48 (0.68)	0.85 (0.80)	4.52 (1.00)	−1.32 (0.09)	1.36 (0.91)	3.36 (1.00)	−0.40 (0.35)
ADF-Fisher	42.91 (0.34)	37.15 (0.51)	38.25 (0.55)	47.06 (0.21)	44.74 (0.28)	28.00 (0.92)	64.81 (0.01)	32.16 (0.81)	26.82 (0.595)	48.65 (0.16)
PP-Fisher	42.64 (0.35)	34.07 (0.65)	49.99 (0.13)	48.74 (0.16)	45.97 (0.24)	29.15 (0.90)	33.61 (0.75)	31.62 (0.83)	20.47 (1.00)	40.53 (0.45)
First-Differences										
LLC	−17.07[a] (0.00)	−16.95[a] (0.00)	−14.03[a] (0.00)	−9.85[a] (0.00)	−9.97[a] (0.00)	−13.73[a]	−12.89[a] (0.00)	−14.79[a] (0.00)	−15.16[a] (0.00)	−10.37[a] (0.00)
IPS	−16.96 (0.00)[a]	−16.56[a] (0.00)	−15.89[a] (0.00)	−9.35[a] (0.00)	−9.44[a] (0.00)	−13.32[a] (0.00)	−11.91[a] (0.00)	−13.79[a] (0.00)	−14.25[a] (0.00)	−10.60[a] (0.00)
ADF-Fisher	295.86 (0.00)[a]	276.72[a] (0.00)	280.79[a] (0.00)	162.82[a] (0.00)	163.76[a] (0.00)	230.38[a] (0.00)	205.54[a] (0.00)	238.38[a] (0.00)	245.46[a] (0.00)	185.84[a] (0.00)
PP-Fisher	352.37 (0.00)[a]	353.70[a] (0.00)	361.91[a] (0.00)	176.17[a] (0.00)	175.52[a] (0.00)	234.01[a] (0.00)	228.05[a] (0.00)	243.45[a] (0.00)	243.43[a] (0.00)	184.55[a] (0.00)

Notes Constant term was included in all regressions; Schwarz information criterion was used for lag selection. Newey-West and Bartlett kernel methods were utilized for bandwidth selection and spectral estimation, respectively. [a] and [b] denote significance at 1% and 5% levels, respectively

where $t = 1, 2, \ldots\ldots\ldots$, T and $i = 1, 2, \ldots\ldots\ldots, N$ denote time and cross-sectional dimensions of the panel while $M = 1, 2, \ldots\ldots\ldots, M$ is the number of independent variables in the regression. The slope coefficients $\beta_{1,t}, \beta_{2,t}, \ldots\ldots\ldots, \beta_{M,t}$ and individual-specific intercept (α_i) are allowed to differ across cross-sectional units of the panel. The panel cointegrating regression in Eq. (3.12) is defined to compute the related panel cointegration test statistics. In case of computations of the *panel* − *p* and *panel* − *t* test statistics, Eq. (3.12) should be redefined in the first-differences of variables and residuals should be estimated from Eq. (3.13)

$$Y_{i,t} = b_{1,i}\Delta X_{1i,t} + b_{2,i}\Delta X_{2i,t} + \cdots\cdots\cdots\cdots b_{M,i}\Delta X_{Mi,t} + \pi_{i,t} \tag{3.13}$$

Using the residuals from the differenced regression, with a Newey and West (1987) estimator, the long-run variance of $\hat{\pi}^2_{i,t}$ that is represented by $\hat{L}^2_{11i}$ is computed as

$$\hat{L}^2_{11i} - \frac{1}{T}\sum_{t=1}^{T}\hat{\pi}^2_{i,t} + \frac{2}{T}\sum_{s=1}^{ki}\left(1 - \frac{s}{k_i + 1}\right)\frac{1}{T}\sum_{t-s+1}^{T}\hat{\pi}_{i,t}\hat{\pi}_{i,t-s} \tag{3.14}$$

For *panel* − *p* and *group* − *p* test statistics, we estimate the regression $\hat{e}_{i,t} = \hat{y}_i\hat{e}_{i,t-1} + \hat{u}_{i,t}$ using the residuals $\hat{e}_{i,t}$ from Eq. (3.11). After that, the long-run variance ($\hat{\sigma}^2_i$) and the contemporaneous variance $\hat{s}^2_i$ of $\hat{u}_{i,t}$ are calculated as follows:

$$\hat{s}^2_i = \sum_{t=1}^{T}\hat{u}_{i,t} \tag{3.15}$$

$$\hat{\sigma}^2_i = \frac{1}{T}\sum_{t=1}^{T}\hat{u}_{i,t} + \frac{2}{T}\sum_{s=1}^{k_i}\left(1 - \frac{s}{k_i + 1}\right)\frac{1}{T}\sum_{t=s+1}^{T}\hat{u}_{i,t}\hat{u}_{i,t-s} \tag{3.16}$$

where k_i is the lag length. Besides, the term, $\lambda_i = \frac{1}{2}(\hat{\sigma}^2_i - \hat{s}^2_i)$, is also computed.

For *panel* − *t* and *group* − *t* test statistics, using the residuals $\hat{e}_{i,t}$ from Eq. (3.12), $\hat{e}_{i,t} = \hat{y}_i\hat{e}_{i,t-1} + \sum_t^k = 1\hat{y}_{ik}\Delta\hat{e}_{it-1} + \hat{u}^*_{i,t}$ is estimated.

Pedroni's seven-panel cointegration test statistics (four-panel test statistics are based on within dimension, and three group test statistics are based on between dimension) are formulated as follows:

1 – ***Panel v stat.:***

$$T^2N^{\frac{3}{2}}Z_{\hat{v}N,T} = T^2N^{\frac{3}{2}}\left(\sum_{i=1}^{N}\sum_{t=1}^{T}\hat{L}^{-2}_{11i}\hat{e}^2_{i,t-1}\right)^{-1} \tag{3.17}$$

2− *Panel ρ stat.:*

$$T\sqrt{N}Z_{\hat{\rho}N,T^{-1}} = T\sqrt{N}\left(\sum_{i=1}^{N}\sum_{t=1}^{T}\hat{L}_{11i}^{-2}\hat{e}_{i,t-1}^{2}\right)^{-1}\sum_{i=1}^{N}\sum_{t=1}^{T}\hat{L}_{11i}^{-2}\hat{e}_{i,t-1}^{2}\left(\hat{e}_{i,t-1}\Delta\hat{e}_{i,t-1} - \hat{\lambda}_i\right) \quad (3.18)$$

3− *Panel t − stat. (non−parametric):*

$$Z_{t_{N,T}} = \left(\tilde{\sigma}_{N,T}^{2}\sum_{i=1}^{N}\sum_{t=1}^{T}\hat{L}_{11i}^{-2}\hat{e}_{i,t-1}^{2}\right)^{-\frac{1}{2}}\sum_{i=1}^{N}\sum_{t=1}^{T}\hat{L}_{11i}^{-2}\hat{e}_{i,t-1}^{2}\left(\hat{e}_{i,t-1}\Delta\hat{e}_{i,t-1} - \hat{\lambda}_i\right) \quad (3.19)$$

4− *Panel t − stat. (parametric):*

$$Z_{t_{N,T}}^{*} = \left(\tilde{s}_{N,T}^{*2}\sum_{i=1}^{N}\sum_{t=1}^{T}\hat{L}_{11i}^{-2}\hat{e}_{i,t-1}^{*2}\right)^{-1/2}\sum_{i=1}^{N}\sum_{t=1}^{T}\hat{L}_{11i}^{-2}\hat{e}_{i,t-1}^{*}\Delta\hat{e}_{i,t}^{*} \quad (3.20)$$

5 − *Groupρ − stat.*:

$$TN^{-1/2}\tilde{Z}_{\hat{\rho}_{N,T}-1} = TN^{-\frac{1}{2}}\sum_{i=1}^{N}\left(\sum_{t=1}^{T}\hat{e}_{i,t-1}^{2}\right)^{-1}\sum_{t=1}^{T}\left(\hat{e}_{i,t-1}\Delta\hat{e}_{i,t} - \hat{\lambda}_i\right) \quad (3.21)$$

6 − *Groupt − stat.(non − parametric)*:

$$N^{-1/2}\tilde{Z}_{t_{N,T}} = N^{-\frac{1}{2}}\sum_{i=1}^{N}\left(\hat{\sigma}_{i}^{2}\sum_{t=1}^{T}\hat{e}_{i,t-1}^{2}\right)^{-\frac{1}{2}}\sum_{t=1}^{T}\left(\hat{e}_{i,t-1}\Delta\hat{e}_{i,t} - \hat{\lambda}_i\right) \quad (3.22)$$

7 − *Groupt − stat.(parametric)*:

$$N^{-\frac{1}{2}}\tilde{Z}_{t_{N,T}}^{*} = N^{-\frac{1}{2}}\sum_{i=1}^{N}\left(\sum_{t=1}^{T}\hat{s}_{i}^{*2}\hat{e}_{i,t-1}^{*2}\right)^{-\frac{1}{2}}\sum_{t=1}^{T}\hat{e}_{i,t-1}^{*}\Delta\hat{e}_{i,t}^{*} \quad (3.23)$$

For the within-dimension statistics (panel test statistics), the null (no-cointegration) and alternative (cointegration) hypotheses are defined as

$$H_0 : \gamma_i = 1\, for\; all\; i$$

$$H_{1:}\gamma_i = \gamma < 1 \, for \, all \, i$$

For the between-dimension statistics (group test statistics), the null (no-cointegration) and alternative (cointegration) hypotheses are defined as

$$H_0 : \gamma_i = 1 \; for \; all \; i$$

$$H_{1:} y_i < 1 \; for \; all \; i$$

Lastly, the appropriate mean and variance adjustment terms are applied to each panel cointegration test statistic to obtain the standard normally distributed statistics as

$$\frac{X_{N,T-\mu\sqrt{N}}}{\sqrt{v}} \rightarrow N(0, 1) \tag{3.24}$$

where $X_{N,T}$ is the standardized form of the test statistic and the values of $\mu \, and \, v$ are functions of moments of the underlying Brownian motion functionals.

As seen in Table 3.4, four out of seven test statistics (Panel PP, Panel ADF, Group PP, and Group ADF) provided evidence of cointegration in all four models. They reject the null hypothesis of no-cointegration at 1% significance level. Therefore, after revealing a cointegrating (long-run) relationship between the related variables defined in Eq. (3.1), the long-run parameters (slope coefficients) of independent variables in the cointegrating vector were estimated, as a next step. To this aim, Pedroni's (2000) fully modified ordinary least squares (FMOLS hereafter) approach was utilized since it controls for endogeneity and serial correlation in regression.

3.4.1.3 Pedroni's FMOLS Estimator

To estimate cointegrating vectors in a dynamic panel data, Pedroni (2000) developed the FMOLS approach, which is able to allow for heterogeneity across individual members of the panel. This approach has an advantage since it corrects for both endogeneity bias and serial correlation to the OLS estimator (Ozcan 2013). The long-run parameters are obtained using the following equation (Pedroni 2000).

$$\beta_{NT}^{*} - \beta = \left[\sum_{i=1}^{N} L_{22i}^{-2} \sum_{t=1}^{T} (u_{it} - \bar{u}_i)^2 \right]^{-1} \sum_{i=1}^{N} \hat{L}_{11i}^{-1} \hat{L}_{22i}^{-1} \left[\sum_{t=1}^{T} (u_{it} - \bar{u}_i)\mu_{it}^{*} - T\hat{\gamma}_i \right] \tag{3.25}$$

where β_{NT}^{*} is standard panel OLS estimator, $\bar{u}$ is the individual specific means.

Table 3.4 Pedroni's cointegration test results

Test stats.	Model 1		Model 2		Model 3		Model 4	
	Statistic	Prob.	Statistic	Prob.	Statistic	Prob.	Statistic	Prob.
Dependent variable: *CO_2 emissions*								
Alternative hypothesis: *Common AR coefs. (within-dimension)*								
Panel v-stat	1.49[c]	0.07	0.17	0.43	0.09	0.46	0.92	0.18
Panel rho-stat	0.56	0.71	1.58	0.94	1.24	0.89	1.14	0.87
Panel PP-stat	−8.97[a]	0	−5.41[a]	0	−5.94[a]	0	−6.59[a]	0
Panel ADF-stat	−9.22[a]	0	−5.11[a]	0	−5.84[a]	0	−6.43[a]	0
Alternative hypothesis: *Individual AR coefs. (between-dimension)*								
Grup rho-stat	3.32	1	3.66	1	3.37	1	3.51	1
Grup PP-stat	−15.12[a]	0	−10.54[a]	0	−11.85[a]	0	−12.17[a]	0
Grup ADF-stat	−8.59[a]	0	−5.73[a]	0	−6.32[a]	0	−6.50[a]	0.00
Dependent variable: *EF*								
Alternative hypothesis: *Common AR coefs. (within-dimension)*								
Panel v-stat	0.35	0.36	1.14	0.13	0.92	0.18	0.93	0.18
Panel rho-stat	1.28	0.9	1.21	0.89	1.27	0.9	1.18	0.88
Panel PP-stat	−8.06[a]	0	−8.93[a]	0	−7.89[a]	0	−8.07[a]	0
Panel ADF-stat	−7.60[a]	0	−8.49[a]	0	−7.82[a]	0	−7.18[a]	0
Alternative hypothesis: *Individual AR coefs. (between-dimension)*								
Grup rho-stat	3.84	1	3.95	1	3.71	1	3.59	1
Grup PP-stat	−16.59[a]	0	−13.15[a]	0	−11.64[a]	0	−18.31[a]	0
Grup ADF-stat	−9.22[a]	0	−8.53[a]	0	−8.74[a]	0	−8.91[a]	0.00

Notes Model 1 includes T.arrivals, Model 2 includes T.expenditures, Model 3 includes T.revenues, and Model 4 includes T.index as tourism variable. [a], [b], and [c] indicate rejection of no-cointegration null hypothesis at 1%, 5% and 10% significance levels, respectively. Constant and trend terms were included in all models. Schwarz information criterion was used for lag selection. For bandwidth selection, Newey-West automatic and for Spectral estimation, Bartlett were employed. Weighted statistics were not provided because in that case Table 3.3 did not fit in page. However, they also provided results favoring cointegration

$$\mu_{it}^{*} = \mu_{it} - \frac{\hat{L}_{21i}}{\hat{L}_{22i}}\Delta_{it},\ \hat{y}_i = \hat{\Gamma}_{21i} + \hat{\Omega}_{21i}^{0} - \frac{\hat{L}_{21i}}{\hat{L}_{22i}}\left(\hat{\Gamma}_{22i} + \hat{\Omega}_{22i}^{0}\right),$$

$\hat{\Omega}_i$ is a scalar long-run variance of the residuals, $\hat{\Omega}_i^0$ is the contemporaneous covariance, $\hat{\Gamma}_i$ is a weighted sum of autocovariance, and $\hat{L}_i$ is a lower triangular decomposition of $\hat{\Omega}_i$. $\hat{L}_{11i} = \hat{\sigma}_u^2 - \hat{\sigma}_{u\varepsilon}^2/\hat{\sigma}_\varepsilon^2$ and $\hat{L}_{22i} = \hat{\sigma}_\varepsilon^2$ are the conditional long-run variances. The estimator of β_{NT}^{*} converges to the true value at rate $T\sqrt{N}$ and is distributed as

$$T\sqrt{N}\left(\hat{\beta}_{NT}^{*}-\beta\right) \rightarrow N(0, v), \text{ where } v=\begin{cases}\frac{2}{6} & \text{if } \bar{u}_i=\bar{y}_1=0 \\ & \text{else}\end{cases} \text{ as } T \rightarrow \infty \text{ and as } N \rightarrow \infty$$

There are two versions of FMOLS estimator: The Pooled FMOLS and the Group Mean FMOLS. In the framework of the panel set, the mean-group FMOLS long-run coefficients are obtained by averaging the group estimates over cross-section units of the panel: $\hat{\beta}_{FMOLS(MG)} = N^{-1}\sum_{i=1}^{N}\hat{\beta}_i$. The related t-statistic converges asymptotically to a standard normal distribution: $t_{FMOLS(MG)} = N^{-1/2}\sum_{i=1}^{N} t_i \rightarrow N(0, 1)$.

The long-run coefficients of pooled FMOLS are computed weighted or unweighted ways. In the weighted case, each group is weighted by the components of the long-run covariance of the group residuals and the right-hand-side variables in differences; in the unweighted case, these components are averaged (Maeso-Fernandez et al. 2006) The weighted statistics need prior knowledge of the estimated parameters, and therefore, to compute a feasible weighted statistic, Pedroni (2000) utilized the values estimated under the null hypothesis.

3.4.2 *Results for Panel FMOLS Estimation*

As a final step of the analysis, the long-run parameters of the cointegrating vector are estimated based on Pedroni's (2000) panel FMOLS approach. Table 3.5 provides the results for the case that CO_2 emissions are dependent variables, while Table 3.6 provides the results for the case that EF is a dependent variable. Four models are estimated for each tourism variable in both the cases of EF and CO_2 emissions. As seen in Table 3.5, all coefficients are significant at 1% level in all models. GDP per capita has a positive and significant effect on CO_2 emissions, while its squared form has a

Table 3.5 FMOLS results (CO_2)

Variables	Model 1		Model 2		Model 3		Model 4	
	Coef.	Prob.	Coef.	Prob.	Coef.	Prob.	Coef.	Prob.
GDP	0.979[a]	0.000	1.166[a]	0.000	1.036[a]	0.000	1.057[a]	0.000
GDP^2	−0.105[a]	0.000	−0.121[a]	0.000	−0.126[a]	0.000	−0.059[a]	0.000
Energy	0.955[a]	0.000	0.929[a]	0.000	0.958[a]	0.000	0.940[a]	0.000
Finance	−0.063[a]	0.000	−0.060[a]	0.000	−0.056[a]	0.000	−0.040[a]	0.000
T.arrivals	−0.044[a]	0.000						
T.expenditures			0.004	0.816				
T.revenues					−0.049[a]	0.004		
T.index							−0.031[a]	0.000
Adjust. RSquare	0.846		0.839		0.744		0.983	
Obs.	460		460		460		460	

Notes Constant term was included in all regression models; the pooled (weighted) option was used against heteroscedasticity problem; and [a] denotes significance at 1% level

Table 3.6 FMOLS results (EF)

	Model 1		Model 2		Model 3		Model 4	
Variables	Coef.	Prob.	Coef.	Prob.	Coef.	Prob.	Coef.	Prob.
GDP	−0.601[a]	0.001	−0.049	0.815	−0.523[b]	0.010	−0.516[a]	0.008
GDP^2	0.036[a]	0.000	0.003	0.771	0.032[a]	0.004	0.035[a]	0.001
Energy	0.721[a]	0.000	0.663[a]	0.000	0.747[a]	0.000	0.772[a]	0.000
Finance	−0.031[a]	0.006	−0.017	0.226	−0.007	0.591	−0.012	0.319
T.arrivals	−0.114[a]	0.000						
T.expenditures			−0.039[a]	0.006				
T.revenues					−0.129[a]	0.000		
T.index							−0.086[a]	0.000
Adjust. RSquare	0.972		0.966		0.970		0.972	
Obs.	399		399		399		399	

Notes Constant term was included in all regression models; the pooled (weighted) option was used against heteroscedasticity problem; [a] and [b] denote significance at 1% and 5% levels, respectively

negative and significant effect. This result confirms the tourism-based EKC hypothesis, assuming an inverted U-shaped relationship between income and environmental degradation. In this sense, air pollution rises until a certain income per capita level beyond which emissions level starts decreasing. In other words, in the early stages of development, countries do not pay special interest in environmental issues because meeting basic needs is the essential task of governments. However, in the later stages of development, governments and residents start paying more attention to environmental problems given that basic needs are met. Confirmation of the EKC hypothesis is not unexpected because the sample of this study includes developed countries, and therefore they are in the later stages of economic development.

Concerning the results for energy consumption per capita, it seems that increase in primary energy consumption contribute to more CO_2 emissions. This result is not a surprise, either because primary energy sources are mostly based on fossil fuels worldwide. 1% increase in energy consumption raises CO_2 emissions between 0.92% and 0.95%. Regarding the financial development variable, the results from all models indicate that financial development helps to reduce CO_2 emissions level. 1% increase in financial development leads to improvements in air pollution between the percentages of 0.4 and 0.6. In this respect, it could be stated that bank credits are channelized to environmentally friendly investments in the panel, including 20 most tourist attractive countries, as a whole. For instance, clean and environmentally friendly technologies might be easily adopted into production processes by firms using bank credits. Finally, three of the four tourism variables indicate that tourism development decreases air pollution in the panel set as a whole. 1% increase in the number of tourist arrivals and tourism receipts reduces CO_2 emissions by 0.04%, while 1% increase in tourism index decreases CO_2 emissions by 0.03%. This result is in line with the results of Dogan and Aslan (2017), Katircioglu (2014), Paramati et al. (2018), and Lee and Brahmasrene (2013). However, tourism expenditures do not have a significant effect on air pollution. The tourism sector, commonly named as a smokeless industry, appears to contribute to the environmental quality of the panel set. In this sense, these 20 countries appear to care about environmental quality to attract more tourist, for instance by making use of renewable energy sources such as hydro and solar for heating and hot water. In this context, it could be stated that human being, in the last century, has cared environmental issues much since last few decades because of global environmental threats such as global warming, drought and floods, melting glaciers, and so on. People choose to visit places that have less harmful effects on environment because they are more educated and conscious about environment.

Furthermore, using another indicator for environmental quality (EF per capita), each model was re-estimated with the FMOLS. As provided in Table 3.6, a different result was obtained relating to the EKC. There exists a U-shaped relationship appears between EF per capita and GDP per capita. As such, in the first stage of development, EF has a declining trend because the environmental pressure of human-being is rather less in an agriculture-based economy; however, this pressure has started to increase with the transformation from an agrarian economy to an industrial economy. This result is not unexpected because EF is a broader indicator than CO_2 emissions. It includes other dimensions of environmental pollution such as land and water pollution along with air pollution. This result (U-shaped EKC) is similar to the results of Al-mulali et al. (2015) in cases of low- and lower-middle-income countries; Al-mulali et al. (2016) for 58 countries; Charfeddine (2017) for Qatar; Charfeddine and Mrabet (2017) in case of nonoil-exporting countries; Caviglia-Harris et al. (2009) for 146 countries; and York et al. (2003) who used EF as a proxy for environmental degradation with regard to 142 countries. However, it is in sharp contrast to those of Al-mulali et al. (2015) in cases of upper-middle and high-income countries; Bello et al. (2018) for Malaysia; Charfeddine and Mrabet (2017) in case of oil-exporting countries; Katircioglu et al. (2018) for ten most tourist destination countries; and Ulucak and Bilgili (2018) who confirmed an inverted U-shaped the EKC for all income-group countries,

Furthermore, 1% increase in energy consumption per capita raises EF per capita between 0.66 and 0.77%. In this context, the more demand for energy a country has, the more CO_2 the country emits into the air; and this situation will result in an increasing level of EF because CO_2 emissions make up a large portion of EF. However, regarding financial development, different results were found than in the case of CO_2. Except Model 1 (consisting of tourist arrivals number), financial development does not have a significant effect on EF. These insignificant results could be due to the fact that financial development has a direct effect on air pollution instead of other dimensions of environmental degradation. Finally, regarding tourism development, all tourism variables have a negative and significant effect on EF per capita. 1% increase in the number of tourist arrivals, tourism expenditures, tourism receipts, and tourism index reduces EF per capita between 0.03 and 0.12%. This result indicates that those 20 countries pay special attention to environmental issues to attract more tourists, e.g. keeping sea and beach clean to get a blue flag and thereby or building bungalow houses amidst forest instead of cutting trees, etc. In this sense, we recently hear a lot about the concept of eco-tourism or sustainable tourism and recognize civilized residents of the world in this century have changed their lifestyles to save the environment and to reduce their burden on nature. This finding is similar to that of Katircioglu et al. (2018) for the ten most tourist destination countries.

3.5 Conclusion and Policy Implications

This chapter analyzed the impacts of tourism development on environmental quality for France, Spain, the United States, China, Italy, Turkey, Mexico, Germany, Thailand, the United Kingdom, Japan, Austria, Greece, Honk-Kong, Malaysia, Russian Federation, Portugal, Canada, Poland and the Netherlands over the period of 1995–2018. As for indicators of the tourism sector, the number of international tourist arrivals, the international tourism expenditures, the international tourism receipts, and a composite index was employed. Environmental quality was represented by CO_2 emissions and EF (in per capita terms) while income, energy consumption, and financial development were included in models as control variables. Based on the panel data models, such as the first-generation panel unit root tests, Pedroni's cointegration tests, and panel FMOLS approach, four models for CO_2 emissions and four models for EF were estimated. The findings indicated that the tourism sector helps to reduce CO_2 emissions and human pressure on the environment. These 20 countries pay special attention to environmental quality to attract more tourists. Regarding income per capita, the results revealed that the EKC hypothesis was confirmed in the case of CO_2 emissions while it was not supported in the case of EF. In this sense, initially CO_2 emissions increase in the early stages of economic growth; however, at the later stages of the growth process, it starts decreasing. Therefore, air quality seems a luxury good meaning that the demand for air quality is only possible beyond a certain income per capita level. In contrast, there is a U-shaped relationship between EF and economic growth meaning that the pressure of humans on the environment is rather less in the early stages of development while this pressure intensifies in the later stages of development. Energy consumption positively affects both CO_2 emissions and EF. The more energy a country consumes the more air pollution or environmental pressure that the country would be exposed to. Finally, financial development appears to reduce CO_2 emissions level while it does not significantly affect the EF level. As such, bank credits are channelized to less energy-intensive investment types, such as renewable energy investment. These environmentally friendly investments emit less CO_2 into air since they utilize advance and sophisticated technologies.

Based on the mentioned results, some crucial policy implications could be suggested for the policymakers of the countries under the examination of this chapter. First, the tourism sector should be supported by both government and private sectors. As such, the government should support the tourism sector by incentives, e.g. financing some part of hotel construction costs, providing bank credits with a low-interest rate and long-term, encouraging employment in the tourism sector by some regulations, or promoting the country in abroad through ads. Additionally, private tourism agencies may collaborate with their counterparts in foreign countries, and they may prepare special holiday packages, including some privileges, etc. Further, renewable energy sources such as hydro and solar can be used more in hotels and tourist places for heating and hot water. Particularly, eco-tourism or sustainable tourism can be considered as alternative tourism ways because their main priority is keeping the environment safe. From the perspective of technology, more sophisticated technologies with environmentally friendly and less energy-intensive can be adapted in the patterns of production and consumption in the tourism sector. By doing these, the tourism sector will inevitably develop and benefit not only for environmental wellness but also for the national government budget. It should not be forgotten that the tourism sector, an important source of national income, compensates for the balance of payment deficit. Regarding other determinants of environmental quality, energy consumption and financial development should be considered by policymakers, as well. In this respect, fossil energy sources still have a high share in the energy mix of countries worldwide. As the main culprit of CO_2 and greenhouse gas emissions, the energy demand for fossil fuels should be reduced by making use of renewables more. Although this is not an easy task, governments, the private sector, and non-governmental organizations should cooperate to reach this target. Finally, a well-developed financial system is of importance, particularly for the air quality. Banks should provide more credits with a low-interest rate and long-term for firms or entrepreneurs that make investments in R&D, environmentally friendly technologies, and renewable energy sectors. In terms of consumers, purchasing of energy-efficient and high technology products should be also facilitated through bank credits.

References

Akadiri, S. S., Lasisi, T. T., Uzuner, G., & Akadiri, A. C. (2020). Examining the causal impacts of tourism, globalization, economic growth and carbon emissions in tourism island territories: bootstrap panel Granger causality analysis. *Current Issues in Tourism*, 23(4), 470–484.

Al-mulali, U., Solarin, S. A., Sheau-Ting, L., & Ozturk, I. (2016). Does moving towards renewable energy cause water and land inefficiency? An empirical investigation. *Energy Policy, 93*, 303–314.

Al-mulali, U., Weng-Wai, C., Sheau-Ting, L., & Mohammed, A. H. (2015). Investigating the environmental Kuznets curve (EKC) hypothesis by utilizing the ecological footprint as an indicator of environmental degradation. *Ecological Indicators, 48*, 315–323.

Azam, M., Alam, M., & Haroon, M. (2018). Effect of tourism on environmental pollution: Further evidence from. *Journal of Cleaner Production, 190*, 330–338.

Bagliani, M., Bravo, G., & Dalmazzone, S. (2008). A consumption-based approach to environmental Kuznets curves using the ecological footprint indicator. *Ecological Economics, 65*(3), 650–661.

Balaguer, J., & Cantavella-Jorda, M. (2002). Tourism as a long-run economic growth factor: the Spanish case. *Applied Economics, 34*(7), 877–884.

Başarir, Ç., & Çakir, Y. N. (2015). Causal interactions between CO2 emissions, financial development, energy and tourism. *Asian Economic and Financial Review, 5*(11), 1227–1238.

Becken, S., Frampton, C., & Simmons, D. (2001). Energy consumption patterns in the accommodation sector—the New Zealand case. *Ecological Economics, 39*(3), 371–386.

Becken, S. (2002). Analyzing international tourist flows to estimate energy use associated with air travel. *Journal of sustainable tourism, 10*(2), 114–131.

Bella, G., & Bella, G. (2018). Estimating the tourism induced environmental Kuznets curve in France Estimating the tourism induced environmental Kuznets curve. *Journal of Sustainable Tourism, 0*(0), 1–10.

Bello, M. O., Solarin, S. A., & Yen, Y. Y. (2018). The impact of electricity consumption on CO(2) emission, carbon footprint, water footprint and ecological footprint: The role of hydropower in an emerging economy. *Journal of Environmental Management, 219,* 218–230.

Berrittella, M., Bigano, A., Roson, R., & Tol, R. S. (2006). A general equilibrium analysis of climate change impacts on tourism. *Tourism Management, 27*(5), 913–924.

Bramwell, B., & Lane, B. (1993). Sustainable tourism: An evolving global approach. *Journal of Sustainable Tourism, 1*(1), 1–5.

Casler, S. D., & Blair, P. D. (1997). Economic structure, fuel combustion, and pollution emissions. *Ecological Economics, 22*(1), 19–27.

Caviglia-Harris, J. L., Chambers, D., & Kahn, J. R. (2009). Taking the "U" out of Kuznets: A comprehensive analysis of the EKC and environmental degradation. *Ecological Economics, 68*(4), 1149–1159.

Charfeddine, L. (2017). The impact of energy consumption and economic development on Ecological Footprint and CO2 emissions: Evidence from a Markov Switching Equilibrium Correction Model. *Energy Economics, 65,* 355–374.

Charfeddine, L., & Mrabet, Z. (2017). The impact of economic development and social-political factors on ecological footprint: A panel data analysis for 15 MENA countries. *Renewable and Sustainable Energy Reviews, 76,* 138–154.

Choi, H. C., & Sirakaya, E. (2006). Sustainability indicators for managing community tourism. *Tourism Management, 27*(6), 1274–1289.

Choi, I. (2001). Unit root tests for panel data. *Journal of International Money and Finance, 20*(2), 249–272.

Danish, & Wang, Z. (2018). Dynamic relationship between tourism, economic growth, and environmental quality. *Journal of Sustainable Tourism, 26*(11), 1928–1943.

Dickey, D. A., & Fuller, W. A. (1981). Likelihood ratio statistics for autoregressive time series with a unit root. *Econometrica, 49*(4), 1057–1072.

Dinda, S. (2004). Environmental Kuznets curve hypothesis: A survey. *Ecological Economics, 49*(4), 431–455.

Dogan, E., & Aslan, A. (2017). Exploring the relationship among CO 2 emissions, real GDP, energy consumption and tourism in the EU and candidate countries: Evidence from panel models robust to heterogeneity and cross-sectional dependence. *Renewable and Sustainable Energy Reviews, 77*(March), 239–245.

Dubois, G., Peeters, P., Ceron, J. P., & Gössling, S. (2011). The future tourism mobility of the world population: Emission growth versus climate policy. *Transportation Research Part A: Policy and Practice, 45*(10), 1031–1042.

Dwyer, L., & Forsyth, P. (2008). Economic measures of tourism yield: What markets to target? *International Journal of Tourism Research, 10*(2), 155–168.

Fereidouni, H. G., & Al-mulali, U. (2014). The interaction between tourism and FDI in real estate in OECD countries. *Current Issues in Tourism, 17*(2), 105–113.

Fisher, R. A. (1932). *Statistical methods for research workers* (4th ed.). Edinburgh: Oliver & Boyd.

Gössling, S. (2002). Global environmental consequences of tourism. *Global Environmental Change, 12*(4), 283–302.

Gössling, S. (2013). National emissions from tourism: An overlooked policy challenge? *Energy Policy, 59,* 433–442.

Grossman, G. M., & Krueger, A. B. (1991). *Environmental impacts of a North American free trade agreement* (No. w3914). National Bureau of economic research.

Hardin, G. (1968). The tragedy of the commons. *Science, 162,* 1243–1248.

Holden, A. (2009). The environment-tourism nexus: Influence of market ethics. *Annals of Tourism Research, 36*(3), 373–389.

Hsieh, C. M., Tsai, B. K., & Chen, H. S. (2017). Residents' attitude toward aboriginal cultural tourism development: An integration of two theories. *Sustainability, 9*(6), 903.

Im, K. S., Pesaran, M. H., & Shin, Y. (2003). Testing for unit roots in heterogeneous panels. *Journal of Econometrics, 115*(1), 53–74. https://doi.org/10.1016/S0304-4076(03)00092-7

Indra, M., & Kumar, S. (2019). Tourist arrivals, energy consumption and pollutant emissions in a developing economy–implications for sustainable tourism. *Tourism Management, 72*(2018), 145–154.

Katircioglu, S. (2009a). Revisiting the tourism-led-growth hypothesis for Turkey using the bounds test and Johansen approach for cointegration. *Tourism Management, 30*(1), 17–20.

Katircioglu, S. (2009b). Tourism, trade and growth: the case of Cyprus. *Applied Economics, 41*(21), 2741–2750.

Katircioglu, S., Gokmenoglu, K. K., & Eren, B. M. (2018). Testing the role of tourism development in ecological footprint quality: evidence from top 10 tourist destinations. *Environmental Science and Pollution Research International, 25*(33), 33611–33619.

Katircioğlu, S. T., & Taşpinar, N. (2017). Testing the moderating role of financial development in an environmental Kuznets curve: Empirical evidence from Turkey. *Renewable and Sustainable Energy Reviews, 68*, 572–586. https://doi.org/10.1016/j.rser.2016.09.127.

Katircioglu, S. (2009c). Testing the tourism-led growth hypothesis: The case of Malta. *Acta Economica, 59*(3), 331–343.

Katircioglu, Salih Turan. (2014). International tourism, energy consumption, and environmental pollution: The case of Turkey. *Renewable and Sustainable Energy Reviews, 36,* 180–187. https://doi.org/10.1016/j.rser.2014.04.058.

Katircioğlu, S. T. (2014). Testing the tourism-induced EKC hypothesis: The case of Singapore. *Economic Modelling, 41,* 383–391.

Katircioglu, Salih Turan, Feridun, M., & Kilinc, C. (2014). Estimating tourism-induced energy consumption and CO 2 emissions: The case of Cyprus. *Renewable and Sustainable Energy Reviews, 29,* 634–640.

Khan, A., Chenggang, Y., Hussain, J., Bano, S., & Nawaz, Aa. A. (2020). Natural resources, tourism development, and energy-growth-CO2 emission nexus: A simultaneity modeling analysis of BRI countries. *Resources Policy, 68*(2019), 101751.

Kim, H. J., Chen, M. H., & Jang, S. (2006). Tourism expansion and economic development: THE case of Taiwan. *Tourism Management, 27*(5), 925–933.

Kuznets, S. (1955). Economic growth and income inequality. *The American Economic Review, 45*(1), 1–28.

Koçak, E., Ulucak, R., & Ulucak, Z. Ş. (2020). The impact of tourism developments on CO2 emissions: An advanced panel data estimation. *Tourism Management Perspectives, 33*(2019).

Lee, J. W., & Brahmasrene, T. (2013). Investigating the influence of tourism on economic growth and carbon emissions: Evidence from panel analysis of the European Union. *Tourism Management, 38,* 69–76.

Levin, A., Lin, C.-F., & James Chu, C.-S. (2002). Unit root tests in panel data: Asymptotic and finite-sample properties. *Journal of Econometrics, 108*(1), 1–24.

Maddala, G. S., & Wu, S. (1999). A comparative study of unit root tests with panel data and a new simple test. *Oxford Bulletin of Economics and Statistics, 61*(S1), 631–652.

Magnani, E. (2001). The Environmental Kuznets Curve: Development path or policy result? *Environmental Modelling and Software, 16*(2), 157–165.

Mathieson, A., & Wall, G. (1982). *Tourism: economic, physical and social impacts.* New York: Longman House.

Maeso-Fernandez, F., Osbat, C., & Schnatz, B. (2006). Towards the estimation of equilibrium exchange rates for transition economies: Methodological issues and a panel cointegration perspective. *Journal of Comparative Economics, 34*(3), 499–517.

Nelson, V. (2013). *An introduction to the geography of tourism.* Lanham, MD: Rowman & Littlefield.

Neto, F. (2003). A new approach to sustainable tourism development: Moving beyond environmental protection. In *Natural resources forum* (Vol. 27, No. 3, pp. 212–222). Oxford, UK: Blackwell Publishing Ltd.

Newey, W. K., & West, K. D. (1987). A simple, positive semi-definite, heteroskedasticity and autocorrelation consistent covariance matrix. *Econometrica, 55*(3), 703–708.

Ostrom, E., & Field, B. (1999). Revisiting the commons: Local lessons, global changes. *Science, 284,* 278–282.

Ozcan, B. (2013). The nexus between carbon emissions, energy consumption and economic growth in Middle East countries: A panel data analysis. *Energy Policy, 62,* 1138–1147.

Perch-Nielsen, S., Sesartic, A., & Stucki, M. (2010). The greenhouse gas intensity of the tourism sector: The case of Switzerland. *Environmental Science & Policy, 13*(2), 131–140.

Paramati, S. R., Alam, S., Keung, C., & Lau, M. (2018). The effect of tourism investment on tourism development and CO 2 emissions: empirical evidence from the EU nations. *Journal of Sustainable Tourism, 0*(0), 1–21. https://doi.org/10.1080/09669582.2018.1489398.

Paramati, S. R., Shahbaz, M., & Alam, M. S. (2017). Does tourism degrade environmental quality? A comparative study of Eastern and Western European Union. *Transportation Research Part D: Transport and Environment, 50,* 1–13. https://doi.org/10.1016/j.trd.2016.10.034.

Pedroni, P. (1999). Critical values for cointegration tests in heterogeneous panels with multiple regressors. *Oxford Bulletin of Economics and Statistics, 61,* 653–670.

Pedroni, P. (2000). Fully modified OLS for heterogeneous cointegrated panels. *Advances in Econometrics, 2000*(15), 93–130.

Pedroni, P. (2004). Panel cointegration: asymptotic and finite sample properties of pooled time series tests with an application to the PPP hypothesis. *Econometric Theory*, 597–625.

Peeters, P., Szimba, E., & Duijnisveld, M. (2007). Major environmental impacts of European tourist transport. *Journal of Transport Geography, 15*(2), 83–93.

Pintassilgo, P., & Silva, J. A. (2007). 'Tragedy of the commons' in the tourism accommodation industry. *Tourism Economics, 13*(2), 209–224.

Rees, W. E. (1992). Ecological footprints and appropriated carrying capacity: What urban economics leaves out. *Environment and Urbanization, 4*(2), 121–130.

Rosselló-Batle, B., Moià, A., Cladera, A., & Martínez, V. (2010). Energy use, CO2 emissions and waste throughout the life cycle of a sample of hotels in the Balearic Islands. *Energy and Buildings, 42*(4), 547–558.

Sadorsky, P. (2012). Information communication technology and electricity consumption in emerging economies. *Energy Policy, 48,* 130–136.

Saenz-de-Miera, O., & Rosselló, J. (2014). Modeling tourism impacts on air pollution: The case study of PM10 in Mallorca. *Tourism Management, 40,* 273–281.

Saint, S., Adewale, A., & Chigozie, A. (2019). The role of globalization, real income, tourism in environmental sustainability target. Evidence from Turkey. *Science of the Total Environment, 687,* 423–432.

Santana-Gallego, M., Ledesma-Rodríguez, F., & Pérez-Rodríguez, J. V. (2011). Tourism and trade in OECD countries. A dynamic heterogeneous panel data analysis. *Empirical Economics, 41*(2), 533.

Sghaier, A., Guizani, A., Ben, S., & Mohammad, J. (2018). Tourism development, energy consumption and environmental quality in Tunisia, Egypt and Morocco: A trivariate analysis. *GeoJournal* (Wttc 2017).

Shaheen, K., Zaman, K., Batool, R., Khurshid, M. A., & Aamir, A. (2019). *Dynamic linkages between tourism, energy, environment, and economic growth : evidence from top 10 tourism-induced countries*. (Bennett 2014).

Sorabella, J. (2003). The grand tour. In *Heilbrunn timeline of art history*. New York, NY: The Metropolitan Museum of Art.

Tamazian, A., Chousa, J. P., & Vadlamannati, K. C. (2009). Does higher economic and financial development lead to environmental degradation: Evidence from BRIC countries. *Energy Policy, 37*(1), 246–253. https://doi.org/10.1016/j.enpol.2008.08.025.

Tang, C. F., & Abosedra, S. (2014). Small sample evidence on the tourism-led growth hypothesis in Lebanon. *Current Issues in Tourism, 17*(3), 234–246.

Towner, J. (1985). The grand tour: A key phase in the history of tourism. *Annals of Tourism Research, 12,* 297–333.

Tsai, K. T., Lin, T. P., Hwang, R. L., & Huang, Y. J. (2014). Carbon dioxide emissions generated by energy consumption of hotels and homestay facilities in Taiwan. *Tourism Management, 42,* 13–21.

Ulucak, R., & Bilgili, F. (2018). A reinvestigation of EKC model by ecological footprint measurement for high, middle and low income countries. *Journal of Cleaner Production, 188,* 144–157.

UNEP & UNWTO. (2005). Making tourism more sustainable: A guide for policy makers. eISBN: 978-92-844-0821-4.

Wackernagel, M., Onisto, L., Bello, P., Callejas Linares, A., Susana Lopez Falfan, I., Mendez Garcia, J., et al. (1999). National natural capital accounting with the ecological footprint concept. *Ecological Economics, 29*(3), 375–390. Retrieved from https://econpapers.repec.org/RePEc:eee:ecolec:v:29:y:1999:i:3:p:375–390.

Wackernagel, M., & Rees, W. (1996). *Our ecological footprint: Reducing human impact on the earth.* The New Catalyst Bioregional Series: New Society Publishers.

Wackernagel, M., Schulz, N. B., Deumling, D., Linares, A. C., Jenkins, M., Kapos, V., et al. (2002). *Tracking the ecological overshoot of the human economy, 99*(14).

York, R., Rosa, E. A., & Dietz, T. (2003). Footprints on the earth: The environmental consequences of modernity. *American Sociological Review, 68*(2), 279–300.

Zaman, K., Shahbaz, M., Loganathan, N., & Ali, S. (2016). Tourism development, energy consumption and Environmental Kuznets Curve: Trivariate analysis in the panel of developed and developing countries. *Tourism Management, 54,* 275–283.

Zhang, L., & Gao, J. (2016). Exploring the effects of international tourism on China' s economic growth, energy consumption and environmental pollution: Evidence from a regional panel analysis. *Renewable and Sustainable Energy Reviews, 53,* 225–234.

Zhang, Y.-J. (2011). The impact of financial development on carbon emissions: An empirical analysis in China. *Energy Policy, 39*(4), 2197–2203.

Chapter 4
The Effects of Tourism, Economic Growth and Renewable Energy on Carbon Dioxide Emissions

Nuno Carlos Leitão and Daniel Balsalobre-Lorente

Abstract This study discusses the relationship between tourism demand and carbon dioxide (CO_2) emissions, as well as the correlation between renewable energy and CO_2 emissions. To achieve this, a robust panel methodology is employed for EU-28 countries. First, preliminary descriptive summary statistics and correlation analysis. Second, unit root tests in panel data (Levin et al. in J Econom, 108:1–24, 2002; ADF–Fisher Chi-square, and Phillips–Perron) to establish stationarity traits of the outlined variables Subsequently, we use Pedroni (Rev Econom Stat 83(4):727–731, 2001, Econom Theory 20(03):597–625, 2004), a panel data Random Effects (RE), DOLS (Panel Dynamic Least Squares) and panel Granger causality test, as econometric methodologies for long-run equilibrium relationship and detection of causality flow, respectively. The econometric empirical results confirm that there exists an Inverted-U linkage between economic growth and environmental degradation, which validates the Environmental Kuznets Curve (EKC) hypothesis for EU-28 countries. Furthermore, empirical results show a negative association between tourist arrivals and CO_2 emissions, making it possible to infer that the tourism sector accentuates climate change. Regarding renewable energy, the results validate the negative relationship between this variable and carbon dioxide emissions, which is in line with previous studies. This result validates the position of the United Nation Sustainable Development Goals (UN-SDG's) of access to clean energy (renewable energy) and mitigation of climate change issues. This empirical study also presents conclusions that are useful for policymakers and stakeholders.

Keywords CO_2 emissions · Tourism arrivals · Renewable energy · Panel data

N. C. Leitão (✉)
Polytechnic Institute of Santarém, Center for Advanced Studies in Management and Economics, Évora University, Évora, Portugal
e-mail: nunocarlosleitao@gmail.com

Center for African and Development Studies, Lisbon University, 1200-781 Lisbon, Portugal

D. Balsalobre-Lorente
Department of Political Economy and Public Finance, Economic and Business Statistics and Economic Policy, University of Castilla-La Mancha, Cuenca, Spain
e-mail: daniel.balsalobre@uclm.es

D. Balsalobre-Lorente et al. (eds.), *Strategies in Sustainable Tourism, Economic Growth and Clean Energy*, https://doi.org/10.1007/978-3-030-59675-0_4

JEL Classification Q20 · Q50 · Z30

4.1 Introduction

Researchers have been concerned over the past few decades to look at the tourism sector as it contributes to the economic growth of tourist receiving economies, as well as stimulating employment and the possibility of establishing business networks.

The empirical studies have been an interest in evaluating the determinants of tourism demand (Tavares and Leitão 2017; Harb and Bassil 2018; Gallego et al. 2019), the impact of tourist arrivals on economic growth, i.e. tourism-led growth hypothesis (TLHG) (e.g. Castro-Nuño et al. 2013; Brida et al. 2016; Mitra 2019) and the relationships between the tourism sector and climate change (e.g. Solarin 2014; Shakouri et al. 2017; Işik et al. 2017; Sharif et al. 2017; Nepal et al. 2019). This present study seeks to advance in the empirical literature, checking the linkage between economic growth, renewable energy use and international tourism on carbon dioxide emissions, which empirical results offer fresh evidence of the effects of the tourism industry and renewable energy use on carbon emissions under the EKC empirical evidence.

As suggested by the literature review, (Al-Mulali et al. 2014; Shakouri et al. 2017; Işik et al. 2017; Sharif et al. 2017; Solarin 2014; Nepal et al. 2019; Etokakpanet al. 2019; Balsalobre-Lorente et al. 2020a, b) tourism sector encourages climate change and stimulate greenhouse gas. However, according to the report developed by the WTO (2018) about tourism trends in EU-28, the tourism sector is essential by the EU because of creating employment and promotes economic growth (UNWTO 2018:7).

In 2017, tourism arrivals in the EU increased by 8%, when compared with 2016. (UNWTO 2018). According to the same source (UNWTO 2018: 23–24), the destinations with the highest growth rates are Spain and Portugal, followed by Slovenia (+12%), Croatia (+8%), Greece (+5%) and Italy (+3%). The EU has encouraged sustainable tourism development developing different programmes to invest in sustainable transnational tourism products (European Commission 2010).

This paper considers the relationship between tourism arrivals, economic growth and climate change (carbon dioxide emissions) for the case of EU-28 countries for the period 2000–2014. The impact of renewable energy consumption on CO_2 emissions is also considered in this research. This study complements the extant literature by investigating the effect of tourism arrivals on climate change for several reasons. The first, the countries of EU-28 are important receptors of international tourism arrivals. Secondly, even some empirical literature has exhibited a direct connection between the tourism sector and environmental degradation (Becken and Patterson 2006; Jones and Munday 2007; Tovar and Lockwood 2008; Scott et al. 2010; Balli et al. 2019). Moreover, several studies have supported a negative connection between the tourism industry and environmental degradation (Paramati et al. 2018). In this context, this study evaluates the connection between tourism arrivals, within the environmental Kuznets curve (EKC) framework renewable energy, and climate change. Besides, it tries to verify the Tourism-Led Growth Hypothesis (TLGH) under a sustainable scenario for selected EU-28 countries, where international tourism contributes to promoting sustainable economic growth.

The study is organized as follows. Section 4.2 offers a literature review. The methodology is presented in Sect. 4.3. The empirical study is considered in Sect. 4.4. The final section discusses the conclusions and some policy recommendations.

4.2 Literature Review

This section aims to present some conclusions of the recent empirical studies that evaluate the relationship between climate change (measured by carbon dioxide emissions) and the tourism arrivals, renewable energy and the assumptions of EKC (income per capita and squared income per capita). The selected studies that we present allow giving support by EU-28 countries. As is well known, there is a EU concern to reduce carbon dioxide emissions as well as greenhouse effects. Besides, the international community has made efforts to minimize the tourism sector (e.g. UNWTO 2008).

EU-28 region is a homogeneous group of countries, where the tourism industry is directly related to income, employment and tax revenues (Paramati et al. 2018). In 2015, the contribution of the tourism industry to the GDP was near US$1610 billion, being around 10% of the total EU's GDP. The tourism industry also generated more than 26 million direct and indirect jobs. In forthcoming years, it is expected that the tourism industry sector will rise by 2.9% per annum over the next (WTTC 2016).

4.2.1 The Effects of Tourism Arrivals on Carbon Dioxide Emissions

Indeed, the empirical studies realized around tourism arrivals, and carbon dioxide emissions are inconclusive, and predominant the studies that show the tourism sector stimulate climate change. Nevertheless, there are studies that to present a negative correlation between the tourism arrivals and carbon dioxide emissions defending that tourism demand contributes to the decrease of carbon dioxide emissions and, consequently, by promoting sustainable development, decreasing greenhouse gas emissions. In this line, the studies by Ben Jebli et al. (2014, 2019), Lee and Brahmasrene (2013) demonstrate that tourist arrivals are negatively correlated with carbon dioxide emissions, thus contributing to more sustainable tourism management. Katircioglu (2014) applied panel dynamic least squares—DOLS econometric technique, which empirical results showed that tourism helps to reduce CO_2 emissions in Singapore between 1971 and 2010. Paramati et al. (2018) found similar results for a panel of EU-28 countries during 1990–2013.

On the other hand, some studies have considered a direct connection between tourism and carbon emissions. For instance, Becken and Patterson (2006) found that tourism industry leads to an increase in carbon emissions for the case of the case study of New Zealand.

Balli et al. (2019) showed that tourism increases carbon emissions in selected Mediterranean countries. Dogan et al. (2015) obtained a similar relationship, using dynamic panel least square—DOLS methodology. Therefore, Yorucu (2016) explored the impact of tourist arrivals on carbon emissions from 1960 to 2010 in Turkey, which empirical results suggested that the ascending number of foreign tourist arrivals would increase the carbon emissions.

The relationship between income per capita, renewable energy, tourism arrivals, trade and foreign direct investment on climate change was analysed by Jebli et al. (2019). The authors considered panel data (panel fully modified least-squares—FMOLS, panel dynamic least squares—DOLS) for Central and South American Countries to cover the period 1995–2010. The study proves that renewable energy, tourism arrivals and foreign direct investment permit to decrease CO_2 emissions. However, the study also demonstrates that income per capita and trade stimulate climate change.

Most empirical studies show that tourism demand accentuates climate change and global warming. If we think that tourism demand inputs contribute to economic growth, we easily accept that tourist inputs contribute to environmental degradation and, consequently, to carbon dioxide emissions. Based on the assumptions of the EKC equation, the studies by Nepala et al. (2019), Shakouri et al. (2017), Işik et al. (2017), Sharif et al. (2017), Solarin (2014) and Al-Mulali et al. (2014) demonstrate that tourist arrivals can contribute to the increase of carbon dioxide emissions, causing environmental degradation.

The empirical study of Nepal et al. (2019) considered tourism arrivals, energy consumption and CO_2 emissions in Nepal. Using an econometric methodology autoregressive distributed lag—ARDL model and Granger causality, the authors found that tourism presents a positive effect on carbon dioxide emissions and in investment. However, energy consumption is negatively correlated with tourism arrivals.

The research proposed by Shakouri et al. (2017) considered panel data applied to the Asian-Pacific region for 1995–2015. The econometric results prove that income per capita presents a positive correlation with CO_2 emissions, and squared income per capita presents a negative effect on CO_2 emissions, showing that this empirical work gives support to environmental Kuznets assumptions. Besides, tourism arrivals have a positive impact on CO_2 emissions in the long term.

The empirical study of Işik et al. (2017) evaluates the cointegration among economic growth, financial development, trade, tourism and CO_2 emissions. The authors considered the Greek experience for the period 1970–2014. Using time-series (autoregressive distributed lag—ARDL model and vector error correction model—VECM), the authors demonstrate that economic growth, trade, and tourism origin an increase of CO_2 emissions in Greece. In this context, the Pakistan experience was considered by Sarif et al. (2017) using the ARDL model, and FMOLS, DOLS estimators, for the period 1972–2013. The panel FMOLS and DOLS estimators show that tourism arrivals, income per capita and foreign direct investment have a positive impact on CO_2 emissions. ARDL model also demonstrates that carbon dioxide emissions, in the long run, present a positive effect, showing that there is climate change.

The relationship between the tourism sector, social distribution and EKC arguments and pollution haven hypothesis (PHH) by G-7 counties were investigated by Anser et al. (2019) for 1990–2015, and the authors used as econometric strategy Random Effects (RE) estimator. To test this link with carbon dioxide emission, Anser et al. (2019) selected as explanatory variables the Gini index, squared Gini, income per capita, squared income per capita, foreign direct investment flows, education expenditures, health expenditures and tourism arrivals. The variables of the Gini index, squared Gini, income per capita and squared income per capita are according to the premises of the EKC. The coefficient of tourism arrivals presents a positive correlation with CO_2 emissions.

Considering the effect of tourism on CO_2 emissions and the EKC, the empirical study of Mikayilov et al. (2019) applied to Azerbaijan using cointegration methods demonstrated that energy and trade present a positive correlation with the ecological footprint. However, the authors concluded that there is no evidence between tourism and EKC hypotheses in Azerbaijan.

4.2.2 The Effects of the Kuznets Environmental Curve on Carbon Dioxide Emissions

Since the 1990s, the relationship between growth and the Kuznets Environmental Curve has been studied (e.g. Grossman and Krueger 1991, 1993, 1995 and Douglas and Selden 1995). In line with Soytas and Sari (2009), we analyse the connection between economic growth, carbon dioxide emissions and energy consumption in a multivariate framework, we have proposed as additional explanatory variable tourism.

As a rule, empirical studies find a positive correlation between per capita income and carbon dioxide emissions, demonstrating that economic growth promotes climate change and greenhouse effects. Regarding the ratio of squared income per capita and CO_2 emissions, studies find a negative association. Thus, from a medium- to long-term perspective, countries achieve higher states of development, which justifies the correlation between squared income per capita and carbon dioxide emissions, i.e. an inverted U-shaped. When conducting a literature survey, it is observed that there are studies that use time series (autoregressive models, ARDL, ARMA, VECM and Granger Causality), which highlight the studies by Esteve and Tamarit (2012), Baek (2015), Shahbaz et al. (2015), Özokcu and Özdemir (2017), Alshehry and Belloumi (2017), and Och (2017). Also, other types of studies such as Kim et al.

Table 4.1 Selection of empirical studies of EKC

Empirical studies	Time period	Econometric strategy	Country
Kim et al. (2016)	2000–2012	Panel Data	Korea
Leitão and Shahbaz (2013)	1970–2009	Time series	Portugal
Shahbaz et al. (2015)	1971–1980	Time series	Portugal
Leitão and Shahbaz (2013)	1980–2010	Panel data	18 countries
Özokcu and Özdemir (2017)	1980–2010	Panel data	26 OECD countries and 52 emerging countries
Alshehry and Belloumi (2017)	1971–2011	Time series	Saudi Arabia
Och (2017)	1981–2012	Time series	Mongolia
Romero et al. (2017)	1995–2009	Panel data	27 EU countries
Apergis (2016)	1960–2013	Panel data	15 OECD countries
Onafowora and Owove (2014)	1970–2010	Time series	Pairs Countries
Zaman et al. (2016)	1965–2011	Panel data	34 developed and developing countries

Source Authors composition and selection

(2016), Balogh and Jambor (2017), Apergis (2016), Leitão and Shahbaz (2013), Leitão (2015), Olale et al. (2018) and Romero et al. (2017) used panel data (Fixed Effects—FE, RE, Generalized Moments System Method—GMM-System or Panel Cointegration).

Zaman et al. (2016) explored the connection between carbon emissions and the tourism sector. They confirmed through DOLS technique the EKC for 34 developed and developing countries during the period 1965–2011.

Table 4.1 presents a selection of the EKC studies, and the selected studies validate the EKC hypotheses.

4.2.3 *The Effects of Renewable Energy on Carbon Dioxide Emissions*

In recent years, researchers have been addressing the link between renewable energy and climate change. Most studies apply the assumptions of the EKC. Thus, the studies by Tiwari (2011), Vasylieva et al. (2019), Zoundi (2017), Acheampong et al. (2019), Liu et al. (2017), Bilan et al. (2019) found a negative association between renewable energies and carbon dioxide emissions, demonstrating that renewable energies reduce carbon dioxide emissions and greenhouse effects.

India experience was investigated by Tiwari (2011), the author used a structural VAR model and concluded that renewable energy causes economic growth, and the use of renewable energy decrease CO_2 emissions. Moreover, Tiwari (2011 also demonstrated that income per capita has a positive effect on carbon dioxide emissions.

Also, ascending international tourism has increased the use of energy use, not only in transportation, construction, or hospitality, where many tourist-related activities increase pollution levels (Becken et al. 2001, 2003; Gössling 2002). Becken and Simmons (2002) found that the tourism industry is a key contributor to energy consumption, leading to hasten environmental degradation. In consequence, the promotion of renewable energy sources will both reduce emission and promote sustainable tourism industry (Etokakpan et al. 2019; Balsalobre-Lorente et al. 2018, 2020a, b).

The empirical study of Zoundi (2017) tested the EKC hypotheses for 25 African countries for the period 1980–2012. Using DOLS, FEs and GMM-System, the study doesn't find in totality EKC assumptions. But income per capita presents a positive impact on CO_2 emissions, and renewable energy is negatively correlated with carbon dioxide emissions.

The ECK in Asian countries (Indonesia, Malaysian, Philippines and Thailand) was investigated by Liu et al. (2017) for the period 1970–2013. The authors used OLS, FMOLS, DOLS estimators as an econometric strategy. The results of this research don't find support on ECK hypotheses. However, the renewable energy and agriculture sector encourage the decrease of CO_2 emissions, and energy consumption (non-renewable energy) is positively correlated with carbon dioxide emissions.

The interconnection between renewable energy, carbon dioxide emissions and growth for candidates or potential candidates of EU memberships was investigated by Bilan et al. (2019). The econometric results with FMOLS and DOLS demonstrate that renewable energy consumption is negatively correlated with CO_2 emissions. Furthermore, renewable energy consumption promotes economic growth; nevertheless, income per capita is positively associated with CO_2 emissions. In this context, Acheampong et al. (2019) considered the association between globalization in Sub-African, using panel data (FE, RE and GMM-System) for the period 1980–2015. The results show that CO_2 emissions present a positive sign in the long term, demonstrating that climate changes increased. Though renewable energy encourages a decrease of CO_2 emissions, and income per capita, and squared, income per capita has support in ECK assumptions. The empirical study of Vasylieva et al. (2019) using a panel cointegration (FMOLS, DOLS) also shows that income per capita presents a positive effect on CO_2 emissions, and squared income per capita present a negative impact on carbon dioxide emissions.

Moreover, the variable of corruption is negatively correlated with CO_2 emissions. In this context, the research of Araby et al. (2019) considered the effect of renewable energy and energy consumption on carbon dioxide emissions applied to Euro Mediterranean countries. For the period 2002–2016, the authors used a FE estimator, and the empirical results show that renewable energy is negatively correlated with CO_2 emissions, indicating that renewable energy allows decreasing climate change. Furthermore, the study exhibited that nuclear and coal electricity encourages climate change. Finally, Araby et al. (2019) found a positive relationship between economic growth and CO_2 emissions. A different perspective has the study of Adams and Nsiah (2019), demonstrating that renewable energy and non-renewable energy present a positive effect on carbon dioxide emissions. The authors utilized a panel cointegration for the period 1980–2014 to sub-Sahara African countries. Nevertheless, in the long run, only non-renewable energy has a positive effect on CO_2 emissions.

Consequently, the impact of renewable and non-renewable energy applied to Pakistan was studied by Zaidi et al. (2018), and the authors used as econometric strategy ARDL model for the period 1970–2016. The empirical results show that non-renewable energy and economic growth stimulate carbon dioxide emissions, and renewable energy presents an insignificant effect on Carbon dioxide emissions (Zaidi et al. 2018).

4.3 Methodology

The effects of tourism arrivals, renewable energy on carbon dioxide emissions are considered in section, utilizing a panel data such as unit root (Levin et al. 2002; Im et al. 2003; ADF–Fisher Chi-square and Phillips–Perron); panel cointegration test suggested by Pedroni (2001, 2004), RE, DOLS and Panel Granger causality test were also considered in this study. The panel unit root test was considered to evaluate the variables used in this research are stationary at the level or cointegrated in differences. The methodologies of Pedroni (2001, 2004), and Kao Residual Cointegration Test (1999) were applied to evaluate the cointegration among variables in the long run. The cointegration test suggested by Pedroni (1999) is evaluated by with-dimension (panel cointegration) and between-dimension (mean panel). The statistics of four groups of tests, namely panel v-statistic (nonparametric, used on the variance ratio), panel-rho statistic, panel P.P.- statistic, and panel ADF- statistic are proposed in with-dimension. Moreover, in between-dimension, the residuals are evaluated by three statistics, specifically group rho-statistic; group PP-statistic; group ADF-statistic. The next step considered is the Kao Residual Cointegration Test (1999) that showed if there exists or not cointegration (the null hypothesis) between variables measured by ADF statistic, this test a complementary test of Pedroni (1999).

Subsequently, the econometric models were estimated by DOLS considered the arguments of Saikkonen (1991) and Stock and Watson (1993) to determine the long-run relationship among the variables.

The study uses as dependent variable carbon dioxide emissions (CO_2), and the independent variables are income per capita (Y), squared income per capita (Y^2), renewable energy (REN) and tourism arrivals ($TOUR$) for the period 2000–2014.[1]

Based on the studies of Leitão and Shahbaz (2013), Lim and Won (2019), Balsalobre-Lorente et al. (2019), Anser et al. (2019), and Acheampong et al. (2019) the following function is considered:

[1]The choice of the data is confined to available of data.

$$CO_2 = f\left(Y, Y^2, REN, TOUR\right) \tag{4.1}$$

Given the general function presented, we formulate two models. All variables are expressed in the logarithmic form:

Model [1]:

$$LogCO_2 = \beta_0 + \beta_1 LogY + \beta_2 LogY^2 + \beta_3 LogREN + \delta_t + \eta_i + \varepsilon_{\text{it}} \tag{4.2}$$

Model [2]:

$$LogCO_2 = \beta_0 + \beta_1 LogY + \beta_2 LogY^2 + \beta_3 LogREN + \beta_4 LogTOUR + \delta_t + \eta_i + \varepsilon_{\text{it}} \tag{4.3}$$

where the dependent variable is CO_2, i.e. carbon dioxide emissions expressed in kte from the World Bank (2020).

The explanatory variables are the following:

Y—represents gross domestic product expressed in US dollars, from the World Bank Indicators (2019).

Y^2—squared income per capita expressed in US dollars, from the World Bank Indicators (2019).

REN—represents a percentage of renewable energy in total final energy consumption. The source of this proxy is the World Bank Indicators (2019).

TOUR—Number of arrivals, overnight visitors from the World Tourism Organization.

δ_t_signifies the common deterministic trend.

η_i—represents the unobserved time.

ε_{it}—denotes random disturbance considered normal and identically distributed. Regarding the literature review (Anser et al. 2019; Bilan et al. 2019; Balsalobre-Lorente et al. 2019, and Jeli et al. 2019; Etokakpan et al. 2019; Balsalobre-Lorente et al. 2020a, b), we formulate the following hypotheses:

H_1: There is a positive correlation between income per capita and CO_2 emissions.

H_2: There exists a negative relationship between squared income per capita and CO_2 emissions.

Empirical studies in the area of energy and environmental economics have been testing the EKC with the highest incidence in the last two decades. Thus, there is an extensive and abundant quantity of studies, both with time series and panel data. The inverted U curve shows that economies can go through different stages of development. Thus, at an early stage, countries are concerned only with economic growth, which causes environmental damage. As can be referred from a given moment, when economies have industrialization, countries become aware of sustainable development and climate and environmental issues.

The hypotheses (H_1 and H_2) are formulated based on assumptions of the EKC. The recent studies of Balogh and Jambor (2017), Acheampong et al. (2019), Balsalobre-Lorente et al. (2019), and Anser et al. (2019) give support to this hypothesis.

H_3: Renewable energy is negatively correlated with carbon dioxide emissions.

The issue of the use of renewable energy and energy efficiency is recent in empirical studies. The researchers have used non-renewable energy consumption more often to test its relationship to climate change. The literature review concludes that the consumption of non-renewable energy accentuates climate change and global warming (e.g. Leitão and Balogh 2020). In turn, the use of cleaner energy, also known as renewable energy, made it possible to meet the requirements set out in the Kyoto Protocol (1997).

The previous studies of Balsalobre-Lorente et al. (2019), Acheampong et al. (2019), Zoundi (2017) give support to our hypothesis.

H_4: Tourism arrivals diminish carbon emissions.

All economic activities promote economic growth, cause environmental degradation and accentuate the carbon footprint. Thus, the formalization of this hypothesis aims to understand whether tourism has environmental impacts in the EU-28, considering the report on climate change and its correlation with the tourism sector (UNWTO 2008). We expect a negative connection between tourism and carbon emissions for the selected panel, in line with results offered by Lee and Brahmasrene (2013), Katircioglu (2014), Ben Jebli et al. (2014), Jebli et al. (2019) or Paramati et al. (2018).

Table 4.2 presents the variables and the expected signs considered by the literature review to explanatory variables.

Table 4.2 Definitions of variables

Dependent variable		Source
LogCO_2—Logarithm of carbon dioxide emissions		World Bank: World Development Indicators (2019)
Explanatory variables	*Expected signs*	*Source*
LogY—Logarithm of income per capita based on purchasing power parity (PPP)	[+]	World Bank: World Development Indicators (2019)
LogY^2—Logarithm of Squared income per capita based on purchasing power parity (PPP)	[-]	World Bank: World Development Indicators (2019)
LogREN—Logarithm of renewable energy	[-]	World Bank: World Development Indicators (2019), and International Energy Agency
LogTOUR—Logarithm of tourism arrivals	[-]	World Bank: World Development Indicators (2019), and World Tourism Organization

Source Authors composition

4.4 Econometric Results

Table 4.3 shows the descriptive statistics for each variable used in this study. The variables of squared income per capita ($LogY^2$), tourism arrivals ($LogTOUR$) and carbon dioxide emissions ($LogCO_2$) present the higher values of means. Furthermore, the variables of squared income per capita ($LogY^2$), tourism arrivals ($LogTOUR$) and carbon dioxide emissions ($LogCO_2$) are the higher values of Maximum.

The correlations between all variables considered in this research are presented in Table 4.4. The explanatory variables have a positive impact on CO_2 emissions. The income per capita ($LogY$) is positively correlated with squared income per capita ($LogY^2$), renewable energy ($LogREN$) and with tourism arrivals ($LogTOUR$). Besides, the variable of squared income per capita ($LogY^2$) has a positive relationship with tourism arrivals ($LogTOUR$) and negatively correlated with renewable energy ($LogREN$).

Table 4.5 shows the unit root test for the variables used in this study, considering the criteria proposed by Levin et al. (2002), ADF–Fisher Chi-square and Phillips–Perron (e.g., Choi 2001). According to Table 4.5, all variables are cointegrated at the first difference, I (1).

The results of the Pedroni panel cointegration test are reported in Table 4.6. We observe that the variables considered in this study are cointegrated (Table 4.7).

Considering results (Kao Residual Cointegration Test), we observe that the variables used in this research are cointegrated in the long run.

In Table 4.8, we present DOLS estimation results:

Table 4.3 Descriptive statistics

Variables	Mean	Std. Dev	Min	Max
$LogCO_2$	4.738	0.842	0.000	5.931
$LogY$	4.400	0.420	0.000	5.005
$LogY^2$	8.859	0.398	7.635	10.011
$LogREN$	1.012	0.455	-1.058	1.726
$LogTOUR$	6.854	0.548	5.772	7.926

Source Authors calculation considering the World Bank Indicators database

Table 4.4 Correlation between variables

Variables	$LogCO_2$	$LogY$	$LogY^2$	$LogREN$	$LogTOUR$
$LogCO_2$	1.000				
$LogY$	0.478	1.000			
$LogY^2$	0.182	0.580	1.000		
$LogREN$	0.144	0.104	0.107	1.000	
$LogTOUR$	0.607	0.047	0.177	0.005	1.000

Source Authors calculation considering the World Bank Indicators database

Table 4.5 Panel unit root test

	(A) Null: unit root (assumes common unit root process)		(B) Null: unit root (assumes individual unit root process)			
	Levin, Lin and Chu t		ADF–Fisher Chi-square		PP–Fisher Chi-square	
	t-Statistic	Prob.	t-Statistic	Prob.	t-Statistic	Prob.
At level						
$LogCO_2$	−5.022***	(0.000)	84.990***	(0.000)	83.145**	(0.010)
$LogY$	12.817	(1.000)	1.933	(1.000)	2.579	(1.000)
$LogY^2$	10.665	(1.000)	2.038	(1.000)	0.349	(1.000)
$LogREN$	7.402	(1.000)	6.319	(1.000)	5.988	(1.000)
$LogTOUR$	7.792	(1.000)	3.708	(1.000)	2.380	(1.000)
At First difference						
$\Delta\ LogCO_2$	−12.698***	(0.000)	209.872***	(0.000)	382.841***	(0.000)
$\Delta LogY$	−6.940***	(0.000)	100.209***	(0.000)	169.289***	(0.000)
$\Delta\ LogY^2$	−6.085***	(0.000)	93.713***	(0.000)	125.778***	(0.000)
$\Delta\ LogREN$	−6.971***	(0.000)	147.208***	(0.000)	256.667***	(0.000)
$\Delta\ LogTOUR$	−10.0105***	(0.000)	169.949***	(0.000)	225.094***	(0.000)

Source Authors calculation considering the World Bank Indicators database. *Note* Statistically significant at 1% level (***)

Table 4.6 Pedroni panel cointegration test

	Statistic	Prob.	Weighted statistic	Prob.
Within-dimension				
Panel v-Statistic	2.960***	(0.001)	−1.627	(0.948)
Panel rho-Statistic	−0.299	(0.382)	0.900	(0.816)
Panel PP-Statistic	−3.004***	(0.001)	−1.462*	(0.071)
Panel ADF-Statistic	−3.004***	(0.001)	−1.477*	(0.069)
Between-dimension				
Group rho-Statistic	1.247	(0.894)		
Group PP-Statistic	−5.573***	(0.000)		
Group ADF-Statistic	−2.737***	(0.003)		

Source Authors calculation considering the World Bank Indicators database. *Note* Statistically significant at 1% (***) and 10% level (*)

Table 4.7 Kao Residual Cointegration Test

	t-Statistic	Prob
ADF	−12.825***	(0.000)
Residual variance	0.169	
HAC variance	0.169	

Note Authors calculation considering the World Bank Indicators database. *Note* Statistically significant at 1% (***)

Table 4.8 Panel Dynamic Least Squares (DOLS)

Variables	Model [1] Coef.	Model [2] Coef.	Expected signs
LogY	1.256*** (0.000)	1.283*** (0.000)	[+]
$LogY^2$	−0.310* (0.044)	−0.504*** (0.002)	[-]
LogREN	−0.033*** (0.000)	−0.104 (0.43)	[-]
LogTOUR		−0.623* (0.066)	[-]
Obs	422	422	
Adj R^2	0.77	0.78	
S.E. of regression	0.40	0.41	
Long- run variance	0.144	0.152	

Source Authors calculation considering the World Bank Indicators database. Dependent variable: logarithm of carbon dioxide emissions ($LogCO_2$). *Note* Statistically significant at 1% (***), and 10% level (*)

Table 4.8 (Fig. 4.1) reports the econometric results with the DOLS estimator. In the model [1], all explanatory variables are statistically significant. The coefficients of income per capita *(LogY)* and renewable energy *(LogREN)* are statistically significant at a 1% level, and the variable of squared income per capita ($LogY^2$) is statistically significant at a 10% level. In model [2], it is possible to observe that income per capita *(LogY),* and squared income per capita ($LogY^2$) are statistically significant at 1% level, and tourism arrivals (*LogTOUR*) are statistically significant at 10% level. The studies of Özokcu and Özdemir (2017), Alshehry and Belloumi (2017), and Och (Och 2017) give support to our results $\beta_1 > 0$; $\beta_2 < 0$, and there are according to U-inverted EKC assumptions.

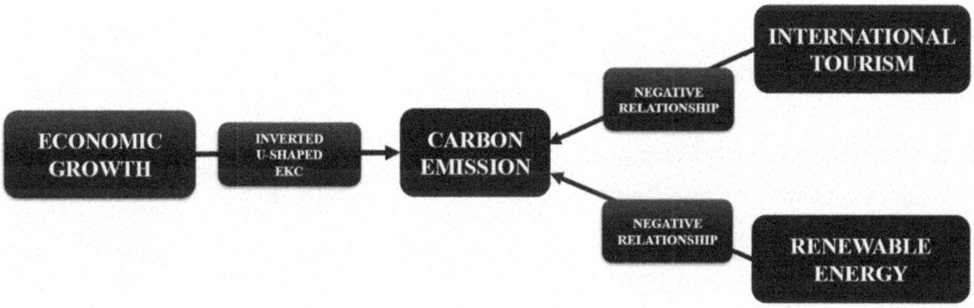

Fig. 4.1 Scheme of empirical results

The variables of income per capita and squared income per capita are according to the hypothesis of EKC. The variable of renewable energy *(LogREN)* presents the expected sign, i.e. a negative impact on CO_2 emissions ($\beta_3 < 0$), demonstrating that renewable energy contributes to the quality of environmental. This result has support in the empirical studies of Tiwari (2011), Vasylieva et al. (2019), Zoundi (2017), Acheampong et al. (2019), Liu et al. (2017), Bilan et al. (2019). Moreover, the empirical studies of Acheampong et al. (2019), Liu et al. (2017), Bilan et al. (2019) also found a negative correlation between renewable energy and carbon dioxide emissions ($\beta_4 < 0$); our result is according to these studies.

In conclusion, tourism arrivals *(LogTOUR)* is negatively correlated with CO_2 emissions. Our empirical results are in line with Katircioglu (2014), Ben Jebli et al. (2014), Jebli et al. (2019) or Paramati et al. (2018).

Finally, Table 4.9 presents the results with the panel Granger causality test.

Figure 4.2 (based in Table 4.9) shows a bidirectional causality between income per capita *(LogY)* and CO2 emissions (Halicioglu 2009; Ghosh 2010; Wang 2011; Amiri and Ventelou 2012; Elmi and Sadeghi 2012), displays that economic activity stimulates the climate change. The variables of renewable energy (*LogREN*) and CO_2 emissions (*LogCO2*) also have a bidirectional causality. This result is consistent with Salim and Rafiq (2012), who found the same relationship for Brazil, China and India. Besides, renewable energy (*LogREN*) and economic growth (*LogY*) present a bidirectional causality (Apergis and Payne 2010).

Our results with the panel Granger causality test prove that there is bidirectional causality between tourism arrivals (*LogTOUR*) and carbon dioxide emissions ($LogCO_2$) in line with (Ben Jebli et al. 2014). The bidirectional causality between economic growth and carbon dioxide validates the feedback hypothesis. Finally, the

Table 4.9 Panel granger causality tests

Null hypothesis:	Obs	F-Statistic	Prob.
Bidirectional causality			
LogY does not Granger Cause $LogCO_2$	395	170.998***	(0.000)
$LogCO_2$ does not Granger Cause *LogY*		7.737***	(0.005)
LogREN does not Granger Cause $LogCO_2$	394	3.967***	(0.004)
$LogCO_2$ does not Granger Cause *LogREN*		42.059***	(0.000)
LogTOUR does not Granger Cause $LogCO_2$	395	172.531***	(0.000)
$LogCO_2$ does not Granger Cause *LogTOUR*		3.253*	(0.072)
LogREN does not Granger Cause *LogY*	394	20.737***	(0.000)
LogY does not Granger Cause *LogREN*		108.781***	(0.000)
Unidirectional causality			
LogTOUR does not Granger Cause *LogY*	395	12.4284***	(0.000)
LogREN does not Granger Cause LogTOUR	366	3.58136*	(0.059)

Source Authors calculation considering the World Bank Indicators database. *Note* Statistically significant at 1% (***), 10% level (*)

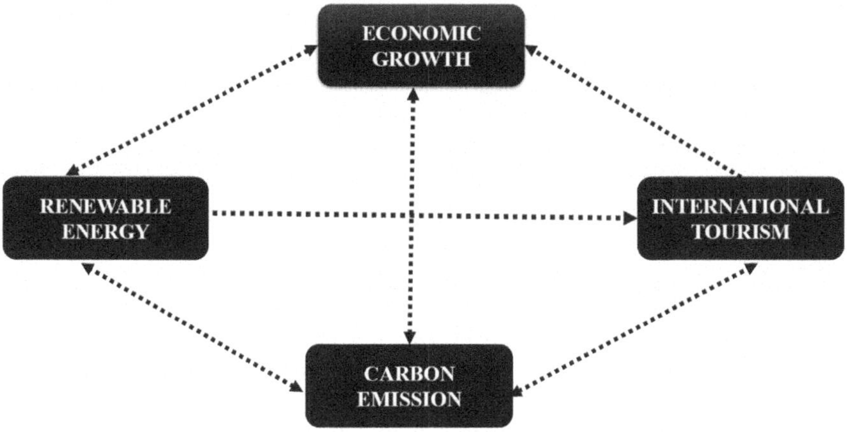

Fig. 4.2 Scheme of causality test

panel Granger causality test demonstrates that there is a unidirectional causality between (*LogTOUR*) and economic growth (*LogY*) showing same results than Lee and Chang (2008), Malik et al. 2010 and Risso et al. (2010), and renewable energy (*LogREN*) and tourism sector (*LogTOUR*).

4.5 Conclusions

This study is constructed in line with the vision 2030 of the sustainable goals with a focus on access to clean energy access and mitigation of climate change issues as a result of anthropogenic activities, which by extensions if properly managed translates into the sustainable economic growth. This study explores the effects of tourism arrivals, renewable energy, economic growth on CO_2 emissions and panel data in the context of EU-28. The results and discussion started with the unit root test analysed by the methodology of Levin et al. (2002); ADF–Fisher Chi-square and Phillips–Perron (e.g. Choi 2001) showing that all variables used in this research are cointegrated at the first difference, I(1). This methodology was complemented by Pedroni panel cointegration (1999) that evaluated the panel cointegration (with-dimension) and mean panel (between-dimension). Kao Residual Cointegration Test (1999) and these tests demonstrated that the variables used in this research are cointegrated in the long run.

The econometric models are tested with DOLS. The empirical results reveal that income per capita and squared income per capita are according to EKC assumptions. Therefore, economic growth (income per capita and squared income per capita) confirms an inverted shaped curve which is consistent with the empirical studies of Tiwari (2011), Vasylieva et al. (2019), Zoundi (2017), Acheampong et al. (2019), Liu et al. (2017), Bilan et al. (2019), Balsalobre-Lorente et al. (2019); the renewable energy consumption is negatively correlated with CO_2 emissions, showing that the renewable energy consumption reduces climate change.

Our study confirms that the tourism industry may reduce environmental damage in EU-28 countries. To reach a sustainable tourism process, it is necessary to promote investments in modern transport or renewable energy use, impacting the environment positively. Hence, EU-28 countries should accelerate tourism-related infrastructure investments, like road or rail transport transportation infrastructures, to preserve ecosystems and biodiversity and reduce pollution levels (Khadaroo and Seetanah 2008).

The panel Granger causality presents an interesting relationship between variables. In this context, we observe that there exists a linkage between income per capita (*LogY*) and carbon dioxide emissions (LogCO_2), referring that economic growth encourages climate change. As in Nepala et al. (2019), Anser et al. (2019), Shakouri et al. (2017), Işik et al. (2017) and Sharif et al. (2017), there is a bidirectional relationship between tourism arrivals (*LogTOUR*) and climate change (CO_2 emissions). However, there is also bidirectional causality among renewable energy (*LogREN*) and CO_2 emissions ($LogCO_2$). Another impressive contribution is that renewable energy (*LogREN*) and economic growth (*LogY*) presented bidirectional causality.

Thus, considering the empirical results in this study, it is possible to present some recommendations. In terms of economic policy recommendations, we think that policymakers should consider the principle of sustainable development, from sustainable tourism advocated by the United Nations World Tourism Organization (UNWTO 2008), which sought to obtain a balance between climate change and the tourism sector. So, as the report of UNWTO (2008:35) refers, strategies should include reducing energy use, improving energy efficiency and increasing renewable energy. It will be necessary to change the paradigm, i.e. a government policy by the Member States of the European Union based on the sustainability of natural resources and where the incentive to ecotourism will significantly reduce carbon dioxide emissions. The international community has long been questioning environmental degradation (Kyoto Protocol 1997) and, more recently, the Paris Agreement (2015).

Moreover, concerning the future work, it will be interesting to introduce other explanatory variables such the index of globalization (KOF globalization index, proposed by Dreher 2006 and revisited by Gygli et al. 2019), and corruption index (Corruption Perceptions Index, suggested by Transparency International) and test the relationship between these proxies and climate change, as well the impact on tourism arrivals and the economic growth to EU-28. Furthermore, it will also be interesting to evaluate the effect of rural tourism and ecotourism on carbon dioxide emissions.

References

Acheampong, A. O., Adams, S., & Boateng, E. (2019). Do globalization and renewable energy contribute to carbon emissions mitigation in Sub-Saharan Africa? *Science of the Total Environment, 667,* 436–446.

Adams, S., & Nsiah, C. (2019). Reducing carbon dioxide emissions; Does renewable energy matter? *Science of the Total Environment, 693*(133288), 1–9.

Al-Mulali, U., Fereidouni, H. G., & Lee, J. Y. M. (2014). Electricity consumption from renewable and non-renewable sources and economic growth: Evidence from Latin American countries. *Renewable and Sustainable Energy Reviews, 30,* 290–298.

Alshehry, A. S., & Belloumi, M. (2017). Study of the environmental Kuznets Curve for transport carbon dioxide emissions in Saudi Arabia. *Renewable and Sustainable Energy Reviews, 75,* 1339–1347.

Amiri, A., & Ventelou, B. (2012). Granger causality between total expenditure on health and GDP in OECD: evidence from the Toda-Yamamoto approach. *Economic Letters, 116*(3), 541–544.

Anser, M. K., Yousaf, Z., Nassani, A. A., Abro, M. M. Q., & Zaman, K. (2019). International tourism, social distribution, and environmental Kuznets curve: evidence from a panel of G-7 countries. *Environmental Science and Pollution Research,* 1–14.

Apergis, N., & Payne, J. E. (2010). Renewable energy consumption and growth in Eurasia. *Energy Economics, 32,* 1392–1397.

Apergis, N. (2016). Environmental Kuznets curves: New evidence on both panel and country-level CO_2 emissions. *Energy Economics, 54,* 263–271.

Araby, L. E., Samak, N., & Ibrahiem, M. D. (2019). The impact of renewable and non-renewable energy on carbon dioxide emission: An empirical analysis for Euro Mediterranean Countries. *International Journal of Energy Economics and Policy, 9*(6), 103–108.

Baek, J. (2015). Environmental Kuznets curve for CO_2 emissions: The case of Arctic countries. *Energy Economics, 50,* 13–17.

Balli, E., Sigeze, C., Manga, M., Birdir, S., & Birdir, K. (2019). The relationship between tourism, CO_2 emissions and economic growth: a case of Mediterranean countries. *Asia Pacific Journal of Tourism Research, 24*(3), 219–232.

Balogh, J. M., & Jambor, A. (2017). Determinants of CO_2 emission: A global evidence. *International Journal of Energy Economics and Policy, 75*(5), 217–226.

Balsalobre-Lorente, D., Driha, O. M., Shahbaz, M., & Sinha, A. (2020a). The effects of tourism and globalization over environmental degradation in developed countries. *Environmental Science and Pollution Research, 27*(7), 7130–7144.

Balsalobre-Lorente, D., Driha, O. M., Bekun, F. V., & Adedoyin, F. F. (2020b). The asymmetric impact of air transport on economic growth in Spain: fresh evidence from the tourism-led growth hypothesis. *Current Issues in Tourism,* 1–17.

Balsalobre-Lorente, D., Gokmenoglu, K. K., Taspinar, N., & Cantos, C. J. M. (2019). An approach to the pollution haven and pollution halo hypotheses in MINT countries. *Environmental Science and Pollution Research, 26,* 23010–23026.

Balsalobre-Lorente, D., Shahbaz, M., Roubaud, D., & Farhani, S. (2018). How economic growth, renewable electricity and natural resources contribute to CO_2 emissions? *Energy Policy, 113,* 356–367.

Becken, S., & Simmons, D. G. (2002). Understanding energy consumption patterns of tourist attractions and activities in New Zealand. *Tourism Management, 23*(4), 343–354.

Becken, S., & Patterson, M. (2006). Measuring national carbon dioxide emissions from tourism as a key step towards achieving sustainable tourism. *Journal of Sustainable Tourism, 14*(4), 323–338.

Becken, S., Frampton, C., & Simmons, D. (2001). Energy consumption patterns in the accommodation sector—The New Zealand case. *Ecological Economics, 39*(3), 371–386.

Becken, S., Simmons, D. G., & Frampton, C. (2003). Energy use associated with different travel choices. *Tourism Management, 24*(3), 267–277.

Ben Jebli, M., Ben Youssef, S., & Apergis, N. (2014). The dynamic linkage between CO2 emissions, economic growth, renewable energy consumption, number of tourist arrivals and trade.

Bilan, Y., Streimikiene, D., Vasylieva, T., Lyulyov, O., Pimonenko, T., & Pavlyk, A. (2019). Linking between renewable energy, CO_2 emissions, and economic growth: Challenges for candidates and potential candidates for the E.U. membership. *Sustainability, 11*, 1528: 1–16.

Blundell, R., & Bond, S. (1998). Initial conditions and moment restrictions in dynamic panel data models. *Journal of Econometrics Review, 87,* 115–143.

Brida, J. G., Lanzilotta, B., & Pizzolon, F. (2016). Dynamic relationship between tourism and economic growth in MERCOSUR countries: A nonlinear approach based on asymmetric time series models. *Economics Bulletin, 36*(2), 879–894.

Castro-Nuño, M., Molina-Toucedo, J. A., & Pablo-Romero, M. P. (2013). Tourism and GDP: A meta-analysis of panel data studies. *Journal of Travel Research, 52*(6), 745–758.

Choi, I. (2001). Unit root tests for panel data. *Journal of International Money and Finance, 20*(2), 249–272.

Corruption perceptions index. https://www.transparency.org/.

Dogan, E., Seker, F., & Bulbul, S. (2015). Investigating the impacts of energy consumption, real GDP, tourism and trade on CO_2 emissions by accounting for cross-sectional dependence: A panel study of OECD countries. *Current Issues in Tourism, 12*(2), 1–19.

Douglas, H., & Selden, T. M. (1995). Stoking the fires? CO_2 emissions and economic growth. *Journal of Public Economics, 57,* 85–101.

Dreher, A. (2006). Does globalization affect growth? Evidence from a new Index of globalization. *Applied Economics, 38*(10), 1091–1110.

Elmi, Z. M., & Sadeghi, S. (2012). Health care expenditures and economic growth in developing countries: panel cointegration and causality. *Middle-East Journal of Science Research, 12*(1), 88–91.

Esteve, V., & Tamarit, C. (2012). Threshold cointegration and nonlinear adjustment between CO_2 and income: The Environmental Kuznets curve in Spain, 1857–2007. *Energy Economics, 34,* 2148–2156.

Etokakpan, M. U., Bekun, F. V., & Abubakar, A. M. (2019). Examining the tourism-led growth hypothesis, agricultural-led growth hypothesis and economic growth in top agricultural producing economies. *European Journal of Tourism Research, 21,* 132–137.

European Commission. (2010). Europe, the world's No 1 tourist destination a new political framework for tourism in Europe. COM (2010) 352 final. Brussels, Belgium: European Commission.

Faustino, H. C., & Leitão, N. C. (2007). Intra-industry trade: A static and dynamic panel data analysis. *International Advances in Economic Research, 13*(3), 313–333.

Gallego, Á., Serrano, R. M. A., & Casanueva, C. (2019). Dynamic panel data models in tourism. *Current Issues in Tourism, 22*(4), 379–399.

Ghosh, S. (2010). Examining carbon emissions economic growth nexus for India: a multivariate cointegration approach. *Energy Policy, 38,* 2613–3130.

Gössling, S. (2002). Global environmental consequences of tourism. *Global Environmental Change, 12*(4), 283–302.

Grossman, G. M., & Krueger, A. B. (1995). Economic growth and the environment. *Quarterly Journal of Economics, 110*(2), 353–377.

Grossman, G. M., & Krueger, A. B. (1991). Environmental impacts of a North American free trade agreement. NBER Working Paper, no. 3914, Washington.

Grossman, G. M., & Krueger, A. B. (1993). Environmental impacts of a North American free trade agreement. In: Garber, P. (Ed.), *The Mexico-U.S. free trade agreement*. Cambridge, MA: MIT Press.

Gygli, Savina, Haelg, Florian, Potrafke, Niklas, & Sturm, Jan-Egbert. (2019). The KOF globalisation index – Revisited. *Review of International Organizations, 14*(3), 543–574.

Halicioglu, F. (2009). An econometric study of CO_2 emissions, energy consumption, income and foreign trade in Turkey. *Energy Policy, 37*(3), 1156–1164.

Harb, G., & Bassil, C. (2018). Gravity analysis of tourism flows and the 'multilateral resistance to tourism. *Current Issues in Tourism, 23*(6), 666–678.

Im, K. S., Pesaran, M. H., & Shin, Y. (2003). Testing for unit roots in heterogeneous panels. *Journal of Econometrics, 115*(1), 53–74.

Işik, C., Kasımati, E., & Ongan, S. (2017). Analyzing the causalities between economic growth, financial development, international trade, tourism expenditure and/ on the CO_2 emissions in Greece. *Energy Sources, Part B: Economics, Planning and Policy, 12*(7), 665–673.

Jebli, M. B., Youssef, S. B., & Apergis, N. (2019). The dynamic linkage between renewable energy, tourism, CO_2 emissions, economic growth, foreign direct investment, and trade. *Latin American Economic Review, 28*(2), 1–19.

Jones, C., & Munday, M. (2007). Exploring the environmental consequences of tourism: A satellite account approach. *Journal of Travel Research, 46*(2), 164–172.

Katircioglu, S. T. (2014). Testing the tourism-induced EKC hypothesis: The case of Singapore. *Economic Modelling, 41*(4), 383–391.

Khadaroo, J., & Seetanah, B. (2008). The role of transport infrastructure in international tourism development: A gravity model approach. *Tourism Management, 29*(5), 831–840.

Kim, S. J., Jung T. Y., & Kang, S. J. (2016). Regional environmental kuznets curves and their turning points for air pollutants in Korea. *Korea and the World Economy, 17*(3):327–349.

Lee, C. C., & Chang, C. P. (2008). Tourism development and economic growth: A closer look at panels. *Tourism Management, 29*(1), 180–192.

Lee, J. W., & Brahmasrene, T. (2013). Investigating the influence of tourism on economic growth and carbon emissions: Evidence from panel analysis of the European Union. *Tourism Management, 38*, 69–76.

Leitão, N. C., & Balogh, J. M. (2020). The impact of energy consumption and agricultural production on carbon dioxide emissions in Portugal. *AGRIS on-line Papers in Economics and Informatics, 12*(1), 49–59.

Leitão, N. C. (2015). Energy consumption and foreign direct investment: A panel data analysis for Portugal. *International Journal of Energy Economics and Policy, 5*(1), 138–147.

Leitão, N. C., & Shahbaz, M. (2013). Carbon dioxide emissions, urbanization and globalization: A dynamic panel data. *The Economic Research Guardian, 3*(1), 22–32.

Levin, A., Lin, C.-F., & Chu, C.-S. J. (2002). Unit root tests in panel data: Asymptotic and finite-sample properties. *Journal of Econometrics, 108*, 1–24.

Lim, J., & Won D. (2019). Impact of CARB's tailpipe emission standard policy on CO_2 reduction among the U.S. States. *Sustainability, 11*, 1202, 1–15.

Liu, X., Zhang, S., & Bae, J. (2017). The impact of renewable energy and agriculture on carbon dioxide emissions: Investigating the environmental Kuznets curve in four selected ASEAN countries. *Journal of Cleaner Production, 164*, 1239–1247.

Malik, S., Chaudhry, I. S., Sheikh, M. R., & Farooqi, F. S. (2010). Tourism, economic growth and current account deficit in Pakistan: Evidence from cointegration and causal analysis. *European Journal of Economics, Finance, and Administrative Sciences, 22*, 21–31.

Mikayilov, J. I., Mukhtarov, S., Mammadov, J., & Azizov, M. (2019). Re-evaluating the environmental impacts of tourism: does EKC exist? *Environmental Science and Pollution Research, 26*, 19389–19402.

Mitra, S. K. (2019). Is tourism-led growth hypothesis still valid?. *International Journal of Tourism Research, 21*(5), 615–624.

Nepala, R., Irsyad, M. I., & Nepal, S. K. (2019). Tourist arrivals, energy consumption and pollutant emissions in a developing economy–implications for sustainable tourism. *Tourism Management, 72*, 145–154.

Och, M. (2017). Empirical investigation of the environmental Kuznets curve hypothesis for nitrous oxide emissions for Mongolia. *International Journal of Energy Economics and Policy, 7*(1), 117–128.

Olale, E., Ochuoodho, O., Lantz, V., & Armali, J. El. (2018). The environmental Kuznets curve model for greenhouse gas emissions in Canada. *Journal of Cleaner Production, 184*, 859–868.

Onafowora, A. O., & Owove, O. (2014). Bounds testing approach to analysis of the environment Kuznets curve hypothesis. *Energy Economics, 44,* 47–62.
Özokcu, S., & Özdemir, O. (2017). Economic growth, energy, and environmental Kuznets curve. *Renewable and Sustainable Energy Reviews, 72,* 639–647.
Paramati, S. R., Alam, M. S., & Lau, C. K. M. (2018). The effect of tourism investment on tourism development and CO_2 emissions: empirical evidence from the E.U. nations. *Journal of Sustainable Tourism, 26*(9):1587–1607.
Pedroni, P. (1999). Critical values for cointegration tests in heterogeneous panels with multiple regressors. *Oxford Bulletin of Economics and Statistics, 61,* 653–670.
Pedroni, P. (2001). Purchasing power parity tests in cointegrated panels. *Review of Economics and Statistics, 83*(4), 727–731.
Pedroni, P. (2004). Panel cointegration: asymptotic and finite sample properties of pooled time series tests with an application to the PPP hypothesis. *Econometric Theory, 20*(03), 597–625.
Peng, Z., & Li, Q. W. M. (2020). Spatial characteristics and influencing factors of carbon emissions from energy consumption in China's transport sector: An empirical analysis based on provincial panel data. *Polish Journal of Environmental Studies, 29*(1), 217–232.
Risso, W. A., Barquet, A., & Brida, J. G. (2010). Causality between economic growth and tourism expansion: empirical evidence from Trentino-Alto Adige. *Tourismos: An International Multidisciplinary Journal of Tourism, 5*(2):87–98.
Romero, M. P. P., Cruz, L., & Barata, E. (2017). Testing the transport energy–Environmental Kuznets curve hypothesis in the E.U. 27 countries. *Energy Economics, 62,* 257–269.
Saikkonen, P. (1991). Asymptotically efficient estimation of cointegration regressions. *Econometric Theory, 7*(1), 1–21.
Salim, R. A., & Rafiq, S. (2012). Why do some emerging economies proactively accelerate the adoption of renewable energy? *Energy Economics, 34,* 1051–1057.
Scott, D., Peeters, P., & Gössling, S. (2010). Can tourism deliver its "aspirational" greenhouse gas emission reduction targets? *Journal of Sustainable Tourism, 18*(3), 393–408.
Shahbaz, M., Dube, S., Ozturk, I., & Jali, Adbul. (2015). Testing the environmental Kuznets curve hypothesis in Portugal. *International Journal of Energy Economics and Policy, 5*(2), 475–481.
Shakouri, B., Yazdi, S. K., & Ghorchebigi, E. (2017). Does tourism development promote CO_2 emissions?, *Anatolia, 28*(3):444–452.
Sharif, A., Afshan, S., & Nisha, N. (2017). Impact of tourism on CO_2 emission: Evidence from Pakistan. *Asia Pacific Journal of Tourism Research, 22*(4), 408–421.
Solarin, A. S. (2014). Tourism arrivals and macroeconomic determinants of CO_2 in Malaysia, Anatolia. *An International Journal of Tourism and Hospitality Research, 25,* 228–241.
Soytas, U., & Sari, R. (2009). Energy consumption, economic growth, and carbon emissions: Challenges faced by an E.U. candidate member. *Ecological Economics, 68*(6):1667–1675.
Stock, J. H., & Watson, M. W. (1993). Simple estimator of cointegrating vectors in higher order integrated systems. *Econometrica, 61*(4), 783–820.
Tavares, J. M., & Leitão, N. C. (2017). The determinants of international tourism demand for Brazil. *Tourism Economics, 23*(4), 834–845.
Tiwari, A. (2011). Tourism, exports and FDI as a means of growth: Evidence from four Asian Countries. *Romanian Economic Journal, 14*(40).
Tovar, C., & Lockwood, M. (2008). Social impacts of tourism: An Australian regional case study. *International journal of tourism research, 10*(4), 365–378.
United Nation World Tourism Organization. (2008). *Climate change and tourism: Responding to global challenges.* UNWTO Spain. (Accessed May 2020).
Vasylieva, T., Lyulyov, O., Bilan, Y., & Streimikiene, D. (2019). Sustainable economic development and greenhouse gas emissions: The dynamic impact of renewable energy consumption, GDP, and corruption. *Energies, 12*(3289), 1–12.
Wang, Y. C. (2011). Health care expenditure and economic growth: Quantile panel-type analysis. *Journal of Economic Modelling, 28*(4), 1536–1549.

WTTC (2016). *Travel & tourism: Economic impact 2016*. European Union. London, United Kingdom: WTTC.

World Tourism Organization. (2018). *European union tourism trends,* UNWTO, Madrid. https://doi.org/10.18111/9789284419470.

Yorucu, V. (2016). Growth impact of CO_2 emissions caused by tourist arrivals in Turkey. *International Journal of Climate Change Strategies and Management.*

Zaidi, S. A. H., Danish, Hou, & Mirza, F. M. (2018). The role of renewable and non- renewable energy consumption in CO_2 emissions: a disaggregate analysis of Pakistan. *Environmental Science and Pollution Research, 25,* 31616–31629.

Zaman, K., Shahbaz, M., Loganathan, N., & Raza, S. A. (2016). Tourism development, energy consumption and environmental Kuznets curve: Trivariate analysis in the panel of developed and developing countries. *Tourism Management, 54,* 275–283.

Zoundi, Z. (2017). CO_2 emissions, renewable energy and the environmental Kuznets curve, a panel cointegration approach. *Renewable and Sustainable Energy Reviews, 72,* 1067–1075.

Chapter 5
Clean India Mission and Its Impact on Cities of Tourist Importance in India

Perfecto G. Aquino Jr., Mercia Selva Malar Justin, and Revenio C. Jalagat Jr.

Abstract Tourism is an essential industry for India for its economic advancement. Tourism contributed to one-tenth of India's GDP and 42 million jobs in 2018. India has launched various measures to promote and push the tourism industry. Amidst the various schemes of the Government of India, Clean India Mission seemed to have a positive impact on the tourism Industry. The study aims to examine the impact of the Clean India campaign on four cities of India, and in turn, the impact on the tourist flows in these cities. The cities identified for this study are the top five tourist destinations of India during 2019: Delhi, Mumbai, Chennai, Agra and Jaipur. Clean India Mission is an effort to deal with all kinds of solid waste, sanitation and hygiene issues and other forms of pollution. Clean India Mission is an initiative to restore India's natural beauty by making the country cleaner and pollution-free.

Keywords Clean india mission · India tourism · Cities of tourist · India

5.1 Introduction

Tourism is an essential industry for economic growth. It is an industry that has an excellent trickling effect and triggering impact on economic growth. Some countries are dependent on tourism for their survival as they have no much other resources. Countries like Croatia and Malta have 15% of their GDP contributed by the tourism industry. Other countries like Thailand, Jamaica and Iceland have as close as 10% of their GDP contributed by tourism. The potential of the tourism industry in bringing

P. G. Aquino Jr. (✉)
Duy Tan University, Da Nang, Vietnam
e-mail: jesusper186@gmail.com

M. S. M. Justin
Xavier Institute of Management and Entrepreneurship, Bengaluru, India
e-mail: mercia@xime.org

R. C. Jalagat Jr.
Al-Zahra College for Women, Muscat, Oman
e-mail: Revenio@zcw.edu.om

D. Balsalobre-Lorente et al. (eds.), *Strategies in Sustainable Tourism, Economic Growth and Clean Energy*, https://doi.org/10.1007/978-3-030-59675-0_5

economic strength to a nation is very high. For India, tourism is essential to improve employment opportunities, the income of individuals and the nation, earn foreign exchange and attract foreign direct investment. India has been working hard to improve its tourism industry and improve the economic benefits that trickle from this industry for quite long. In this regard, India has made many attempts to improve the tourists flow into India. The facts presented below in Table 5.1 would stand evidence to the fact that India has been improving slowly but steadily on its economic benefits from the tourism industry. Various factors positively influence tourism. Certainly, cleanliness and hygiene add value to the place of tourist attraction. With cleanliness and hygiene, any tourist location would draw more tourists, both domestic and foreign. Failing which even the best tourist attractions can miss the potential and opportunity to attract tourists. Indian cities under the Clean India Mission (CIM) had the benefit and the opportunity to continuously improve their cleanliness, hygiene and reduce their pollution levels.

The essentiality of the Clean India Mission had been felt like a tourism booster not only by the Indian government but also their constituents considering it as a critical factor in economic development and employment generation. Many believed that emphasis on cleanliness and proper hygiene had impacted the increasing trend of tourism in the country in both domestic and international arena and the evidence of tourism potential's full realization. The common notion that the first impression lasts a lifetime means a lot in developing the image of India into a tourism hub in Asia. However, others are also pessimistic about the programme's effectiveness and sustainability. Mixed responses were obtained from various stakeholders in determining whether the Clean India Mission holistically indeed contributed to the tourism growth and well-being of the country that prompted the authors to conduct its investigation to obtain a clearer perspective and evidence to support whether or not these are affirmable. Thus, the primary objective of this study is to investigate the impact of the Clean India Mission, specifically in for cities in India as well as its influence on the tourist flows as identified in this paper. Examination of recent and relevant literature was endeavoured to evaluate the status of tourism in India, cities of tourism, listed Indian cities under the top 100 global destinations, Top 5 cleanest cities.

For the past 5 years, statistics showed the growth of tourism in India (See Table 5.1). This was evidenced by many indicators such as the number of foreign tourists' arrivals, number of deistic tourists, foreign exchange earnings from foreign tourists, etc., although the percentage annual growth of Foreign Tourist Arrivals (FTA) was not consistently growing as shown in the Table 5.1.

Table 5.1 reveals that India's tourism has been growing over the last 5 years. The number of foreign tourists' arrival (FTA) a key indicator has grown for 5 years from 2014 to 2019 from 7.68 million to 10.56 million (72.72%) The percentage of annual growth of FTA has not shown consistent growth. The highest percentage growth of FTA has been during 2018 with 14% annual growth. The number of deistic tourists has increased from just 1290.12 million in 2014 to 1854.9 million in 2019, showing an increase of approximately 69.55% over the 5 years. The highest annual growth has been during 2017 with 12.68% since 2014.

Table 5.1 India tourism: vital statistics

India tourism statistics	2019	2018	2017	2016	2015	2014
No. of foreign tourist arrivals in India	10.56 million (P)	10.04 million	8.8 million	8.89 million	8.03 million	7.68 million
Annual growth rate	5.20%	14.00%	9.70%	10.70%	4.50%	10.20%
No. of domestic tourist visits to all States/Uts	1854.9 million(R)	1652.49 million	1613.55 million		1432 million	1290.12 million
Annual growth rate	11.90%	2.30%	12.68		11.6	12.9
Estimated foreign exchange earnings from tourism						
(i) In INR terms	1,94,892 crore (# 2)	Rs. 177,874 crore (#2)	154,146 crore	135,193 crore	123,320 crore	(i) In INR terms
Annual growth rate	9.60%	15.40%	14%		9.60%	14.50%
(ii) In US$ terms, US$	28.585 billion (#2)	US$ 27.31 billion (#2	US$ 22.92 billion		26.07 billion	20.24 billion
Annual growth rate	4.70%	19.10%	8.80%		4.10%	9.70%
Share of India in international tourist arrivals	1.24%	1.17%	1.18%		0.68%	0.68%
• India's rank in world tourist arrivals	25th	26th	25th		40th	41.00%
• Share of India in international tourism receipts (US$ terms)	1.97%	2.05%	1.88%		1.71%	1.62%
• India's rank in world tourism receipts	13th	13th	13th		14th	15th

Source Compiled by the authors

Table 5.2 Indian cities listed in Global 100 tourist destinations

Sl. No	Cities	Rank 2019*	Rank 2017**	Rank 2016*	Rank 2015*
1	New Delhi	11	3	48	28
2	Mumbai	14	2	27	30
3	Agra	26	–	–	45
4	Jaipur	34	–	–	52
5	Chennai	36	1	30	43
6	Kolkata	74	4	62	–
7	Bengaluru	100	6	–	–
8	Pune	–	5	91	–

Source Compiled by the author (Based on MasterCard Index for the specific years). *Note* *Global rank, **Indian rank

There has been a consistent growth in the foreign exchange earnings from foreign tourists over the 5 years in both rupee terms and US$ terms. The average annual growth of foreign exchange earnings in terms of both rupee and dollars is approximately 9%. India has consistently improved its share of foreign tourist arrivals in terms of number and tourism receipt over the 5 years. It has been improving its rank as well.

The chapter addresses the question if *Clean India Mission* has resulted in cleaner cities in India, drawing and pulling more tourists. Dutta and Baskar (2017) said that Clean India Mission has not only improved India's rural and urban sanitation scenario but seems to have had a positive effect on India's ranking in the Travel and Tourism Competitive Index as well. India moved from rank 52 to 40 in a year of implementation of Clean India Mission that focused on urban and rural hygiene, sanitation and cleanliness. The cities chosen for the study are Jaipur, Cochin, Coimbatore and Indore (Table 5.2).

Some Indian cities have been ranked in the top 100 global destinations according to the MasterCard Index. The Index indicates that Indian cities have reached a better ranking over a period of 5 years from 2015 to 2019. Indian cities are becoming globally desired destinations over the last 5 years. Some of them have topped the ASEAN top 10—Mumbai during 2017 and Chennai during 2016. Thus, Indian cities are becoming globally desired destinations of tourist importance. Many government initiatives have brought about a positive impact like 'Incredible India', e-visa, etc. Apart from all these initiatives focused on international tourists, India also has been working on internal development initiatives—one such is Swachh Bharat Abhiyan or Clean India Programme. Clean India Programme is expected to have made positive impacts on the cleanliness, hygiene and pollution levels of the cities.

5.1.1 Clean India Mission

Clean India Mission was a nation-wide campaign in India for the period 2014–2019 that aimed to clean up the street, roads and infrastructure of cities towns, urban and rural cities and area in India. The campaign's official name is in Hindi and translates to "Clean India Mission" in English. The campaign aimed to have a clean India presentable and appealing to international tourists has been in place since the early days. The campaign has been started by earlier governments and has existed in different names over the years. Under the Narendra Modi government, it was termed as Swachh Bharath and was given the thrust it required. Adding to the urgency and significance of it was Mahatma Gandhi's 100th birth anniversary that was aligned with the completion of the campaign. Mahatma Gandhi stated that 'sanitation is more important than independence'.

Table 5.3 presents the results of the Swachh Bharath survey during the past 5 years. It is a clear indicator of the fact that none of the cleanest cities, according to the survey, are the top Indian destinations for international tourists.

Under the Clean India Mission, it looks like India has given too much emphasis for building toilets and making India Open Defecation Free country. It looks like the other aspects of cleanliness like garbage piles being attended to through solid waste management, safe drinking water provision, sanitation facilities, air–water–land pollution have not been attended to through Swachh Bharat Abhiyan.

As Table 5.4 presents the news articles from two significant newspapers on Clean India Mission shows that it has focused too much on ODF rural areas, cities and states and therefore on building toilets. It is also found that the claims of Clean India Mission on its ODF and toilets achievements were not realistic. The other important issues for cleanliness and hygiene like sanitation, safe drinking water, solid waste management and air pollution, etc., were not given serious consideration. Clean India Mission does not seem to have made the impact it should have made in terms of cleanliness and hygiene across the country.

It can be seen from Table 5.5, that most of the cleanest railway stations are present in the state of Rajasthan (70%) and most of the dirtiest railway stations are present in Tamil Nadu (60%). If noted carefully, all the Tamil Nadu railway stations mentioned in the dirtiest list are in the suburbs of Chennai, one of the top five international

Table 5.3 Top five cleanest cities

Rank	Cities 2019	Cities 2018	Cities 2017	Cities 2016	Cities 2015
1	Indore	Indore	Indore	Mysore	Mysore
2	Ambikapur	Bhopal	Bhopal	Indore	Tiruchirappalli
3	Mysore	Chandigarh	Vishakhapatnam	Ujjain	Navi Mumbai
4	Ujjain	Vijayawada	Surat	Ahmedabad	Kochi
5	New Delhi	Mysore	Mysore	Tiruchirappalli	Hassan

Source Compiled by author

Table 5.4 Newspaper articles on Clean India Mission

Newspaper	News based on ODF	Remarks
Economic Times 02 Dec 2019	Rural India achieved 100 pc ODF status: Govt.	Government of India officially declared India as an open defecation free country
Economic Times 15 Sep 2019	How Swachh Bharat Mission helped millions live with dignity	Swachh Bharat Mission has helped women and children
Economic Times 14 Aug 2019	94 local bodies stand in the way of Open Defecation Free India	Except for these 94 local bodies, others seemed to have achieved ODF
Economic Times 02 Oct 2019	A Swachh Bharat milestone: Google Maps show 57,000 public toilets in India	Google could recognize the increased number of public toilets in India
Economic Times 31 Jan 2019	Over 9 crore toilets constructed under Swachh Bharat: Kovind	Total count of toilets constructed under the Clean India Mission
Economic Times 05 Jun 2018	Swachh Bharat Abhiyan: Where progress on paper hits quicksand of ground reality	There is criticism that the progress under the scheme is only paper-oriented and not realistic
Economic Times 04 Jul 2019	Many states became open defecation free, achieved 100 pc toilet coverage since Swachh Bharat Mission launch: Economic Survey	Most states have been declared ODF under the Clean India Mission
Economic Times 03 Feb 2019,	Over 4,000 urban cities declared open defecation free: Govt	Another account of the success of Clean India Mission with cities becoming ODF
LiveMint 07 Oct 2019	In 5 yrs, 600 mn people have changed their sanitation habits: Parameswaran Iyer	The person who heads the project of Clean India Mission suggests it to be a success as 600 million people have been transformed
LiveMint 09 Jan 2019	Swachh Bharat Abhiyan: Why India's toilet data is too good to be true	Doubts about the authenticity of the claims on Swachh Bharat Abhiyan
LiveMint 04 Jul 2019	Swachh Bharat Mission improved health and nutrition: Economic Survey	Economic Survey 2018–2019 highlights Clean India Mission as a reason for improved health and nutrition at households
LiveMint 06 Jun 2017	The real challenge for Swachh Bharat Abhiyan	Requires more investment in waste management
LiveMint 21 Nov 2017	WaterAid report on India's sanitation 'factually incorrect': Govt	Doubts on the Clean India Mission impact

Source Compiled from different newspapers

tourist destinations of India. This tends to indicate that there is no impact caused by cleanliness on tourism. Further, if Swachh Bharat Abhiyan has been working effectively, there must have been no railway stations under the head dirtiest railway stations (Tables 5.6 and 5.7).

Table 5.5 India's cleanest and dirtiest railway stations (2019)

Ranks	Cleanest railway stations	Dirtiest railway stations
1	Jaipur (Rajasthan)	Perungalathur (Tamil Nadu)
2	Jodhpur (Rajasthan)	Guindy (Tamil Nadu)
3	Durgapura (Rajasthan)	Delhi Sadar Bazar
4	Jammu Tawai	Velacheri (Tamil Nadu)
5	Gandhinagar-jp (Rajasthan	Guduvancheri (Tamil Nadu)
6	Suratgarh (Rajasthan)	Singaperumalkoil (Tamil Nadu)
7	Vijayawada	Ottappalam (Kerala)
8	Udaipur City (Rajasthan)	Pazhavanthangal (Tamil Nadu)
9	Ajmer (Rajasthan)	Araria Court (Bihar)
10	Haridwar	Khurja (Uttar Pradesh)

Source Compiled from LiveMint, 3 October 2019

Table 5.6 Newspaper articles on positive aspects of Indian tourism

Newspaper	Positive news
LiveMint, 11 September 2019	South India tops as favourites destination among travellers: Report Chennai, Bengaluru, Hyderabad, Kochi in the list of top 10 in the first half of 2019
LiveMint, 2 January 2020	Foreign tourist arrivals in India rose 3% in Jan–Nov 2019
LiveMint, 30 December 2019	Bonanza for foreign tourists visiting India: New destinations, reduced visa fee
LiveMint, 1 February 2020	Budget 2020: Tourism industry cheers FM's ₹2,500 crore budget boost
LiveMint, 27 February 2019	India tops international overnight visitors to Dubai in 2018
Business Wire, 30 April 2019	Tourism's direct contribution to GDP is expected to grow from $98 billion in 2018 to $106.9 billion in 2019

Source Compiled from different newspapers

5.2 Literature Review

5.2.1 Tourism's Economic Impact

Dar (2019) analysed the destination image of India among foreign tourists and revealed that the overall destination image of India was significantly positive. Various destination attributes, such as history and heritage, culture and traditions, connectively, spiritual atmosphere, etc., impacted tourists' opinions positively and developed a better image of India among foreign travellers according to Dar. Sanjeev and Birdie (2019) presented an overview of recent developments in the tourism and hospitality industry with some statistics and trends relating to prospects for the

Table 5.7 Newspaper articles on negative aspects of Indian tourism

Newspaper	Negative news
The Economic Times, 31 January 2020	Foreign tourist arrivals in India slowed down in Jan–Oct 2019: Eco Survey
Aljazeera, 29 December 2019	India's tourism industry hit hard by citizenship law protests
The Economic Time, 13 February 2020	Growth in foreign tourist arrivals, earnings slowest in 10 years
LiveMint, 15 April 2019	We need more tourists but not an insurge that overwhelms us
LiveMint, 17 January 2020	Kerala Tourism draws flak for beef post on Twitter
LiveMint. 7 November 2019	Delhi's toxic air adds to India's woes as the economy goes from bad to worse

Source Compiled from different sources

tourism and hospitality industry in India. They examined as to what Indian tourism and hospitality managers should focus on to stay competitive in the coming decade. The study found a strong growth predicted for the tourism and hospitality industry. It also identified some of the underpinning issues that will influence the competitiveness of Indian tourism such as the role of social media, business model innovations, risk management, talent management, valuation models, the influence of information technology, employee loyalty and design thinking in hospitality higher education.

It was also interesting to note that travellers in air polluted cities in China tended to spend more money. Mishra et al. (2011) used popular time series models for the period spanning from 1978 to 2009, provided the evidence of long-run unidirectional causality from tourism activities to the economic growth of the country. Mathew and Sreejesh (2017) attempted to examine the correlation between GDP and foreign exchange earnings (FEE) from tourism and found that both variables co-integrated yet causality was missing. Rout et al. (2019) examined the causality between tourism and economic growth. The study confirmed the tourism-led economic growth in the long run and the growth-led tourism in the short run to be existent.

5.2.2 *Tourism, Environment and Sustainability*

The dynamic linkages between tourism development, energy consumption, environmental degradation and economic growth in the context of the Indian economy was studied by Mishra et al. (2019). The short-run findings establish a chain-link between tourism development, economic growth, energy consumption and environmental degradation. Foreign tourist arrivals positively contribute to economic growth which in turn increases per capita energy use thereby raising CO2 emissions—a significant cause of environmental degradation and consequential adverse effects on tourism development. George (2019) identified the positive impacts of responsible tourism initiatives on environmental sustainability. Integrating tourism and green growth concept minimized the negative impact of tourism development on the environment.

Venkatesh and Raj (2016) emphasized the importance of tourism industry impact on the Indian economy in terms of GDP and employment. They highlighted the need for government involvement in making tourism a flourishing and prosperous industry. It was found that as the residents of the local community, perceived responsible tourism plays a pivotal role in the formulation of perceived destination sustainability, which in turn impacts their perceived quality of life. Panigrahi (2019) examined the value of eco-tourism for tribals in Orissa and concluded that the eco-tourism was beneficial for the tribals attracting more tourists and generating more significant revenue for the local inhabitants. Basariya and Ahmed (2019) examined adventure tourism impact on eco-tourism development of Tamil Nadu. The study concluded that the impact is positive on the tourism revenue of Tamil Nadu and also on eco-tourism.

5.2.3 International Tourism and Environment

Dong et al. (2019) analysed data from 274 Chinese cities during the period 2009–2012. They found that air pollution significantly reduced the international inbound tourism: an increase of PM 10 (particulate matter smaller than 10 μ m) by 0.1 mg/m 3 will cause a decline in the tourism receipts-to-local gross domestic product (GDP) ratio by 0.45 percentage points. It was also established that air quality could potentially influence inbound tourists' city destination choices. Collins and Millar (2019) studied the impact of aggressive street behaviour on tourists' perception of safety and security and on destination image. The study revealed that there was a slight impact on safety and security perception and not much on destination image. Post hoc analyses suggest that there could be a link between these factors specifically for leisure tourists. Kim and Bachman (2019) examined the impact of restroom cleanliness on restaurant satisfaction and customer intent to return. Restroom cleanliness impacted customer satisfaction and did not vary across age groups. Restroom appearance had the greatest impact on cleanliness, followed by personal hygiene items. Ginting and Halim (2019) presented how the limitations of tourist facilities caused a lag in the tourist flow in a village Lumban Suhi-Suhi of Indonesia. They suggested that environment-based tourist facilities could attract tourists into the destination. Alazaizeh et al. (2019) studied the tourist experience at Petra Archeological Park, Jordan. They formulated cultural and natural scenic value, crowding, attractions accessibility, vendor persistence and odour of animal's waste are potential indicators for the tourism experience at the park. Kubickova (2017) conducted a study where seven countries of the Central American region were analysed over 18 years. The results indicated that not all government decisions impact destination competitiveness in the same way, as some may have more influence than others.

Goffi et al. (2019) concluded that sustainability factors are positively associated with competitiveness indicators used as dependent variables in the regression model. Their study supported the hypothesis that sustainability plays a crucial role in fostering tourism destination competitiveness. These results indicate that a new model of cleaner tourism that favourably affects the economy, environment and society is required. Nadalipour et al. (2019) conducted a literature review to create a framework for sustainable competitiveness for tourism. They realized that such a model framework required considering economic, sociocultural and ecological dimensions on the one hand and considering all stakeholders participating in the tourism process on the other hand.

5.2.4 Indian Cities, Clean India Mission and Tourism

Juneja et al. (2019) studied the tourist experience after visiting the heritage sites of Delhi. The study revealed that the majority of the tourists were satisfied with the cleanliness and washroom facilities at all three sites. Deficient was felt in the availability of audio guides in the preferred language. Tourists were satisfied with the clarity of the labels and descriptors, but the internal directional signs were a disappointment. Ramps and pathways for walking were at the satisfactory level for the majority of the tourist but on the other hand, resting and sit down places were unsatisfactory due to the less number of seating at the sites. The study has helped in identifying the areas which require immediate attention and can be an area of focus while the government implements the new schemes. Basu and Punjabi (2020) examined the Advanced Locality Management (ALM) programme of Mumbai and its contribution to solid waste management. It evolved as a local association partnership with the Municipal Corporation of Greater Mumbai in 1997 and has worked at the middle class and high-class locality. It has not become a people's movement for its missing out on the locations of the sparse population.

Mohanty et al. (2019) studied the issues of heritage tourism in Bhubaneswar and suggested a holistic approach to solving the issues. The suggestions include a priority to environmental sustainability, social worthiness, economic feasibility and cultural vivacity, of the place, thus paving the way for holistic heritage tourism which is sustainable. Mohapatra et al. (2015) proposed a technology-oriented system to bring in cleanliness in cities. The proposed system "SUCHITRA" is conceptualized to empower citizens to rate the cleanliness at public places in India, make aware the responsible local authorities of places requiring immediate attention for cleaning. It will also assist the local authorities in resources optimization and rank their administrative areas to identify uncleaned spots and assess the success of their cleanliness initiatives.

Satchi and Venkatesan (2017) studied the awareness and compliance of the Clean India Mission among students of higher secondary schools in Chennai and found that only 40% were aware of the requirements and, of the 40, 98% were compliant to the requirements. Beena and Raju (2020) present Chennai as the Medical-Hub of Asia.

The study identified the factors that draw international medical tourists to hospitals in Chennai. Senthilkumar (2019) studied the role of budget hotels in Chennai and found that they are catering to 70% of the business and recreation tourists, enjoying good occupancy all through the year. Raja (2019) studied the foreign tourist arrival to Tamil Nadu as compared to other states of India. The study highlights the cultural tourism and medical tourism of Tamil Nadu earning foreign exchange.

Vasavada et al. (2019) studied Ahmedabad as a cultural city and preservation of Indian culture & heritage in the city. They have proposed the 3Rs—Reduce, Recycle and Reuse. The paper concludes that sustainability practices would lead to sustainable tourism. Jain and Thakkar (2019) examined the holistic approach towards tapping the intangible knowledge using experiential tourism as a tool. It examined the craft experiential tourism models developed by Design Innovation and Craft Resource Centre (DICRC), CEPT University, India, and suggested how such models can be the forerunners to promote craft and cultural tourism in India. It discussed at length multiple activities like mapping craftspeople, developing connections, conducting contextual programmes (craft-design innovation and community participation), knowledge dissemination (through craft awareness programmes and exhibitions), developing infrastructure: with a core vision of sustaining the intangible knowledge for future generations.

5.2.5 *Indian Cities of Tourist Attraction*

New Delhi, as a city has been receiving the highest number of international tourists in 2015 and 2019. In between, it has fluctuated to different positions among the Indian cities. In the international scenario, it has climbed up the ranks from 28 (2015) to 11 (2019). This forward movement of India's capital city is an encouraging indicator for India's tourism industry. New Delhi's cleanliness seemed to have improved over 5 years. New Delhi appeared as the fifth-ranked cleanest city in 2019. Despite the ranking, there are several negative remarks about the cleanliness of the capital city in the print media. In 2019, itself several news articles critically viewed the impact of Clean India Mission on the cleanliness and hygiene of the city. However, it is interesting to see the city appear on the cleanest city rank at fifth position. The news articles also extensively talked about the poor air quality and unsafe drinking water of Delhi all through 2019.

Mumbai, the commercial capital of the country, is said to be the second most attractive destination of India for international visitors. 'Indian slum tour becomes country's most popular tourist attraction', reported The Telegraph on 23 June 2019. The news article speaks on Dharavi slum experience being rated higher than the Taj Mahal experience. Mumbai's Dharavi topped the ten destinations of must-visit in India and made it to '10 Travellers' Choice Experiences 2019 in Asia' list. In yet another positive news on Mumbai, Asian Review reported that Mumbai surged in the international list of destination cities along with Singapore and Delhi. Gunasekar et al. (2018) studied International Tourist Arrival in India: Impact of Mumbai 26/11

Terror Attack. The analysis was done using the vector autoregression (VAR) model, where the foreign tourists arriving in India is a function of the economic condition prevailing in the country, captured here by real gross domestic product of India and the terror attack dummy variable. The results indicate that post-26/11 there has been a significant decline in the number of foreign tourists arriving in India. They further analysed the disaggregated airport level data, where similar significant negative impacts were found for Mumbai and Delhi airports.

Chennai Tamil Nadu's capital and ranks third among the Indian cities as an international tourist destination. Chitra and Arun (2016) in their research titled 'Tourist Preference on Chennai Tourism', suggested that Central and State Governments must coordinate to promote tourism as it is of significant economic and environmental importance. Beena & Raju explored the Medical Tourism in Chennai city through the case study method. The study focused on the factors attracting foreign patients as medical tourists to Chennai, the satisfaction level of international patients and the comparison of the cost of medical treatment at international destinations versus Chennai. Vimitha and Shobana (2017) studied medical tourism in Chennai city. They found that Chennai medical treatment was 10% less costly than the USA. Chennai had international standards of medical treatment, with low cost on medical expenses that attracted international patients as medical tourists to Chennai. Raja (2019) examined the growth and expansion of tourism in Tamil Nadu identified that Chennai has a high capacity to attract international tourists as it has opportunities for various forms of tourism like Medical Tourism, Spiritual Tourism, Eco-Tourism, Cultural Tourism, Business and Commercial visits by corporate people for business meetings and conferences. Verma and Rajendiran (Verma and Rajendran 2017) studied the impact of historic nostalgia on tourists' destination loyalty intention in Mahabalipuram. They found the antecedent role of historical nostalgia and suggested that tourism managers to use historical monuments and structures to evoke historical nostalgia in order to attract heritage.

Agra is in the state of Uttar Pradesh. Agra is internationally known for Taj Mahal and tourists from all over the world come to Agra to see and experience the beauty and magnificence of Taj Mahal. Sahu and Kaurav (2019) studied the impact of human resource shortage on tourism in Agra. Human resources of all categories are essential to support tourism in the city. Thus, it focuses on promoting the necessary human resources for the industry. There is a study with author unknown on the three As of Agra tourism—Attractions, Accommodation and Accessibility. The study closes with a SWOT analysis of Agra tourism. The weaknesses mentioned are a lack of infrastructure and maintenance. Roads and electricity are significant concerns. The threat pointed out is increasing the communal tension. Srivastava (2019) analysed the reasons for Agra not being not economically benefitted by tourism and identified that lack of the following: airport, excellent accommodation, welcoming citizens towards tourists and an auditorium to exhibit the different cultures across Uttar Pradesh and if possible some adventure experience. 'Tourists footfall in Uttar Pradesh more than its population in 2017' reported Business Today on 5 April 2018. The sustainable tourism development plan of Agra—An attempt to tackle the pulsar effect during the Taj festival.

Jaipur is also known as Pink City. It is a popular tourist destination in Rajasthan. The city is known for its architectural beauty of the heritage monuments, colourful festivals, vibrant culture and exotic cuisine. Jaipur is said to be the cornerstone of the tourism of the state of Rajasthan. Rajasthan experienced higher base of 17% growth in 2016; the arrivals increased by 10.50% to 45.91 million visits in 2017, riding on the rush of domestic and foreign tourists. 'Rajasthan sees a 21% rise in tourism with 5.19 crore visitors in 2018', reported India Today on 11 July 2019. Over 5.19 crore domestic and international tourists visited the state with a royal essence in 2018, an increase of nearly 90,000 visitors from 2016, it said. Outlook reported that Jaipur despite joining the UNESCO heritage cities in 2019 is poorly connected internationally—only three countries, viz. Thailand, the United Arab Emirates and Malaysia. 'Jaipur tourism industry cheerful over record arrival of tourists', reported The Pink City Post on 28 December 2019. FICCI reported on 29 April 2019 Mr. D. B. Gupta, Chief Secretary, Government of Rajasthan has said that tourism infrastructure will get high priority in Rajasthan. Mr Gupta said that in 2018–2019, as many as 83 development works related to tourism infrastructure had been undertaken. In the fiscal year 2019, the hotel occupancy rate in the north-western city of Jaipur amounted to over 67%, an increase from the previous year. An exponential increase in the occupancy rate across the city was seen from the financial year 2010, with an exception of financial years 2012 and 2013, according to Statista. Vijayvargiya et al. (2017) conducted a study to evaluate the before and after the impact of "Clean India Campaign". It was revealed that there is a positive gap due to the Clean India campaign in tourism development of Jaipur City.

5.3 Methods

The study has been carried out with literature review, secondary data and primary data. The literature review was focused on tourism and its impact on India, Global tourism and sustainability and Indian cities and tourism. Newspaper articles were reviewed to understand the positives and negatives of Clean India Mission and India's tourism. Primary data was collected from 307 students from across India on Clean India Mission and its impact on the cleanliness of Indian cities. The respondents were chosen by the snowball method. A structured questionnaire was used. About 500 questionnaires were distributed, and only 307 questionnaires were retrieved entirely for data analysis.

5.4 Results

Table 5.8 indicates that majority of the respondents do not agree to the statement that SBA/CIM had made India cleaner than it was before the project began. This is supported by most of the news articles. Clean India Mission is a noble initiative that

Table 5.8 Impact of SBA/CIM on cleaner India

SBA/CIM causing cleaner India	Number of respondents	Percentage of responses
Yes	95	31
No	212	69
Total	307	100

Source Primary data

has a great goal. Yet the goal has not been achieved for various reasons like population, poor civic sense of the citizens, more new sources of waste and garbage but no equivalent methods to address the issues, slow action of the local administrations towards waste management and cleanliness issues.

Table 5.9 portrays that most respondents agreed that CIM did not have a holistic approach (74%). Only a minimum number of respondents (24%) felt that CIM was an effort with a holistic approach. As is can be seen from the drive initiatives, the focus was more on ODF. Solid waste management, clean water and sanitation, air pollution control, etc., received much lesser attention.

Table 5.10 presents the fact that tourist destinations have not become appealing because of CIM. There is something missing in CIM that does not make it a reason for the appeal of tourist destinations in India. Else something more that matters to the appeal of the cities as far as tourism is concerned. The tourist appeal of Indian cities come more from the cultural heritage, art and architecture, etc., rather than their cleanliness or sustainable environment.

Table 5.11 shows the fact that CIM does not adequately address the substantial waste management issue of cities. A reasonable number of respondents (30%) agreed that CIM partially addresses the issue of solid waste management. A vast majority of respondents (70%) felt that CIM did not address substantial waste management

Table 5.9 Holistic approach of SBA/CIM

Holistic approach of SBA/CIM	Number of respondents	Percentage of responses
Yes	79	26
No	228	74
Total	307	100

Source Primary data

Table 5.10 Tourist destinations made more appealing with SBA/CIM

Tourist destinations made more appealing	Number of respondents	Percentage of responses
Yes	87	28
No	220	72
Total	307	100

Source Primary data

Table 5.11 Solid waste management addressed by SBA/CIM

Solid waste management addressed by SBA/CIM	Number of respondents	Percentage of responses
Fully	0	0
Partially	93	30
Not at all	214	70
Total	307	100

Source Primary data

Table 5.12 Sanitation addressed by SBA/CIM

Sanitation addressed by SBA/CIM	Number of respondents	Percentage of responses
Fully	0	0
Partially	86	28
Not at all	221	72
Total	307	100

Source Primary data

issues. Indian cities have a serious concern to address in this regard. Solid waste management is a huge problem that needs immediate addressing.

Table 5.12 presents the picture of the impact of CIM on the sanitation of cities. None of the respondents has a favourable opinion towards the full or complete contribution of CIM towards sanitation of cities. About 28% of respondents have agreed that CIM contributed partially to the sanitation of cities. A vast majority of 72% felt that CIM has not at all contributed to the sanitation of cities. Sanitation facility in Indian cities needs to improve significantly.

Table 5.13 depicts the contribution of safe drinking water under the CIM. It is seen that none of the respondents agrees that safe drinking water is fully ensured by CIM. A small segment of the respondents (23%) agreed that CIM partially ensured the safe water to Indian cities. A vast majority of the respondents felt that CIM did not assure of safe drinking water to Indian cities. Safe drinking is a concern in most cities. Availability of water and safe drinking water needs to be addressed by the government for the benefit of the citizens and the tourist arrivals.

Table 5.13 Safe drinking water ensured by SBA/CIM

Safe drinking water ensured by SBA/CIM	Number of respondents	Percentage of responses
Fully	0	0
Partially	72	23
Not at all	221	77
Total	307	100

Source Primary data

Table 5.14 Positive contribution to air quality of Indian Cities

Safe drinking water ensured by SBA/CIM	Number of respondents	Percentage of responses
Fully	0	0
Partially	0	0
Not at all	307	100
Total	307	100

Source Primary data

Table 5.15 Contribution of CIM towards the sustainability of Indian tourist cities

Contribution of CIM towards sustainability of Indian cities of tourism importance	Number of respondents	Percentage of responses
Yes	28	9
No	279	91
Total	307	100

Source Primary data

Table 5.14 exhibits the respondents' opinion on CIM's contribution to the air quality of Indian cities. None of the respondents felt that CIM neither fully nor partially contributed to the air quality of the Indian cities. All the respondents felt that CIM has not contributed to the air quality of Indian cities. Air quality in most of the cities is at dangerous levels. Improving air quality must be the primary concern of the government through the Clean India Mission.

Table 5.15 presents the opinion of the respondents on the contribution of CIM towards the sustainability of Indian cities. A very insignificant number of respondents (9%) agreed to the fact that CIM contributed to the sustainability of Indian cities. A significant majority of respondents (91) did not find CIM to be contributing to the sustainability of Indian cities.

5.5 Findings and Conclusion

Impact of Clean India Mission on the cleanliness of Indian cities was found to be minimal or not very significant. It is unfortunate that the respondents found the Clean India Mission to have minimal impact on the cleanliness of Indian cities. It was also found that the Clean India Mission did not found to have a holistic approach, according to the respondents of the study. Clean India Mission instead was lopsided or unduly focused on a fragment of the cleanliness. The opinion of the respondents is against the idea that Indian cities have become more appealing with Clean India Mission—the majority of the respondents did not agree that Indian cities have become more appealing and attractive due to Clean India, the purpose it must have served. Solid waste management, sanitation, safe drinking water and sustainability were not adequately addressed by Clean India Mission. These four aspects that need a total transformation all across India were not fully addressed or totally focused upon under Clean India Mission. Thus, Clean India Mission seems missed out to make Indian cities cleaner and more appealing to the tourists to enjoy their visit and presence in these cities.

The newspaper articles and the review of the literature confirm the primary data analysis. Clean India Mission has been successful in making India Open Defecation Free. Unfortunately, that alone is not sufficient for the cleanliness of the cities and the nation. India's Clean India Mission needs to advance further to make Indian cities and villages cleaner from a holistic perspective. A holistic perspective addressing comprehensive solid waste management, sanitation, clean and safe water and high-quality air would make India cleaner, attractive and appealing for tourists. It is also noteworthy to emphasize that based on the respondents' responses, there are more convincing factors that can shape tourism's attractiveness which includes art and architecture, cultural heritage, etc., than cleanliness and other elements of CIM.

Furthermore, the results of this study provide an avenue for many interpretations. Firstly, the implementation of CIM can be revisited and evaluated to determine areas of improvement in the future. The less impact of the programme to resolve issues on solid waste management, sanitation, safe drinking water, air quality, sustainability and the likes pose a strong signal that the programme is not reflective of the real condition of tourism, the issues currently faced and the certainty of how these issues will be adequately and appropriately addressed. Spending a larger budget on this area that eventually leads to future negative consequences would waste time, efforts and resources. Secondly, future efforts to be initiated by the government related to the tourism agenda can be subjected to consensus and series of trial runs to ensure the feasibility and likelihood of an effective implementation considering that initiatives like CIM require large outlay and failure is not a likely option. The government should seek more collaborative effort with all concerned stakeholders from the idea conception of the programme until its proposed implementation of how the stakeholders respond, and their contribution counts when it comes to collaboration and cooperation. Third, adopting a strategic and most workable model also plays a guiding role in implementing any initiatives for improvement. Obviously, models were adopted by the government prior to its implementation; however; determining which model is suitable is another consideration to ensure implementation effectiveness.

Indeed, Tourism in India has not been significantly influenced by the Clean India Mission. Clean India Mission does not seem to have had a holistic approach towards clean cities in India. The benefit to Indian cities in terms of a clean environment, high level of sanitation, safe and quality drinking water, high-quality air, etc., seems to have been missed out by Clean India Mission. The study does not give a clear indication if the Clean India Mission has contributed to tourism improvement in the selected cities. As the analysis indicates, there seems to be a minimal contribution from the Clean India Mission to tourism in India. Unless Clean India Mission scales up and intensifies its focus on a holistic approach, Indian cities will not benefit from Clean India Mission nor would their tourist inflow. The major tourism concerns were not thoroughly addressed as clearly indicated from the responses obtained. However, this study mainly concentrated in the four cities in India, where the result of this study is a small fraction of the broader scope of CIM implementation in India. Hence, it is recommended that extensive investigation and studies on a broader scope can be undertaken to evaluate further and justify these findings.

References

Alazaizeh, M. M., Hallo, J. C., Backman, S. J., Norman, W. C., & Vogel, M. A. (2019). Giving voice to heritage tourists: indicators of quality for a sustainable heritage experience at Petra, Jordan. *Journal of Tourism and Cultural Change, 17*(3), 269–284.

Basariya, S. R., & Ahmed, R. R. (2019). The influence of 'adventure tourism activities' in promoting tourism business in mountain stations. *Journal of Hospitality and Tourism Management, 8*(2), 1–10.

Basu, A. M., & Punjabi, S. (2020). Participation in solid waste management: lessons from the advanced locality management (ALM) programme of Mumbai. *Journal of Urban Management, 9*(1), 93–103.

Beena, K. S., & Raju, D. V. (2020). Enchanting Tamil Nadu-experience yourself a pull factor for medical tourism. *Our Heritage, 68*(30), 5765–5772.

Chitra, M., & Arun, A. (2016). Tourist preference on Chennai tourism. *International Journal of Advanced Research in Management and Social Sciences, 5*(7), 74–86.

Collins, M.D., & Millar, M. (2019). Tourists' perceptions of destination image, safety, and aggressive street behavior. *International Journal of Hospitality & Tourism Administration* 1–18.

Dar, H. (2019). Analysis of destination image among foreign tourists: A case of India. *Our Heritage, 67*(10), 299–313.

Dong, D., Xu, X., & Wong, Y. F. (2019). Estimating the impact of air pollution on inbound tourism in China: An analysis based on regression discontinuity design. *Sustainability, 11*(6), 1682.

Dutta, S., & Baskar, S. (2017). Swachh Bharat Abhiyan 2017 year in review: India's sanitation coverage goes up in third year of Swachh Bharat Abhiyan. https://swachhindia.ndtv.com/swachh-bharat-abhiyan-2017-year-in-review-indias-sanitation-coverage-goes-up-in-third-year-of-swachh-bharat-abhiyan-16104/

George, A. (2019). Impact of responsible tourism in environmental sustainability issues: A case study in kerala, India. Travel and Tourism Research Association: Advancing Tourism Research Globally. https://scholarworks.umass.edu/ttra.

Ginting, N., & Halim, S. (2019). Environment approach on optimizing tourism facilities at tourism village in Indonesia. *Environment-Behaviour Proceedings Journal, 4*(10), 109–116.

Goffi, G., Cucculelli, M., & Masiero, L. (2019). Fostering tourism destination competitiveness in developing countries: The role of sustainability. *Journal of Cleaner Production, 209,* 101–115.

Gunasekar, S., Patri, R., & Narayanan, B. (2018). International tourist arrival in india: Impact of mumbai 26/11 terror attack. *Foreign Trade Review, 53*(1), 12–28. https://doi.org/10.1177/0015732516681882

Jain, R., & Thakkar, J. (2019). Experiencing craft and culture: An emerging cultural sustainable tourism model in India. In: Stankov, U., Boemi, S. N., Attia, S., Kostopoulou, S., Mohareb, N. (Eds). *Cultural sustainable tourism. Advances in science, technology & innovation (IEREK Interdisciplinary Series for Sustainable Development)* (pp. 29–35). Springer, Cham.

Juneja, M., Guide, C., & Uniyal, M.C. (2019). To study the tourist experience after visiting the World Heritage Sites of Delhi. *World,* 2(6).

Kim, H., & Bachman, J. R. (2019). Examining customer perceptions of restaurant restroom cleanliness and their impact on satisfaction and intent to return. *Journal of foodservice business research, 22*(2), 191–208.

Kubickova, M. (2017) The impact of government policies on destination competitiveness in developing economies. *Current Issues in Tourism, 22*(6), 619–642. https://doi.org/10.1080/13683500.2017.1296416.

Mathew, P. V., & Sreejesh, S. (2017). Impact of responsible tourism on destination sustainability and quality of life of community in tourism destinations. *Journal of Hospitality and Tourism Management, 31*, 83–89.

Mishra, P. K., Rout, H. B., & Kestwal, A. K. (2019). Tourism-energy-environment-growth Nexus: Evidence from India. *Journal of Environmental Management and Tourism, 10*(5), 1180–1191.

Mishra, P. K., Rout, H. B., & Mohapatra, S. S. (2011). Causality between tourism and economic growth: Empirical evidence from India. *European Journal of Social Sciences, 18*(4), 518–527.

Mohanty, S., Mishra, S., & Mohanty, S. (2019). A Stakeholder's perspective on Holistic Heritage tourism—A special reference to the temple city Bhubaneswar, India. *African Journal of Hospitality, Tourism and Leisure, 8*(5), 1–17.

Mohapatra, R. et al (2015). SUCHITRA (System for Urban, Clean, Healthy India Transformation through Rating)–A cleanliness rating tool for cities to empower citizens. *IOSR Journal of Humanities and Social Science, 20*(2), 7–11.

Nadalipour, Z., Khoshkhoo, M. H. I., & Eftekhari, A. R. (2019). An integrated model of destination sustainable competitiveness. *Competitiveness Review: An International Business Journal, 29*(4), 314–335.

Panigrahi, N. (2019). Development of eco-tourism in tribal regions of orissa: potential and recommendations. *Culture Mandala,13*(3), 1–17.

Raja, P. (2019). Growth and spread of tourism in Tamil Nadu. *International Journal of Social Science and Economic Research, 4*(5), 3819–3838.

Rout, H. B., Mishra, P., & Pradhan, B. (2019). Empirics of tourism-led growth in India, 1995 to 2016. *Journal of Environmental Management and Tourism, 9*(6), 1190–1201.

Sahu, S. K., & Kaurav, R. P. S. (2019). Antecedents of HR challenges in tourism industry with reference to Agra. In *Proceedings of 10th International Conference on Digital Strategies for Organizational Success*. Retrieved from SSRN: https://ssrn.com/abstract=3315078 or http://dx.doi.org/10.2139/ssrn.3315078.

Sanjeev, G. M., & Birdie, A. K. (2019). The tourism and hospitality industry in India: emerging issues for the next decade. *Worldwide Hospitality and Tourism Themes, 11*(4), 355–361.

Satchi, N. S., & Venkatesan, L. (2017). Level of awareness and compliance regarding Clean India Campaign among school children. *University Journal of Nursing Sciences*, 1(1).

Senthilkumar, M. P. (2019). Role of budget hotels metropolitan cities with special reference to Chennai City. *Research Review International Journal of Multidisciplinary, 4*(6), 147–152.

Srivastava, S. B. (2019). Relationship between growth and tourism: An empirical investigation from India. *Global Journal of Enterprise Information System, 11*(2), 14–21.

Vasavada, D. M., Shah, D. S., & Gajjar, M. N. A. (2019). A conceptual study on sustainability and preservation of Indian culture and heritage with reference to Heritage City Ahmedabad. *International Journal for Research in Management and Pharmacy, 8*(4), 11–23.

Venkatesh, M., & Raji, S.J.M. (2016). Impact of tourism in india. *International Journal of Scientific Engineering and Applied Science, 2*(1), 167–184.

Verma, A., & Rajendran, G. (2017). The effect of historical nostalgia on tourists' destination loyalty intention: An empirical study of the world cultural heritage site–Mahabalipuram, India. *Asia Pacific Journal of Tourism Research, 22*(9), 977–990.

Vijayvargiya, L., Arrawatia, M. A., & Sharma, N. (2017). Impact of clean India campaign on tourism: A study based on Jaipur city. *Inspira- Journal of Modern Management & Entrepreneurship, 7*(4), 165–170.

Vimitha & Shobana. (2017). Current scenario of medical tourism in Chennai its aspects and strategies. *International Journal of Science and Research, 6*(7), 1158–1561.

Chapter 6
The Effects of Globalization and Terrorism on Tourist Arrivals to Turkey

Zübeyde Şentürk Ulucak and Ali Gökhan Yücel

Abstract Turkey, with rich tourism diversity and destinations, has experienced considerable changes in the tourism sector as one of the top ten most visited countries over two decades. Research shows that increasing mutual interaction and integration among countries and different cultures encourage visitors and contribute to tourism sector development. In this respect, in the light of available data, globalization performance in Turkey has been ongoing above the world average since 1970. On the other hand, the country has suffered terror attacks that are a significant deterrent factor for tourism. Hundreds of bombings and armed assaults occurred until very recently. However, empirical evidence on the role of terrorism and globalization in the tourism sector is not sufficient to clearly understand tourist behaviors and to provide new insights into the literature. Considering the probable effects of globalization and terrorism, Turkey is an excellent case to investigate tourism sector development within this framework. Therefore, this study aims to investigate how tourist arrivals to Turkey react to globalization level and terror attacks by using advanced time-series analysis covering the period 1980–2018. Results reveal that globalization and terrorism are essential determinants of tourist arrivals in Turkey.

Keywords Tourism sector · Globalization · Terrorism · Tourist arrivals · Turkey

6.1 Introduction

Tourism sector has gained significant momentum as a result of economic development, decreases in air transportation costs, and technological advances. According to recent research by the World Travel and Tourism Council (WTTC 2019a), the

Z. Ş. Ulucak (✉)
Department of Public Finance, Erciyes University, FEAS, Kayseri, Turkey
e-mail: zsulucak@erciyes.edu.tr

A. G. Yücel
Department of Economics, Erciyes University, FEAS, Kayseri, Turkey
e-mail: agyucel@erciyes.edu.tr

D. Balsalobre-Lorente et al. (eds.), *Strategies in Sustainable Tourism, Economic Growth and Clean Energy*, https://doi.org/10.1007/978-3-030-59675-0_6

contribution of the tourism sector to global gross domestic product is 10.4% and 319 million jobs, which corresponds to 10% of the total employment in 2018.

Table 6.1 presents the world's top ten countries in the number of international tourist arrivals and tourism receipts. The total number of tourists reached 1,460 billion in 2019, growing by 3.6% as compared to 2018. Nine out of the top ten countries enjoyed favorable growth rates in the number of tourist arrivals. France remained the world's most popular tourist destination hosting more than 90 million foreign tourists, followed by Spain, the United States, China, and Turkey. Alongside these countries, Mexico, Thailand, Germany, and the United Kingdom also managed to enter the top ten tourist destinations worldwide. International tourism receipts reached $1,479 billion in 2019, growing 1.5% compared to the previous year. The United States is the top tourism earner by far, followed by Spain, France, Thailand, and the UK.

A striking insight from Table 6.1 is that while the United States ranks third in the number of tourist arrivals, it earns the most revenue from tourism by far. The share of the United States in tourist arrivals is around 5.5%, while the percentage of its tourism receipts is 14.5%. Another aspect worth mentioning is that the improvement in the number of tourist arrivals to Turkey is not in line with its tourism receipts. While Turkey ranks as the 6th most popular tourist destination, it ranks only 13th in terms of tourism receipts. To make a comparison, Turkey hosted 51.2 million tourists with a tourism receipt of $29.8 billion, while Germany hosted 39.6 million tourists with a receipt of $41.6 billion. In other words, Germany generated $10 billion more in tourism revenue than Turkey despite hosting 10 million fewer visitors than Turkey.

Table 6.1 Top ten countries in tourist arrivals and tourism receipts

Country	Tourist Arrivals (million, 2019)	Share (%)	Change (%) (2019/2018)	Country	Tourism receipts (USD billion)	Share (%)	Change (%) (2019/2018)
France	90.3*	6.19	2.4*	US	214.1	14.5	−0.3
Spain	83.7	5.74	1.1	Spain	79.7	5.4	3.2
US	79.3	5.53	−0.6	France	63.8	4.3	1.9
China	65.7	4.36	4.5	Thailand	60.5	4.1	3.2
Italy	64.5	4.27	4.8	UK	49.9	3.4	7.4
Turkey	51.2	3.17	11.9	Italy	49.6	3.4	6.2
Mexico	45.0	2.87	9.0	Japan	46.1	3.1	8.0
Thailand	39.8	2.70	4.2	Australia	45.7	3.1	9.1
Germany	39.6	2.65	1.8	Germany	41.6	2.8	2.2
UK	37.5	2.52	3.2	Macao	39.5	2.7	−2.9
TOP-10	*596.6*	*40.86*	*4.2*	*TOP-10*	*616.9*	*46.6*	*3.8*
WORLD	*1,460*	*100*	*3.6*	*WORLD*	*1,479*	*100*	*7.1*

Source World tourism barometer (UNWTO 2020b); *Projected

As a result of increased wealth and life expectancy, lower transportation costs, better pensions and advertising, worldwide tourism has been growing since the 1950s. Between 1950 and 2019 the number of international tourist arrivals has grown massively from 25 million to 1.5 billion. On the other hand, the short-term tourism sector is highly volatile due to domestic and foreign shocks. Among the various shocks affecting the tourism sector, diseases and terrorism are the most harmful ones. Broke out in Wuhan, China, at the end of 2019, Covid-19 hit tourism sector so hard that UNWTO (2020a) projects that the number of international visitors will drop by 60–80% in 2020, contrary to the previous forecast of 3–4% growth. The tourism sector also takes a massive hit by terror attacks. Since the beginning of the twenty-first century, terrorism has been on the rise. There has been a five-fold increase in terror-related deaths since 2000. Terrorist attacks of September 11, 2001, in New York, Bali bombings in 2002, 2004 Jakarta bombings, 2005 London bombings, 2006 Egypt bombings, 2008 Mumbai attacks, 2013 Nairobi armed assault, 2015 November Paris attacks, 2016 March Brussel attacks, 2016 Nice attack, March 2017 London attack, 2019 Sri Lanka attacks are only some of the terrorist incidents that the world has witnessed.

Buckelew (1984, p. 18) defines terrorism as "Violent, criminal behaviour designed primarily to generate fear in the community, or in a substantial segment of the community, for political purposes." Based on Buckelew's definition, it could be said that the fundamental purpose of terrorism is to inspire fear so that people change their behavior. In most of the terrorist attacks, tourist destinations are the main target. The reason behind this is to use tourists as a political tool to attract more media coverage.

A report issued by WTTC (2019b) shows that it takes 13 months on average for the tourism sector to recover from a terrorist attack. As for the epidemics, it takes relatively longer, 21 months, to bounce back from an epidemic. However, the situation might be different for Turkey as the country has suffered several attacks in the last two decades. Tourism losses have a substantial impact on Turkey's economy, where tourism constitutes a significant share (20%) of total exports. Given the importance of tourism in Turkey's economy, analyzing the determinants of tourism demand in Turkey is an appealing research area. Turkey has several distinct features that make the country an essential venue for study. Firstly, Turkey is an excellent case to analyze as the country hosted 51.2 million tourists in 2019, which gave it a sixth spot in the world rankings. Understanding the dynamics of tourism demand in Turkey plays an important role in achieving sustainable tourism. The results obtained from this study may also provide hints for other countries.

Secondly, among the many determinants of tourism demand, globalization and terrorism are the leading ones. Therefore, we analyze the determinants of tourism demand in Turkey in a multivariate framework incorporating globalization and terrorism. Turkey is the only country in the top ten tourist destinations consistently and severely suffering from terrorism/violence. In his book published in 1984, Buckelew stated that Turkey had managed to bring endemic terrorism under control. As opposed to the author's statement, the country has been fighting terrorism for nearly 40 years. On the other hand, globalization has become the main driver of the tourism sector since it integrates countries through economic, societal, and cultural aspects and enables them to get in touch more closely (Ulucak et al. 2020). Given the pivotal role of globalization in the tourism sector, one may consider Turkey as one of the most globalized countries because the globalization performance of the country has been over the world average globalization level since the 1970s (Bilgili et al. 2020).

Thirdly, the imbalance between the number of international tourist arrivals and tourism receipts of Turkey may be better understood by revealing the quantitative impacts of each variable. If the elasticities of each variable are revealed, then the tourism receipts may be boosted by increasing the price of inelastic determinants of tourism demand as well as decreasing the price of elastic determinants of tourism demand.

Fourthly, we used annual data of Turkey covering the period of 1980–2018. We analyze the stationarity properties of the variables and cointegration among the variables through a recently developed *Fourier* approach, which provides more robust results. In the last stage of the empirical analysis, we used fully modified ordinary least squares (FMOLS) and dynamic ordinary least squares (DOLS) estimators to obtain the long-run parameters. Overall, the findings are expected to provide valuable insights into both researchers and policymakers.

The remainder of this study is organized as follows. The next section provides a brief literature review. Section three presents the model, data, and methodology. Empirical results and discussion are given in section four. The last section concludes the study.

6.2 Literature Review

Since the influential paper of Enders and Sandler (1991), several researchers have studied the empirical relationship between terrorism and tourism. There is a consensus in the literature that terrorism harms tourism demand (Bhattarai et al. 2005; Drakos and Kutan 2003; Llorca-Vivero 2008; Thompson 2011; Ulucak et al. 2020; Yap and Saha 2013; Yaya 2009, among others). However, the magnitude of the impacts of terror incidents varies from case to case.

Advanced econometric techniques became an essential tool for identifying the impacts of terrorism on tourism. In their pioneering study, Enders and Sandler (1991) examined the effects of terrorist activities by ETA, a separatist group, on Spain's tourism using monthly data from 1970 to 1988. The authors found that terrorist events had a substantial impact on the number of tourists visiting Spain. Enders and Sandler also found that there exists unidirectional causality running from terrorism to tourism. The authors further argued that a typical terror incident was estimated to scare away 140,000 tourists.

In another study by Enders and friends (1992), the authors employ autoregressive integrated moving average (ARIMA) covering the period from 1974 to 1988 for Austria, Greece, and Italy. The authors conclude that visitors change their country preferences for tourism to minimize the risk of witnessing terror incidents. Enders et al. (1992) estimate that terrorist attacks led to a loss of 16 billion SDRs.

Aly and Strazicich (2000) investigate the effects of terrorism on two countries with a highly developed tourism industry. Using data from 1955 to 1997 for Egypt and 1971–1997 for Israel and employing two-break Lagrange Multiplier (LM) unit root tests, the authors conclude that the effects of terrorism on tourism in these countries are temporary, not permanent.

In an appealing study carried out by Pizam and Fleischer (2002), the authors examined the monthly tourist arrivals to Israel, covering 1991:05–2001:05. The results of the study confirm the existence of the hypothesis that "the frequency of acts of terrorism cause a larger decline in international tourist arrivals than the severity of these acts." An important implication of the study is that tourism destinations could recover from even severe terror incidents so long as the incidents are not repeated.

Drakos and Kutan (2003) developed a consumer-choice model to investigate the regional effects of terrorism. This study employed monthly data from 1991:01 to 2000:12 for Greece, Israel, and Turkey. In addition to the harmful effects of terrorism on tourism, the authors found evidence of substitution between countries as tourism destinations if one appears safer than the other. An essential finding of the study regarding Turkey is that Turkey's tourism market share dropped by 5.21% as a result of terror incidents between 1991 and 2001.

Llorca-Vivero (2008) analyzed the impacts of terrorist attacks on international tourist arrivals. Using a panel gravity equation for tourism from the G7 countries to a sample of 134 countries covering 2001–2003, Llorca-Vivero found that both domestic and international terror incidents affect tourist arrivals. The author also argues that the impact of terrorism is more severe in developing countries.

In a similar study, Yaya (2009) examined the effect of terrorism on the number of tourist arrivals to Turkey using monthly data 1997:01–2006:12. The findings of the study show that there exists a negative but relatively small impact of terrorism on tourism. The author further argues that terror incidents in Turkey accounted for a reduction of 6 billion visitors over the last 9 years. The economic cost, on the other hand, was estimated to be around $700 billion in 2006.

An important study by Thompson (2011) compared the effects of terrorism on the tourism sector of developing and developed countries. The findings obtained from cross-sectional data of 60 countries suggest that the effect of terror incidents on tourism is more significant in developing countries than in developed countries. The author suggests that developing countries should pay special attention to prevent terrorism.

Feridun (2011) examined the relationship between terrorism and tourism in Turkey for the period between 1986 and 2006. The results of the ARDL bounds test suggest that there is a long-run relationship between tourism and terrorism. The findings also prove the existence of a negative unidirectional causality running from terrorism to tourism.

Employing a fixed-effects panel data analysis for 139 countries over the period 1999–2009, Yap and Saha (2013) found that political instability, corruption and terror incidents have negative impacts on tourism demand, and the effects are substantial.

For instance, a one-unit increase in terror incidents reduced tourist arrivals by 16.2% and tourism revenue by 17.8%. An interesting finding of the study is that terror incidents have less impact compared to political instability.

Buigut (2018) employed a dynamic panel model to make a comparison between the effects of terrorism on developed and emerging country demand for tourism in Kenya. Spanning 2010Q1–2013Q4, the author found that the intensity of terrorist attacks significantly decrease tourist inflows from developed countries but not from emerging countries. Buigut (2018) argues that a 1% increase in fatality decreases tourist arrivals from developed countries by 0.082%, which corresponds to 2487 visitors per year.

In a recent study, Montes and Bernabé (2020) examine the tourist arrivals to Rio de Janeiro, which is the only city in Brazil among the 100 most visited in the world. Employing a panel data analysis covering the period of 2003–2016, the authors conclude that violence reduces the tourist arrivals. The authors also suggest that tourists from developed countries are more sensitive to violence than in developing countries. More specifically, every death as a result of violence in Rio de Janeiro scares away almost four tourists from developed countries and three tourists from developing countries.

In another recent study, Seabra et al. (2020) analyzed the connections between terrorist attacks and tourist arrivals in a multivariate time series analysis between 2002 and 2016 for Portugal. The main finding of the study suggests that terror incidents have a substantial impact on tourist arrivals. The authors also confirm the existence of terrorism spillover meaning that terrorist attacks in other countries also have consequences for Portuguese tourism.

From the globalization perspective, the current literature underlines a mutual relationship between globalization and tourism, which takes globalization into account as a leading factor affecting the dynamics of the tourism sector (Song et al. 2018). Using alternative globalization proxies such as trade openness, international affairs or cooperations, KOF globalization measurements, available studies confirmed the positive relationship between tourism and globalization (Chung et al. 2019; Hjalager 2007; Sugiyarto et al. 2003; Ulucak et al. 2020)

Despite being not limited to those mentioned above, studies investigating the relationship between terrorism and tourism for the case of Turkey are limited. As terrorism and other dynamics of tourism demand continue to evolve, it deserves special attention to analyze the relationship using more recent data and advanced econometric methods. This study identifies the gap and employs robust tests that examine the dynamics of tourism demand while paying special attention to terrorism and globalization.

6.3 Model, Data, and Methodology

The current literature on the determinants of tourism demand indicates that there are many socioeconomic variables employed to explain the number of tourist arrivals for a hosting country except tourism destinations worth seeing and cheap tourism opportunities. In a broad spectrum, these variables can be listed as domestic price level, per capita income, household debt level/debt crisis, nominal/real exchange rate, relative prices between visiting and hosting country, transportation costs, accommodation costs, crime rates, terror incidents, the rule of law, security, justice, globalization, corruption perception, etc. (Dogru et al. 2017; Fourie et al. 2019; Khalid et al. 2019). Although many leading factors explain tourist arrivals and can be employed in a stochastic framework; given the limitations of time series econometric applications and data availability, this study investigates the impacts of terrorism and globalization on the number of tourist arrivals visiting Turkey by constructing the following model:

$$lnTA_t = \beta_0 + \beta_1 lnFI_t + \beta_2 lnRP_t + \beta_3 lnREX_t + \beta_4 lnTer_t + \beta_5 lnGlob_t + \varepsilon_t \tag{6.1}$$

where $TA, FI, RP, REX, Ter, Glob$ and ε stand for tourist arrivals, foreign income level based on industrial production of advanced economies, relative prices calculated through proportion foreign price index and domestic price index (CPI_F/CPI_D), relative exchange rate, number of terror attacks, KOF globalization index and the stochastic error term reflects the effects of undefined factors in the model on tourist arrivals, respectively. The variables in Eq. (6.1) cover the period 1980–2018 based on the annual data for a single country—Turkey—which is suitable for using time series econometric analyses in the estimation of β coefficients. To this end, before proceeding the parameter estimation, the first step is to check the stationarity of variables since time series may have a unit root that may lead to a spurious regression relationship in the estimation of model parameters (Engle and Granger 1987). Then the analysis is maintained by verifying a long-run equilibrium relationship between study variables if the estimation model has non-stationary variables (Enders and Lee 2012a).

To check the stationarity properties of time series, there are some issues to be considered for reliable outcomes. For instance, traditional approaches such as augmented Dickey–Fuller (ADF 1981), Phillips and Perron (PP 1988), Kwiatkowski et al. (KPSS 1992), and Ng and Perron (NP 2001) tests may result in non-stationarity if there is a structural break affecting the state of affairs for the related variable (Perron 1989). On the other hand, the number of breaks is another problem in more recent methodologies which take structural breaks into account in stationarity check such as those proposed by Lumsdaine and Papell (1997), Zivot and Andrews (1992), Lee and Strazicich (2003) since it is difficult to capture the correct number and magnitude of multiple breaks (Prodan 2008). Therefore, a third option in controlling unit root properties of variables is to use Fourier approximation that allows capturing known and unknown breaks through the use of low-frequency components (Enders and Lee 2012a, b; Rodrigues and Taylor 2012). In this process of Fourier approximation, known and unknown structures of breaks are regarded through Eq. (6.2) by expanding the unit root test equation with sine and cosine waves that occur over time in the series.

$$d(t) = a_0 + \sum_{r=1}^{z} \delta_{1,r}\sin(2\pi kt/T) + \sum_{r=1}^{z} \delta_{2,r}\cos(2\pi rt/T);\ z \leq T/2 \tag{6.2}$$

Equation (6.2) includes the number of frequencies (z) and a particular frequency (r) over the T period, and its inclusion into the unit root equation enables capturing known and unknown breaks in the unit root checks. Due to their flexibility structures, this study performs Fourier unit root tests proposed by Enders and Lee (2012a, b), Rodrigues and Taylor (2012). The deterministic term $d(t)$ defined by Eq. (6.2) is also proposed by Banerjee et al. (2017) to check stationary combination of non-stationary variables, which means the presence of a long-run equilibrium–cointegration relationship that is the second step of non-stationary time-series analyses, between study variables.

$$\Delta x_{1,t} = d(t) + \gamma_1 x_{1,t} + \gamma_2 x_{2,t-1} + \gamma_3 \Delta x_{2,t} + \varepsilon_t \tag{6.3}$$

where x_1 and x_2 are the dependent and explanatory variables, respectively, and lagged values of their differenced terms in Eq. (6.3) are useful to control probable serial correlation of the residuals. As defined in Eq. (6.2), $d(t)$ is the deterministic term that considers non-linear behavior of time. Having performed the procedure in Eq. (6.3), a cointegration relationship is determined based on the null and alternative hypothesis through t-statistics

$$H_0 : \gamma_1 = 0 \rightarrow \text{no cointegration}$$

$$H_A : \gamma_1 < 0 \rightarrow \text{presence of cointegration}$$

Fourier approximation procedure to check the stationarity of variables and cointegration relationship between them provides a consistent basis before the estimation of long-run parameters in Eq. (6.1). Another critical issue is to carry out an appropriate estimator for the cointegrated variables. Because various problems such as serial correlation, heteroskedasticity and endogeneity may lead to unreliable results, the ordinary least squares (OLS) estimator is conducted. To this end, fully modified OLS (FMOLS) and dynamic OLS (DOLS) estimators are capable in consideration of those issues in the long-run parameter estimations and produce more robust results for cointegration equations.

6.4 Results and Discussions

To check the stationarity properties of the variables, Fourier unit root tests proposed by Enders and Lee (2012a, F-ADF hereafter; 2012b, F-LM hereafter, and Rodrigues and Taylor (2012, F-GLS hereafter) were applied and three statistics were obtained for each variable with different number of Fourier frequencies. Unit root test results in Table 6.2 indicate that number of tourist arrivals (*TA*), foreign income (*FI*), relative prices (*RP*), relative exchange rate (*REX*), number of terror incidents (*TER*) and globalization data follow a non-stationary process over the period 1980–2018. According to the results at their level values, test statistics calculated for each one shown in column 3 is greater than critical values for 5% significance levels, which implies the null hypothesis of stationarity should be rejected.

Having rejected stationarity for the level values, it is useful to confirm the order of integrations level, i.e., I(1), I(2) for each variable since the second step investigates the linear combination of study variables is I(0) or I(1). So, unit root tests were rerun for first differenced data of each variable, and results in the fourth column of Table 6.2 were obtained, which each one is lower than their critical values of 5%. So, one might conclude that all the variables turn out to be stationary when their first differences are taken, meaning that all the variables are I(1). Such a circumstance

Table 6.2 Fourier unit root test results

		Test Statistics (level)	Test Statistics (First Difference)	Number of Fourier	Critical value (5%)
TA	F-ADF	−2.652	−5.145	1	−4.35
	F-LM	−3.854	−6.768	1	−4.10
	F-GLS	−3.456	−5.951	1	−4.17
FI	F-ADF	−3.245	−6.857	2	−4.05
	F-LM	−2.358	−5.386	2	−3.57
	F-GLS	−1.978	−4.158	2	−3.64
RP	F-ADF	−3.157	−6.518	2	−4.35
	F-LM	−3.054	−6.024	2	−4.10
	F-GLS	−4.015	−7.156	1	−4.17
REX	F-ADF	−3.752	−6.785	2	−4.05
	F-LM	−3.854	−5.963	1	−4.10
	F-GLS	−3.571	−5.487	1	−4.17
TER	F-ADF	−2.419	−5.611	2	−4.05
	F-LM	−3.625	−6.452	1	−4.10
	F-GLS	−2.984	−5.325	2	−3.64
GLOB	F-ADF	−3.547	−6.547	1	−4.35
	F-LM	−2.990	−4.982	1	−4.10
	F-GLS	−3.276	−6.541	2	−4.05

Table 6.3 Fourier cointegration test result

$\hat{k}$	t-stat	Critical value (5%)	Lags for [Δy]	[Δx]	Result
2	−5.715	−3.80	1	1	Cointegration

that all variables of study model in Eq. (6.1) are I(1) requires to check whether a cointegration relationship between those variables exists or not before proceeding with the estimation of long-run coefficients. Therefore, a Fourier unit root test proposed by Banerjee et al. (2017) was applied and the results in Table 6.3 were obtained, which verifies a cointegration/long-run equilibrium relationship for Eq. (6.1) since the calculated t-statistic depicted in column 2 is lower than the 5% critical value in column 3. Hence, it falls into the rejection region for the null hypothesis of no cointegration.

After confirmation of the cointegration relationship, cointegration estimators such as FMOLS and DOLS should be employed to avoid possible endogeneity problem that leads to biased and inconsistent results. FMOLS and DOLS estimators are also robust against serial correlation and heteroskedasticity problems. Therefore, to get long-run cointegration parameters of Eq. (6.1), we applied these cointegration estimators whose results are noted in Table 6.4.

Long-run coefficients estimated through FMOLS and DOLS are statistically significant, and the explanatory power of regressors for variabilities in tourist arrivals is sufficiently high with R-squared values of 90% and 91%. According to the results, an increase in the income level of visitors increase the number of tourist arrivals by 2.258%, which confirms the theoretical expectation that tourists respond to income rise by increasing their demand for tourism. Similarly, rises in relative price and relative exchange rate as well, which are essential determinants for both the purchasing powers of foreign visitors and destination preferences, increase the number of tourist arrivals. However, one may conclude that terrorism matters for the tourism sector in Turkey since it has a detrimental impact on tourist arrivals, and the coefficient of terrorism with a negative sign is higher than those of other regressors in the estimation. Results on the role of terrorism in the tourism sector are in line with findings of

Table 6.4 Long-run cointegration parameters

Variable	FMOLS			DOLS		
	Coeff.	s.e.	t-stat	Coeff.	s.e.	t-stat
Constant	10.253	3.140	3.265***	12.785	2.546	5.021***
FI	2.258	1.107	2.039**	2.457	0.945	2.600**
RP	0.825	0.256	3.222***	1.126	0.254	4.433***
REX	1.568	0.457	3.431***	1.235	0.154	8.019***
TER	−2.871	1.245	−2.306**	−2.687	0.763	-3.521***
GLOB	0.954	0.475	2.008**	1.521	0.368	4.133***

***, **, *denote statistical significance level at 1%, 5%, and 10%, respectively

Bhattarai et al. (2005), Drakos and Kutan (2003), Llorca-Vivero (2008), Thompson (2011), Ulucak et al. (2020), Yap and Saha (2013), Yaya (2009), among others.

On the other hand, some discussion points arise for Turkey to underline threats in the tourism sector in terms of terror attacks. Accordingly, the well-known hypothesis recalls that "the frequency of acts of terrorism cause a larger decline in international tourist arrivals than the severity of these acts." Turkey is almost subjected to daily terror attacks as can be verified by daily data of the Global Terrorism Database. Although a vast majority of them targeted military or police powers in the southeast region, which is far from major tourist destinations of the country, these attacks may create a considerable uncertainty or may be a robust dissuading factor for visitor decisions in choosing Turkey. One may claim that tourist arrivals to Turkey have increased over time despite terror attacks, but this claim should be justified by revealing and comparing potential and available tourist numbers. In parallel with this argument, Yaya (2009) examined the effect of terrorism on the number of tourist arrivals to Turkey using monthly data 1997:01–2006:12. The findings of the study show that there exists a negative but relatively small impact of terrorism on tourism. However, our results show that terror attacks are the most influenced factor affecting tourist arrivals.

Moreover, tourist arrivals may continue to increase since other factors such as global interaction, cheap holiday opportunities based on relative prices/exchange rates, and the absence of alternatives may have dominated the adverse effect of terrorism. However, countries may not increase tourism receipts in parallel with the increase of arrivals since tourists may prefer not to get out of hotels with full service during their vacations which are made especially for summer holidays due to the probability of any victimization. So, this may be an alternative explanation of why Turkey is not ranked within the top ten countries by tourism receipts even though it always takes part in the most visited ten countries (see Table 6.1). This discussion is in line with the findings of Montes and Bernabé (2020), showing that tourists from developed countries are more sensitive to violence than developing countries. More specifically, they found that every death as a result of violence in Rio de Janeiro scares away almost four tourists from developed countries and three tourists from developing countries. Considering this situation for Turkey, developed countries such as Germany, France, the United Kingdom, Netherlands, Belgium, Greece, Poland, Switzerland, and the United States are among the most visitor sending countries to Turkey.

On the other hand, Russia, Georgia, Iran, Iraq, Bulgaria, Ukraine, Azerbaijan, Saudi Arabia, Romania, and Kazakhstan are the most tourist-sending countries with lower and middle income relatively. Therefore, terror attacks may have different impacts on tourists who come from developed and developing countries to Turkey as well. Overall, empirical findings and all these discussion points may shed new light on the impact of terrorism on tourist arrivals in Turkey, and the situation noted down in Table 6.1. Thus, struggling with terrorism matters for Turkey to increase tourist arrivals and receipts.

Our results for globalization and other explanatory variables are also consistent with the theoretical expectations. Globalization variable is another focal point of the study and results indicate that tourist arrivals increase as the globalization level of the country rises, thus the country can increase benefits from the tourism sector through economic, societal, and cultural integration to the world.

6.5 Conclusion

The tourism sector is a major source of employment, government revenue, and foreign currency for many economies since it provides a considerable employment opportunity and creates a huge demand for domestic products. However, it is affected by many variables that increase or decrease the number of visitors and/or tourism receipts and needs to be well managed to get its potential benefits and increase economic added values from tourism. To this end, the impacts of driving factors and their roles in shaping tourism policies should be well investigated and revealed in detail. Turkey is an emerging economy in which the tourism sector has a significant share and needs to improve gains from tourism, given its tourism potential. On the other hand, it suffers from a chronic terror problem that should be emergently solved, which constrains the further expansion of the tourism sector. To reveal how tourist arrivals react to terror attacks in Turkey, this study investigates the effects of globalization and terrorism on tourist arrivals by using yearly data covering the period 1980–2018 through time series econometric methods. Firstly, stationarity conditions of time series data used in the study were tested through Fourier unit root tests that allow known and unknown breaks, and then the cointegration relationship was confirmed by the Fourier cointegration approach. Finally, FMOLS and DOLS estimators were carried out to get long-run cointegration parameters that show how 1% change in the related variable would affect the response variable. Empirical findings indicate that terror attacks have a detrimental effect on tourist arrivals, while globalization, relative prices, and exchange rate, foreign income have a positive influence on the tourism sector. Combining the facts and empirical results, the study provides valuable insights and discussion points for the tourism sector in Turkey. For instance, Turkey cannot reach the potential tourism gains despite its visitor numbers, and visitors are highly sensitive to terror attacks, as is verified by our findings. Also, research reveals that visitors who may spend more money from advanced or high-income countries are more vulnerable to terror attacks than other visitors coming from developing or low-income countries. Besides, it is accepted that the frequency of terror attacks has a more significant negative impact on tourist arrivals than the severity of these acts. Therefore, terrorism is one of the most important problems in the tourism sector in Turkey. Policymakers should focus on not only military or security purposes in combating terrorism but also, they realize the potential economic loss from the tourism sector due to terror attacks and spend more efforts to convince all agents to struggle with violence and terrorism firstly. On the other hand, results suggest that global integration helps shape the tourism sector, and it leads to an increase in the number of tourist arrivals that may also result in more gains from the tourism sector. So, policymakers should expand the integration level by also designing good relationships with other countries to get more benefits and gains from the sector.

References

Aly, H. Y., & Strazicich, M. C. (2000). Terrorism and tourism: Is the impact permanent or transitory? Time-series evidence from some MENA countries. Ohio State University.

Banerjee, P., Arčabić, V., & Lee, H. (2017). Fourier ADL cointegration test to approximate smooth breaks with new evidence from crude oil market. *Economic Modelling, 67,* 114–124. https://doi.org/10.1016/j.econmod.2016.11.004.

Bhattarai, K., Conway, D., & Shrestha, N. (2005). Tourism, terrorism and turmoil in Nepal. *Annals of Tourism Research, 32*(3), 669–688. https://doi.org/10.1016/j.annals.2004.08.007.

Bilgili, F., Ulucak, R., Koçak, E., & İlkay, S. Ç. (2020). Does globalization matter for environmental sustainability? Empirical investigation for Turkey by Markov regime switching models. *Environmental Science and Pollution Research, 27*(1), 1087–1100. https://doi.org/10.1007/s11356-019-06996-w.

Buckelew, A. F. (1984). *Terrorism and the American response.* Mira Academic Press.

Buigut, S. (2018). Effect of terrorism on demand for tourism in Kenya: A comparative analysis. *Tourism and Hospitality Research, 18*(1), 28–37. https://doi.org/10.1177/1467358415619670.

Chung, M. G., Herzberger, A., Frank, K. A., & Liu, J. (2019). International tourism dynamics in a globalized world: A social network analysis approach. *Journal of Travel Research, 57*(8), 999–1011. https://doi.org/10.1177/0047287519844834.

Dickey, B. Y. D. a, & Fuller, W. a. (1981). Likelihood ratio statistics for autoregressive time series with a unit root. *Econometrica, 49*(4), 1057–1072.

Dogru, T., Sirakaya-Turk, E., & Crouch, G. I. (2017). Remodeling international tourism demand: Old theory and new evidence. *Tourism Management, 60,* 47–55. https://doi.org/10.1016/j.tourman.2016.11.010.

Drakos, K., & Kutan, A. M. (2003). Regional effects of terrorism on tourism in three mediterranean countries. *Journal of Conflict Resolution, 47*(5), 621–641. https://doi.org/10.1177/0022002703258198.

Enders, W., & Lee, J. (2012a). The flexible fourier form and Dickey-Fuller type unit root tests. *Economics Letters, 117*(1), 196–199.

Enders, W., & Lee, J. (2012b). A unit root test using a fourier series to approximate smooth breaks. *Oxford Bulletin of Economics and Statistics, 74*(4), 574–599. https://doi.org/10.1111/j.1468-0084.2011.00662.x.

Enders, W., & Sandler, T. (1991). Causality between transnational terrorism and tourism: The case of Spain. *Terrorism, 14*(1), 49–58. https://doi.org/10.1080/10576109108435856.

Enders, W., Sandler, T., & Parise, G. F. (1992). An econometric analysis of the impact of terrorism on tourism. *Kyklos, 45*(4), 531–554. https://doi.org/10.1111/j.1467-6435.1992.tb02758.x.

Engle, R. F., & Granger, C. W. J. (1987). Co-integration and error correction: Representation, estimation, and testing. *Econometrica, 55*(2), 251–276. http://links.jstor.org/sici?sici=0012-9682%28198703%2955%3A2%3C251%3ACAECRE%3E2.0.CO%3B2-T.

Feridun, M. (2011). Impact of terrorism on tourism in Turkey: Empirical evidence from Turkey. *Applied Economics, 43*(24), 3349–3354. https://doi.org/10.1080/00036841003636268.

Fourie, J., Rosselló-Nadal, J., & Santana-Gallego, M. (2019). Fatal attraction: How security threats hurt tourism. *Journal of Travel Research, 58,* 1–11. https://doi.org/10.1177/0047287519826208.

Hjalager, A.-M. (2007). Stages in the economic globalization of tourism. *Annals of Tourism Research, 34*(2), 437–457. https://doi.org/10.1016/J.ANNALS.2006.10.006.

Khalid, U., Okafor, L. E., & Shafiullah, M. (2019). The effects of economic and financial crises on international tourist flows: A cross-country analysis. *Journal of Travel Research, 58,* 004728751983436. https://doi.org/10.1177/0047287519834360.

Kwiatkowski, D., Phillips, P. C. B., Schmidt, P., & Shin, Y. (1992). Testing the null hypothesis of stationarity against the alternative of a unit root. How sure are we that economic time series have a unit root? *Journal of Econometrics, 54*(1–3), 159–178.

Lee, J., & Strazicich, M.C. (2003). Minimum lagrange multiplier unit root test with two structural breaks. *Review of Economics and Statistics, 85*(4), 1082–1089. https://doi.org/10.1162/003465303772815961.
Llorca-Vivero, R. (2008). Terrorism and international tourism: New evidence. *Defence and Peace Economics, 19*(2), 169–188. https://doi.org/10.1080/10242690701453917.
Lumsdaine, R. L., & Papell, D. H. (1997). Multiple trend breaks and the unit-root hypothesis. *Review of Economics and Statistics, 79*(2), 212–218.
Montes, G. C., & Bernabé, S. de P. (2020). The impact of violence on tourism to Rio de Janeiro. *International Journal of Social Economics, 47*(4), 425–443. https://doi.org/10.1108/IJSE-09-2019-0590.
Ng, S., & Perron, P. (2001). LAG length selection and the construction of unit root tests with good size and power. *Econometrica, 69*(6), 1519–1554.
Perron, P. (1989). The great crash, the oil price shock, and the unit root hypothesis. *Econometrica, 57*(6), 1361–1401.
Phillips, P. C. P. C. B., & Perron, P. (1988). Testing for a unit root in time series regression. *Biometrika, 75*(2), 335–346. https://doi.org/10.1093/biomet/75.2.335.
Pizam, A., & Fleischer, A. (2002). Severity versus frequency of acts of terrorism: Which has a larger impact on tourism demand? *Journal of Travel Research, 40*(3), 337–339. https://doi.org/10.1177/004728750204000301l.
Prodan, R. (2008). Potential pitfalls in determining multiple structural changes with an application to purchasing power parity. *Journal of Business & Economic Statistics, 26*(1), 50–65. http://www.tandfonline.com/doi/abs/10.1198/073500107000000304.
Rodrigues, P. M. M., & Robert Taylor, A. M. (2012). The flexible fourier form and local generalized least squares de-trended unit root tests*. *Oxford Bulletin of Economics and Statistics, 74*(5), 736–759. https://doi.org/10.1111/j.1468-0084.2011.00665.x.
Seabra, C., Reis, P., & Abrantes, J. L. (2020). The influence of terrorism in tourism arrivals: A longitudinal approach in a Mediterranean country. *Annals of Tourism Research, 80*. https://doi.org/10.1016/j.annals.2019.102811.
Song, H., Li, G., & Cao, Z. (2018). Tourism and economic globalization: An emerging research agenda. *Journal of Travel Research, 57*(8), 999–1011. https://doi.org/10.1177/0047287517734943.
Sugiyarto, G., Blake, A., & Sinclair, M. T. (2003). Tourism and globalization: Economic impact in Indonesia. *Annals of Tourism Research, 30*(3), 683–701. https://doi.org/10.1016/S0160-7383(03)00048-3.
Thompson, A. (2011). Terrorism and tourism in developed versus developing countries. *Tourism Economics, 17*(3), 693–700. https://doi.org/10.5367/te.2011.0064.
Ulucak, R., Yücel, A. G., & İlkay, S. Ç. (2020). Dynamics of tourism demand in Turkey: Panel data analysis using gravity model. *Tourism Economics*, 135481662090195. https://doi.org/10.1177/1354816620901956.
World Tourism Organization. (2020a). *International tourism numbers could fall 60–80% in 2020 [Press Release -7 May 2020]*. Retrieved from https://www.unwto.org/news/covid-19-international-tourist-numbers-could-fall-60-80-in-2020.
World Tourism Organization. (2020b). *UNWTO World Tourism Barometer* (Vol. 18, No. 3, pp. 14–15). Madrid: UNWTO. https://doi.org/10.18111/wtobarometereng.
World Travel and Tourism Council. (2019a). *Travel & tourism economic impact 2019 world*.
World Travel and Tourism Council. (2019b, 4 November). Travel & tourism industry is more resilient than ever. Retrieved from https://wttc.org/News-Article/Travel-Tourism-Industry-is-More-Resilient-Than-Ever-According-to-New-Research-by-WTTC-and-Global-Rescue.
Yap, G., & Saha, S. (2013). Do political instability, terrorism, and corruption have deterring effects on tourism development even in the presence of Unesco heritage? A cross-country panel estimate. *Tourism Analysis, 18*(5), 587–599. https://doi.org/10.3727/108354213X13782245307911.

Yaya, M. E. (2009). Terrorism and tourism: The case of Turkey. *Defence and Peace Economics, 20*(6), 477–497. https://doi.org/10.1080/10242690903105414.

Zivot, E., & Andrews, D. W. K. (1992). Further evidence on the great crash, the oil-price shock, and the unit-root hypothesis. *Journal of Business & Economic Statistics, 10*(3), 251–270.

Chapter 7
Testing the Dynamic Relationship Among CO_2 Emissions, Economic Growth, Energy Consumption and Tourism Development. Evidence for Uruguay

Juan Gabriel Brida, Bibiana Lanzilotta, and Fiorella Pizzolon

Abstract Carbon dioxide (CO_2) emission is directly linked with energy usage and plays an essential role in the debate on sustainable tourism development and environmental protection. Some authors argue that tourism faces the problem of being "addicted" to growth, which is incompatible with sustainable goals. The literature shows that the environmental Kuznets curve (EKC) hypothesis induced by tourism is verified, with some differences between developed and developing economies. Previous country-studies usually apply linear cointegration techniques and Granger causality tests in order to test the EKC hypothesis. This study explores the linkages between CO_2 emissions, economic growth, energy consumption and tourism development for Uruguay without imposing—a priori—any parametric model, in order to investigate the presence of nonlinearity in the relation, as postulated by the EKC hypothesis. This paper examines the dynamic long-run relationship among these variables, using data from 1960 to 2014. We test the existence of nonlinear cointegration relationship and the causality applying nonparametric tests. We find that this methodology provides a more suitable way to represent linkages between the variables under study for Uruguay. Nevertheless, the evidence regarding the causality between tourism growth and CO_2 emissions is weak. Finally, we discuss policy implications, limitations, and future research.

Keywords Tourism-induced EKC hypothesis · Nonlinear cointegration · Nonparametric causality tests · Uruguay

JEL Codes C30 · E43 · L83

J. G. Brida (✉) · B. Lanzilotta · F. Pizzolon
Facultad de Ciencias Económicas y de Administración, Research Group in Economic Dynamics (GIDE), Universidad de la República, Montevideo, Uruguay
e-mail: gbrida@ccee.edu.uy

B. Lanzilotta
e-mail: bibiana.lanzilotta@gmail.com

F. Pizzolon
e-mail: fiorella.pizzolon@gmail.com

D. Balsalobre-Lorente et al. (eds.), *Strategies in Sustainable Tourism, Economic Growth and Clean Energy*, https://doi.org/10.1007/978-3-030-59675-0_7

7.1 Introduction

Many studies explore the relationship between energy consumption, environmental pollution—usually proxied by carbon dioxide (CO_2) emissions, and economic growth. However, less attention has been paid to this relationship concerning the tourism sector, a particular sector of the economy that deserves attention.

Tourism is a large and dynamic economic sector. Its total contribution accounts for 10.4% of global gross domestic product (GDP), 9.9% of total employment, and 4.5% of total investment, in 2017. In recent years, it has registered a steady growth rate of 4% annually (World Travel and Tourism Council 2018). Several studies have confirmed tourism development as an engine of economic growth in some countries. Brida et al. (2014) conducted an exhaustive review of approximately 100 peer-reviewed published papers on the tourism-led growth hypothesis (TLGH) and found that with a few exceptions, the empirical findings suggest that overall international tourism drives economic growth. More recently, Risso (2018) finds evidence in support of the TLGH at a worldwide level.

However, with growth arise sustainability challenges. Like any other industry, tourism can induce pressure on the environment. The development of this sector also contributes to the development of other segments, leading to a higher demand of energy (Becken et al. 2001, 2003 and Gössling 2002), which can be a source of environmental degradation (Xuchao et al. 2010). Therefore, an exploration of the relationship among CO_2 emissions, economic growth, energy consumption, and tourism development is of great interest to both policymakers and practitioners in tourism and would be a contribution to tourism literature.

The empirical link between environmental pollution (proxied by CO_2 emissions) and energy consumption has been extensively investigated and validated in the energy economics literature (Alam et al. 2011; Ang 2008; Soytas et al. 2007; Xing-Ping and Xiao-Mei 2009); even though, the direction of causality between them remains unclear. The literature focusing on the impact of tourism on CO_2 emissions has been growing in the last decade (Neto 2003; Holden 2009; Lin 2010; Perch-Nielsen et al. 2010; Dubois et al. 2011; Gössling 2013; Saenz-de-Miera and Rosselló 2014; Solarin 2014; Tsai et al. 2014). Regarding country panel analysis, Lee and Brahmasrene (2013) find that a long-run equilibrium relationship exists among tourism, CO_2 emissions, economic growth, and foreign direct investment (FDI) in European Union countries. Furthermore, they find that tourism, CO_2 emissions, and FDI have high significant positive effects on economic growth. Economic growth, in turn, shows a highly significant positive impact on CO_2 emissions while tourism and FDI incur a highly significant negative impact on CO_2 emissions. Gössling (2013) finds that while emissions from tourism are significant in all 22 countries studied,[1] they may, in some countries, exceed "official" emissions as calculated on the basis of

[1]New Zealand, Norway, Sweden, Germany, Australia, Switzerland, Maldives, The Netherlands, Anguilla, Antigua and Barbuda, Bahamas, Barbados, Belize, Dominica, Dominican Republic, Grenada, Jamaica, St. Kitts and Nevis, St. Lucia, St. Vincent and the Grenadines, Suriname, and Turks and Caicos.

guidelines for national emission inventories under the Kyoto Protocol. Regarding individual country analysis, Katircioglu (2014b) finds that tourism development in Turkey has resulted not only in considerable increases in energy use but also considerable increases in climate change. Katircioglu et al. (2014) find that tourism development is a catalyst for increases in energy consumption and carbon emissions in the case of Cyprus.

Other studies have tested the validity of the Environmental Kuznets Curve (EKC) hypothesis, which investigates the relationship between environmental pollution and real income growth (Grossman and Krueger 1991; Shafik and Bandypadhyay 1992; Coondoo and Dinda 2002; Dinda 2004; Stern 2004; Luzzati and Orsini 2009). Empirical evidence on the EKC hypothesis is anything but convergent. While some studies find a linear relationship between environmental degradation and economic growth (Shafik and Bandyopadhyay 1992; Akbostanci et al. 2009), others have provided evidence in support of an inverted U-shaped relationship in line with the EKC prediction (Lindmark 2002) though discrepancies have emerged in terms of the exact "turning point". The benchmark study in this literature remains that by List and Gallet (1999) who showed that over the period 1929–1994, in the US, an inverted U-shaped EKC characterized the relationship between per capita emissions and per capita income at the state level. Other studies still have found an "N-shaped relationship" (for example, Friedl and Getzner 2003) which suggests that any decline of environmental degradation is only limited to the short term (He and Richard 2010). Moreover, many studies have integrated the investigation of the relationship between energy consumption and output growth within the EKC framework (for example, Ozturk and Acaravci 2010). Indeed, the evidence of the impact of energy use and economic growth on CO_2 emissions is so overwhelming, that both of these variables are now routinely integrated into the analysis.

To sum up, the increasing importance of the tourism sector and its potential threat to the environment, made researchers focus on the relationship between tourism growth and environmental sustainability. This is a key question for policymakers, who face a dilemma of stimulating economic development via tourism-led growth while protecting the environment, and a crucial concept on the subject, as mentioned above, is the EKC hypothesis. However, as stated before, there are not many studies that explore the EKC hypothesis induced by tourism. Katircioglu (2014a) and De Vita et al. (2015) for individual countries, and Dogan et al. (2015) and Ozturk et al. (2015) for a panel of countries, are the first ones to examine the relationship between tourism development and environmental degradation—measure as CO_2 emissions in the first three studies and the ecological footprint in the last one[2]—within an EKC framework. Katircioglu (2014a) finds that tourism development and carbon emissions are in long-term equilibrium relationship; tourist arrivals have a negatively significant effect on carbon dioxide emission levels both in the long term and the short term and verifies

[2]Ozturk et al. (2015) point that most of the existing literature employed environmental indicators such as CO_2 emissions, sulfur dioxide (SO_2), dark matter (fine smoke), and suspended particle matter (SPM) as the endogenous variables. Nevertheless, these indicators only represent a small portion of the environmental degradation caused by tourism.

the tourism-induced EKC hypothesis in the case of Singapore. De Vita et al. (2015) find that international tourist arrivals into Turkey alongside income, squared income and energy consumption, cointegrate with CO_2 emissions. They find that tourist arrivals, growth, and energy consumption exert a positive and significant impact on CO_2 emissions in the long run. The results of this study provide empirical support to EKC hypothesis showing that at exponential levels of growth, CO_2 emissions decline. Dogan et al. (2015) analyze the long-run dynamic relationship of CO_2 emissions, real GDP, energy consumption, trade and tourism under an EKC model for the OECD countries, and find that the EKC hypothesis cannot be supported. Finally, Ozturk et al. (2015) analyze a panel of 144 countries and find that the number of them that have a negative relationship between the ecological footprint and its determinants (GDP growth from tourism, energy consumption, trade openness, and urbanization) is more existent in the upper-middle and high-income countries. Moreover, they notice that the EKC hypothesis is more present in the upper-middle and high-income countries than the other income countries. More recently, Paramati et al. (2017) analyze 26 developed economies and 18 developing economies and find that the impact of tourism on CO_2 emissions is reducing much faster in developed economies than in developing economies, providing evidence of the EKC hypothesis on the link between tourism growth and CO_2 emissions.

In Uruguay, the tourism sector's total contribution accounted for 10.6% of the GDP, 10.2% of total employment, and 7.7% of total investment, in 2017. In recent years, it has registered a volatile behavior, but since 2015 it has verified positive growth rates. (World Travel and Tourism Council 2018). Brida et al. (2013a, b) confirm that the TLGH is valid for Uruguay; tourism is the locomotive sector of the economy.

The rest of the article is structured as follows. First, we define the methodological framework of the present study; then, we introduce the data used; the following section presents the empirical results, and the final section contains the conclusions.

7.2 Methodological Framework

In order to test and estimate the relationship between CO_2 emissions, economic growth, energy consumption, and tourism development for Uruguay, we follow the procedure suggested by Breitung (2001), Holmes and Hutton (1990), and Ye Lim et al. (2011). This procedure has the following steps: (i) testing nonparametric unit root, (ii) testing the existence of cointegration by using nonparametric tests, (iii) testing linearity, and (iv) performing the rank-causality test. A quick reference of these tests is presented in what follows.

7.2.1 Nonparametric Unit Root Test

Breitung (2002) proposes a statistic test that does not require the specification of the short-run dynamic; such approach is called "model-free" or "nonparametric" because the asymptotic properties of the test do not depend on the short-run dynamics or the nuisance parameters. As a result, the test proposed by Breitung is robust against possible misspecification. Following Davidson (2002), Breitung takes a definition of integration that is not restricted to a specific time series model:

A time series y_t is integrated of order one (I(1)) if, as $T \to \infty$,

$$T^{-1/2} y_{[aT]} \underset{T\to\infty}{\Rightarrow} \sigma W(a) \tag{7.1}$$

where the symbol $\underset{T\to\infty}{\Rightarrow}$ means weak convergence concerning the associated probability measure, $\sigma > 0$ is a constant, [.] represents the integer part, and $W(a)$ is a Brownian motion defined on C[0,1]. Breitung (2002) proposes the variance ratio statistic to test the null hypothesis that y_t is I(1), against the alternative hypothesis y_t is I(0). Critical values are available in Breitung (2002).

The Q_T is the variance ratio of the partial sums and the original series. The variance ratio statistic is defined as

$$\widehat{Q_T} = \frac{T^{-1}\sum_{t=1}^{T} \widehat{U}_t^2}{\sum_{t=1}^{T} \widehat{u}_t^2} \tag{7.2}$$

where $\widehat{U}_t = \widehat{u}_1 + \cdots + \widehat{u}_t$ and $\widehat{u}_t = y_t - \widehat{\delta}' z_t$ are the ordinary least square (OLS) residuals from the regression of the data y_t on (i) $z_t = 0$, let $\widehat{u} = y_t$, with no deterministic term, (ii) $z_t = 1$, with an intercept, or (iii) $z_t = (1, t)'$, with an intercept and linear trend, respectively. The variance ratio statistic is a left tailed test, where the hypothesis of a unit root process is rejected if the test statistic value is smaller than the critical value.

7.2.2 Rank Test for Cointegration

Breitung (2001) introduces a nonparametric test procedure to test the hypothesis of a cointegration relationship and to identify whether this link is nonlinear. Breitung's procedure proposes a rank transformation for the series involved and checks whether the ranked series move together over time toward a linear or nonlinear long-term cointegrating equilibrium. The procedure starts checking the cointegration by using the rank test. If cointegration is accepted, the technique follows with examining linearity in the cointegration relationship, by using a scoring test.

Let $f(x_t) \sim I(1)$ and $g(y_t) \sim I(1)$ be nonlinear increasing functions of x_t and y_t, and $\mu_t \sim I(0)$. Let us suppose that a nonlinear cointegration relationship between x_t and y_t is given by

$$\mu_t = g(y_t) - f(x_t) \tag{7.3}$$

The rank statistic is constructed by replacing $f(x_t)$ and $g(y_t)$ by the ranked series

$$R_T[f(x_t)] = R_T(x_t) \tag{7.4}$$

and

$$R_T[g(y_t)] = R_T(y_t) \tag{7.5}$$

Given that the sequence of ranks is invariant under monotonic transformations of the variables, if x_t or y_t are random walk process then $R_T\ [f(x_t)]$ and $R_T\ [g(y_t)]$ behaves like the ranked random walks as $R_T(x_t)$ and $R_T(y_t)$.

The rank test procedure is based on two "distance measures" between the sequences of $R_T(x_t)$ and $R_T(y_t)$. The cointegration test is based on the difference between the sequences on the ranks that can be detected by the bivariate statistics K_T^*: and ξ_T^* :

$$K_T^* = T^{-1} \max_t |d_t| / \hat{\sigma}_{\Delta d} \tag{7.6}$$

$$\xi_T^* = T^{-3} \sum_{t=1}^{T} d_t^2 / \hat{\sigma}_{\Delta d}^2, \tag{7.7}$$

where

$$d_t = R_T(y_t) - R_T(x_t), \tag{7.8}$$

for $R_T(y_t) = \text{Rank}$ [of y_t among $y_1, \ldots, y_T$] and $R_T(x_t) = \text{Rank}$ [of x_t among $x_1, \ldots, x_T$]. The $\max_t |d_t|$ is the maximum value of $|d_t|$ over t = 1, 2,…, T and

$$\hat{\sigma}_{\Delta d}^2 = T^{-2} \sum_{t=2}^{T} (d_t - d_{t-1})^2 \tag{7.9}$$

adjusts for possible correlation between the series of interest.

7.2.3 Rank Test for (Neglected) Nonlinearity

If cointegration is not neglected in the first step, then we test the linearity of the cointegration relationship. For a convenient representation of the alternative and null hypothesis Breitung (2002) follows Granger (1995) and represents the nonlinear relationship as.

$$y_t = \gamma_0 + \gamma_1 x_t + f^*(x_t) + u_t, \tag{10}$$

where $\gamma_0 + \gamma_1 x_t$ is the linear part of the relationship. Only when $f^*(x_t) = 0$ there is a linear relationship between the variables. In this test, the multiple of the rank transformation is used instead of using $f^*(x_t)$.

Under the assumption that x_t is exogenous and u_t is a white noise with $u_t \sim N(0, \sigma^2)$ a *score test* is obtained as the T^*R^2 statistic of the OLS:

$$\tilde{u}_t = c_0 + c_1 x_t + c_2 R_t(x_t) + e_t. \tag{7.11}$$

Breitung (2001) generalizes the score test for the error correction model representation and applies it to contrast the null hypothesis of linear cointegration against the alternative hypothesis of nonlinear cointegration. To compute the score statistic, the following two multiple regressions are run, consecutively:

$$y_t = \alpha_0 + \sum_{i=1}^{p} \alpha_{1i} y_{t-i} + \alpha_2 x_t + \sum_{i=-p}^{p} \alpha_{3i} \Delta x_{t-i} + u_t \tag{7.12}$$

$$\tilde{u}_t = \beta_0 + \sum_{i=1}^{p} \beta_{1i} y_{t-i} + \beta_2 x_t + \sum_{i=-p}^{p} \beta_{3i} \Delta x_{t-i} + +\theta_1 R_T(x_t) + +\tilde{v}_t, \tag{13}$$

where $\beta_0 + \sum_{i=1}^{p} \beta_{1i} y_{t-i} + \beta_2 x_t + \sum_{i=-p}^{p} \beta_{3i} \Delta x_{t-i}$ is the linear part of the relationship and it involves the ranked series $R_T(x_{jt})$.

Under the null hypothesis, it is assumed that the coefficients for the ranked series are equal to zero, $\theta_1 = 0$. The appropriate value of p is selected based on Akaike Information Criterion, such that serial correlation $\tilde{u}_t$ and possible endogeneity are adjusted based on Stock and Watson (1993). The *score statistic* $T \cdot R^2$ is distributed asymptotically as a χ^2 distribution, where T is the number of observations and R^2 is the coefficient of determination of the second equation. The null hypothesis may be rejected in favor of nonlinear relationship if the score statistic value exceeds the χ^2 critical values with one degree of freedom (when two variables are involved).

7.2.4 Causality Rank Test

Conventional Granger causality test uses Vector Autoregression (VAR) or Vector Error Correction Model (VECM). However, results from the conventional parametric tests are limited by the augmenting hypothesis of the specific functional forms of the variables and the assumptions of homoscedasticity and normality of the error terms. As pointed by Ye Lim et al. (2011), violation of these conditions can cause spurious causality conclusions. For these cases, Holmes and Hutton (1990) proposed a multiple rank F-test, more robust than the standard Granger causality test. In case that the conditions of Granger estimations are satisfied, the multiple rank F-test results are alike the Granger results.

Holmes and Hutton (1990) analyze the small sample properties of the multiple rank F-test, and show that with non-normal error distributions, the test has significant power advantages both in small and in large samples. This is valid for both weak and strong relationships between the variables.

The Holmes and Hutton (1990) multiple rank F-test is based on the rank ordering of each variable. In this test, the causal relationship between y_t and x_t involves a test of a subset of q coefficients in the Autoregressive Distributed Lag (ARDL) model. The multiple rank F-test in the ARDL (p, q) model can be written as

$$\mathrm{R}(y_t) = a_0 + \sum_{i=1}^{p} a_{1i}\mathrm{R}(y_{t-i}) + \sum_{i=1}^{q} a_{2i}\mathrm{R}(x_{t-i}) + e_t \tag{7.14}$$

$$\mathrm{R}(x_t) = b_0 + \sum_{i=1}^{p} b_{1i}\mathrm{R}(x_{t-i}) + \sum_{i=1}^{q} b_{2i}\mathrm{R}(y_{t-i}) + \varepsilon_t \tag{7.15}$$

where $\mathrm{R}(\cdot)$ represents a rank order transformation and each lagged values of the series in each model are treated as separate variables when calculating their ranks, for example, $\mathrm{R}(Y_t)$ and $\mathrm{R}(Y_{t-1})$. The residuals, e_t and ε_t are assumed to be serially uncorrelated, and p and q may differ in each equation. When choosing p and q, two things have to be considered: the significance of the estimated coefficients and the serial correlation of resulting residuals.

From (7.14), rejection of the null hypothesis ($a_{2i} = 0$) implies causality from X to Y; whereas in (7.15), rejection of the null hypothesis ($a_{2i} = 0$) implies the reverse causality from Y to X. The null hypothesis is rejected if the F-test statistic is significant with respective q's value and N−K (K $= p + q + 1$) degrees of freedom.

7.3 Data

Tourism demand is represented by the number of passengers arrived (T). To measure economic growth, per capita gross domestic product (GDPpc) is considered. In order to account for emissions, we take the indicator of CO_2 emissions (metric tons per capita, CO_2pc) and we represent energy consumption by the per capita consumption

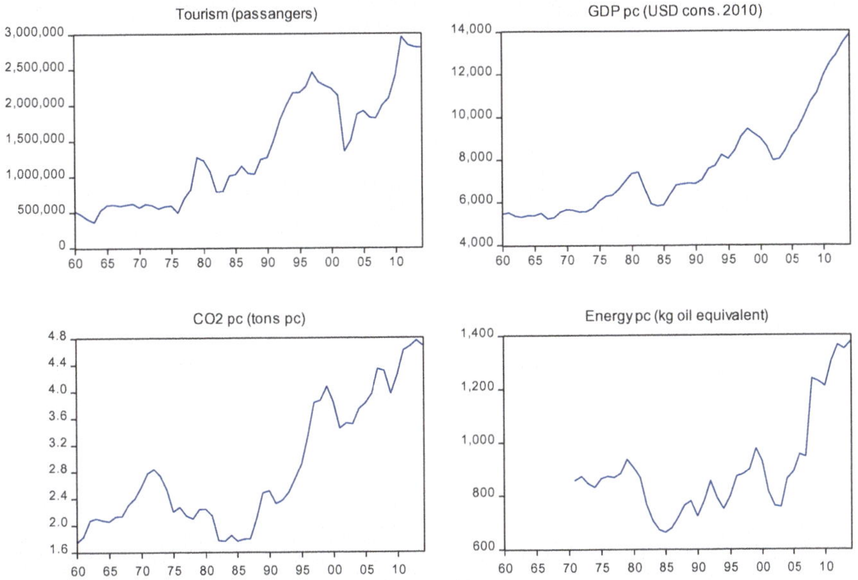

Source: World Bank.

Fig. 7.1 Tourism, GDP pc, CO_2 emissions, and Energy consumption in Uruguay. 1960–2014. Source World Bank

of kilograms of oil equivalent (*Epc*). For the empirical work, variables are considered in their logarithmic transformation ($t = \log(T)$; $gdp = log(GDPpc)$; $co_2 = log(CO_2pc)$; $ene = \log(Epc)$).

All the variables have the same source: World Bank database. The sample under analysis is 1960 to 2014, except for energy consumption, that is only available since 1971. Figure 7.1 shows the evolution of the four variables.

Unit root tests results, reported in Table 7.1, show that all the variables are non-stationary I(1) in the sample analyzed.

7.4 Results

As we explain above, in order to explore the long-run linkages between CO_2 emissions, economic growth, energy consumption and tourism cointegration and causality are tested by running free-model tests.

We consider three alternative sets of variables. The first one consisting of the four aforementioned variables. Each one of the remaining sets consists of only three variables: CO_2 emissions, energy consumption and tourism (the second one), and CO_2 emissions, economic growth, and tourism the third one.

Results of nonparametric cointegration test for each alternative are presented in Table 7.2.

Table 7.1 Nonparametric unit root test results (variables in logs)

	Variables	No deterministic	Mean Adjusted	Trend Adjusted
Level	Tourism	0.33039	0.09085	0.00520
	GDPpc	0.33354	0.08739	0.01008
	CO_2pc	0.24369	0.07133	0.01540
	Energy	0.33829	0.04556	0.01686
First difference	Tourism	0.01498**	0.00082***	0.00082***
	GDPpc	0.026042*	0.00741**	0.00123***
	CO_2pc	0.01980**	0.00268**	0.00214**
	Energy	0.00410***	0.00742**	0.00109***

Notes The hypothesis of a unit root process is rejected if the test statistic falls below the respective critical values
*, ** and *** denote significance at 10, 5%
Source Authors calculations

Table 7.2 Results of nonparametric cointegration test

	Test Statistics	
	$\Xi^*_T[2]$	$T \bullet R^2$
$[t, co_2, ene, gdp]$	0.0010***	9.1150*
Significance Level	Critical values	
10%	0.0160	6.2500
5%	0.0137	7.8100
1%	0.0100	11.3400
$[t, co_2, ene]$	0.0008***	16.9860**
$[t, co_2, gdp]$	0.0014**	7.4910*
Significance level	Critical values	
10%	0.0197	4.6052
5%	0.0165	5.9915
1%	0.0119	9.2104

Notes The hypothesis of no cointegration is rejected if the rank statistic, $\Xi^*_T[2]$, is below the respective critical value and the hypothesis of linearity is rejected if the score statistic, $T \cdot R^2$, exceeds the χ^2 critical values
* and ** denote significance at 10, 5%, according to the grades of freedom of each estimation

According to the results, in all the cases, we can reject non-cointegration hypothesis and linearity. This means that the long-run relationship that links these variables in the three sets would be nonlinear, and the ECK hypothesis can be confirmed.

Finally, we examine causality between the variables of main interest, specifically from tourism to CO_2 emissions, controlling by economic growth and energy consumption. To this end, we follow the nonparametric procedure proposed in

Table 7.3 Results of nonparametric causality test between

H–H causality test	Uruguay		
	df	Pr	NC
$d(t) -> d(co_2)$	(4, 22)	0.30	A
$T -> co_2$	(4, 23)	0.001	R (1%)

Notes F-statistic, df (degrees of freedom), NC: H0: non-causality

Holmes and Hutton (1990). As explained before, this test is more robust than conventional parametric tests usually applied. Results are shown in Table 7.3.

The results of these tests confirm the causality from tourism (represented with arrivals of passengers) to CO_2 emissions when the test is performed in levels (i.e., for the long run). The sign of the effect is positive in the regression of the ranked variables (see Appendix 1).

However, the evidence does not allow accepting causality from tourism to emissions in the short run, that is when the Holmes and Hutton causality test is run in first differences.

In summary, the empirical results show that tourism development, CO_2 emissions, energy consumption, and per capita GDP are in the long-term equilibrium relationship between 1960–2014 for Uruguay. This relationship, as postulated by the EKC hypothesis, is nonlinear. Additionally, results of nonparametric causality tests show that in the long run, there is a causal relationship from tourism to CO_2 emissions, so tourism-induced EKC hypothesis may be verified for Uruguay.

7.5 Concluding Remarks

Differing from other case studies, this paper explores the linkages between CO_2 emissions, economic growth, energy consumption, and tourism development within a time series framework (as suggested by Dinda 2004) but without imposing—a priori—any parametric model.

With this aim, we applied the methodology proposed by Breitung (2001), Holmes and Hutton (1990), and Ye Lim et al. (2011), to the period 1960–2014 in Uruguay. This methodology allows testing the presence of nonlinearities in the relationship between the selected variables.

Empirical results for Uruguay show that tourism development, CO2 emissions, energy consumption, and per capita GDP have nonlinear long-term equilibrium relationship, as postulated for the EKC hypothesis. Since the results do not allow to specify the form of nonlinearity, they open the possibilities to any nonlinear shape of the relationship between contamination, energy consumption and tourism development in Uruguay, not only the U-shaped curve specifically postulated by the EKC hypothesis.

With regard to the causality analysis, results of nonparametric test show that in the long run, there is a causal relationship from tourism to CO_2 emissions. Nevertheless, the evidence with respect to the causality between tourism growth and CO_2 emissions is not verified in the short run.

To similar findings arrived Ozturk et al. (2015) using a different indicator of contamination, (ecological footprint) and different estimation techniques (panel techniques). Nevertheless, as long as we know, this is the first work that empirically tests this relation without imposing a priori any model, within the framework of a case study analysis.

The evidence reached in this paper suggests that tourism policies have to pay attention to the development of more effective environmentally friendly tourism programs.

The study also reveals an important impact of tourism for the Uruguayan economy that also suggests the need for public policies that support tourism development initiatives of the numerous touristic attractions that the country has and that increases the international and domestic tourism demand. Uruguay has the opportunity to learn from previous experiences in the rest of the world, positive and negative, to correct mistakes of other destinations and promote initiatives that minimize the impacts of tourism development.

Further research may extend this methodology to study other Latin-American countries or expanding the perspective to Latin America, through other methodologies such as panel data and other pollution variables. Even if panel data methodologies may have some limitations (Dinda 2004), it has a number of advantages, namely: variability in the data, less collinearity between the variables, a greater number of degrees of freedom, and more ability to identify and measure effects, among others. Additionally, considering other variables to represent contamination, not only atmospheric indicators (as the majority of the literature on the EKC hypothesis, see Sarkodie and Strezov 2019). This may contribute to the limited literature limited on EKC hypothesis which employs, e.g., land indicators, oceans, seas, coasts and biodiversity indicators, and freshwater indicator, ecological footprint, and may give robustness to the empirical findings of the effect of tourism development to the environment.

Finally, our results show that tourism development, CO_2 emissions, energy consumption and per capita GDP have nonlinear long-term equilibrium relationship for the case of Uruguay. However, the study does not specify the shape of the nonlinearity. Ergo, results do not prove that the relationship is U-shaped (as the EKC postulate) and this subject remains open ended. Future research can be performed in that direction and to explore the identity of these nonlinearities in the specific case of Uruguay.

Acknowledgments Our research was supported by the CSIC—UDELAR research project "Grupo de Investigación en Dinámica Económica—GIDE".

Appendix 1: Causality Test (Holmes and Hutton)

See Tables 7.4 and 7.5.

Table 7.4 Regression in levels

Dependent Variable: YRANK Method: Stepwise Regression Sample (adjusted): 1973–2014 Included observations: 42 after adjustments	Number of always included regressors: 1 Number of search regressors: 16 Selection method: Uni-directional Stopping criterion: t-stat = 1.6

Note Final equation sample is larger than stepwise sample (rejected regressors contain missing values)

Variable	Coefficient	Std. error	t-statistic	Prob.*
C	−2.006686	1.974842	−1.016125	0.3170
YRANK_1	0.807544	0.099289	8.133246	0.0000
WRANK_1	0.248516	0.178274	1.394011	0.1726
WRANK_4	−0.539879	0.277002	−1.949010	0.0598
YRANK_4	0.190414	0.105445	1.805809	0.0801
XRANK_3	−0.414428	0.181975	−2.277386	0.0294
ZRANK_2	−0.342754	0.094477	−3.627921	0.0010
XRANK_2	0.393924	0.183517	2.146525	0.0393
WRANK_3	0.590995	0.352657	1.675836	0.1032
R-squared	0.944538	Mean dependent var		31.14286
Adjusted R-squared	0.931093	S.D. dependent var		16.13143
S.E. of regression	4.234532	Akaike info criterion		5.911832
Sum squared resid	591.7315	Schwarz criterion		6.284190
Log likelihood	−115.1485	Hannan-Quinn criter		6.048316
F-statistic	70.25031	Durbin-Watson stat		2.034893
Prob(F-statistic)	0.000000			

Selection summary

Removed: YRANK_3, XRANK_1, ZRANK_3, XRANK_4, ZRANK_1, YRANK_2, WRANK_2, ZRANK_4

Note p-values and subsequent tests do not account for stepwise selection. ($y = co_2$; x = passengers; z = energy pc; w = gdppc)

Table 7.5 Regression in first differences

Dependent Variable: DYRANK Method: Stepwise Regression Sample (adjusted): 1965 2014 Included observations: 50 after adjustments	Number of always included regressors: 1 Number of search regressors: 16 Selection method: Uni-directional Stopping criterion: t-stat = 1.6

Note Final equation sample is larger than stepwise sample (rejected regressors contain missing values)

Variable	Coefficient	Std. error	t-statistic	Prob.*
C	25.98089	5.336542	4.868488	0.0000
DWRANK_4	−0.310329	0.144300	−2.150581	0.0367
DYRANK_1	0.344648	0.135480	2.543897	0.0143
R-squared	0.179113	Mean dependent var		27.18000
Adjusted R-squared	0.144182	S.D. dependent var		15.86125
S.E. of regression	14.67331	Akaike info criterion		8.268062
Sum squared resid	10,119.38	Schwarz criterion		8.382784
Log likelihood	−203.7016	Hannan-Quinn criter		8.311749
F-statistic	5.127573	Durbin-Watson stat		1.990116
Prob(F-statistic)	0.009675			

Selection summary

Removed: DZRANK_3, DZRANK_1, DWRANK_1, DXRANK_1, DWRANK_2, DXRANK_4, DYRANK_2, DYRANK_3, DXRANK_3, DZRANK_4, DYRANK_4, DZRANK_2, DXRANK_2, DWRANK_3

Note p-values and subsequent tests do not account for stepwise selection

References

Akbostanci, E., Türüt-Asik, S., & Tunç, G. I. (2009). The relationship between income and environment in Turkey: Is there an environmental Kuznets curve? *Energy Policy, 37*(3), 861–867.

Alam, M. J., Begum, I. A., Buysse, J., Rahman, S., & Huylenbroeck, G. V. (2011). Dynamic modeling of causal relationship between energy consumption, CO_2 emissions and economic growth in India. *Renewable and Sustainable Energy Reviews, 15,* 3243–3251.

Ang, J. B. (2008). Economic development, pollutant emissions and energy consumption in Malaysia. *Journal of Policy Modeling, 30,* 271–278.

Becken, S., Frampton, C., & Simmons, D. G. (2001). Energy consumption patterns in the accommodation sector-the New Zealand case. *Ecological Economics, 39*(3), 371–386.

Becken, S., Simmons, D. G., & Frampton, C. (2003). Energy use associated with different travel choices. *Tourism Management, 24,* 267–277.

Breitung, J. (2001). Rank test for nonlinear cointegration. *Journal of Business & Economics Statistics, 19*(3), 331–340.

Breitung, J. (2002). Nonparametric tests for unit roots and cointegration. *Journal of Econometrics, 108,* 343–363.

Brida, JG, Lanzilotta, B., Pereyra, J.S. & Pizzolon, F. (2013a) The tourism-led growth hypothesis: A comparative study for the MERCOSUR countries. *Revista de Economia Mundial* 75–96.

Brida, J. G., Lanzilotta, B., Pereyra, J. S., & Pizzolon, F. (2013). A nonlinear approach to the tourism-led growth hypothesis: The case of the MERCOSUR. *Current Issues in Tourism, 18*(7), 647–666.
Brida, J. G., Cortes-Jimenez, I., & Pulina, M. (2014). Has the tourism-led growth hypothesis been validated? A literature review. *Current Issues in Tourism., 19,* 1–37. https://doi.org/10.1080/13683500.2013.868414
Coondoo, D., & Dinda, S. (2002). Causality between income and emission: A country group specific econometric analysis. *Ecological Economics, 40,* 351–367.
Davidson, J. (2002). Establishing conditions for the functional central limit theorem in nonlinear and semiparametric time series processes. *Journal of Econometrics, 106*(2), 243–269.
Dinda, S. (2004). Environmental Kuznets curve hypothesis: A survey. *Ecological Economics, 49*(4), 431–455.
De Vita, G., Katircioglu, S., Altinay, L., Fethi, S., & Mercan, M. (2015). Revisiting the environmental Kuznets curve hypothesis in a tourism development context. *Environmental Science and Pollution Research, 22*(21), 16652–16663.
Dogan, E., Seker, F., & Bulbul, S. (2015). Investigating the impacts of energy consumption, real GDP, tourism and trade on CO_2 emissions by accounting for cross-sectional dependence: A panel study of OECD countries. *Current Issues in Tourism, 20*(16), 1701–1719.
Dubois, G., Peeters, P., Ceron, J. P., & Gössling, S. (2011). The future tourism mobility of the world population: Emission growth versus climate policy. *Transportation Research Part A, 45,* 1031–1042.
Friedl, B., & Getzner, M. (2003). Determinants of CO_2 emissions in a small open economy. *Ecological Economics, 45*(1), 133–148.
Gössling, S. (2002). Global environmental consequences of tourism. *Global Environmental Change, 12*(4), 283–302.
Gössling, S. (2013). National emissions from tourism: An overlooked policy challenge? *Energy Policy, 59*(2013), 433–442.
Granger, C. W. J. (1995). Modelling nonlinear relationship between extended-memory variables. *Econometrica, 63,* 265–279.
Grossman, G.M. & Krueger, A.B. (1991). Environmental impacts of a North American free trade agreement. *NBER Working Papers*. 3914. National Bureau of Economic Research, Inc.
He, J., & Richard, P. (2010). Environmental Kuznets curve for CO_2 in Canada. *Ecological Economics, 69*(5), 1083–1093.
Holden, A. (2009). The environment-tourism nexus influence of market ethics. *Annals of Tourism Research, 36,* 373–389.
Holmes, J. M., & Hutton, P. A. (1990). On the Causal Relationship Between Government Expenditures and National Income. *the Review of Economics and Statistics, 72*(1), 87–95.
Katircioglu, S. T. (2014a). Testing the Tourism-Induced EKC Hypothesis: The Case of Singapore. *Economic Modeling, 41,* 383–391.
Katircioglu, S. T. (2014b). International tourism, energy consumption, and environmental pollution: The case of Turkey. *Renewable & Sustainable Energy Reviews, 36,* 180–187.
Katircioglu, S. T., Feridun, M., & Kilinc, C. (2014). Estimating tourism-induced energy consumption and CO_2 emissions: The case of Cyprus. *Renewable & Sustainable Energy Reviews, 29,* 634–640.
Lee, J. W., & Brahmasrene, T. (2013). Investigating the influence of tourism on economic growth and carbon emissions: Evidence from panel analysis of the European Union. *Tourism Management, 38,* 69–76.
Lin, T. P. (2010). Carbon dioxide emissions from transport in Taiwan's national parks. *Tourism Management, 31*(2), 285–290.
Lindmark, M. (2002). An EKC-pattern in historical perspective: Carbon dioxide emissions, technology, fuel prices and growth in Sweden 1870–1997. *Ecological Economics, 42,* 333–347.
List, J. A., & Gallet, C. A. (1999). The environmental Kuznets curve: Does one size fit all? *Ecological Economics, 31*(3), 409–423.

Luzzati, T., & Orsini, M. (2009). Natural environment and economic growth: Looking for the energy-EKC. *Energy, 34,* 291–300.

Neto, F. (2003). A new approach to sustainable tourism development: Moving beyond environmental protection. *Nat Res Forum, 27,* 212–222.

Ozturk, I., & Acaravci, A. (2010). CO2 emissions, energy consumption and economic growth in Turkey. *Renewable and Sustainable Energy Reviews, 14*(9), 3220–3225.

Ozturk, I., Al-Mulali, U., & Saboori, B. (2015). Investigating the environmental Kuznets curve hypothesis: The role of tourism and ecological footprint. *Environmental Science and Pollution Research International, 23*(2), 1916–1928.

Paramati, S. R., Alam, M. S., & Chen, C. F. (2017). The effects of tourism on economic growth and CO_2 emissions: A comparison between developed and developing economies. *Journal of Travel Research, 56*(6), 712–724.

Perch-Nielsen, S., Sesartic, A., & Stucki, M. (2010). The greenhouse gas intensity of the tourism sector: The case of Switzerland. *Environmental Science and Pollution, 13,* 131–140.

Risso, Wiston Adrian (2018). Tourism and Economic Growth: A Worldwide Study. *Tourism Analysis* Volume 23, Number 1, 2018, pp. 123–135(13).

Saenz-de-Miera, O., & Rosselló, J. (2014). Modeling tourism impacts on air pollution: The case study of PM10 in Mallorca. *Tourism Management, 40,* 273–281.

Shafik, N., & Bandyopadhyay, S. (1992). *Economic growth and environmental quality: Time series and cross-country evidence.* The World Bank, Washington, DC: Background Paper for the World Development Report.

Sarkodie, S. A., & Strezov, V. (2019). A review on environmental Kuznets curve hypothesis using bibliometric and meta-analysis. *Science of the Total Environment, 649,* 128–145.

Solarin, S.A. (2014). Tourist arrivals and macroeconomic determinants of CO_2 emissions in Malaysia. *Anatolia: An International Journal of Tourism and Hospitality Research, 25*(2), 228–241.

Soytas, U., Sari, R., & Ewing, B. T. (2007). Energy consumption, income, and carbon emissions in the United States. *Ecological Economics, 62,* 482–489.

Stern, D. I. (2004). The rise and fall of the environmental Kuznets curve. *World Development, 32*(8), 1419–1439.

Stock, J. H., & Watson, M. W. (1993). A simple estimator of cointegrating vector in higher order integrated systems. *Econometrica, 61,* 783–820.

Tsai, K. T., Lin, T. P., Hwang, R. L., & Huang, Y. J. (2014). Carbon dioxide emissions generated by energy consumption of hotels and home stay facilities in Taiwan. *Tourism Management, 42,* 13–21.

World Travel and Tourism Council. (2018). Travel & Tourism Economic Impact 2018.

Xing-Ping, Z., & Xiao-Mei, C. (2009). Energy consumption, carbon emissions, and economic growth in China. *Ecological Economics, 68,* 2706–2712.

Xuchao, W., Priyadarsini, R., & Eang, L. S. (2010). Benchmarking energy use and greenhouse gas emissions in Singapore's hotel industry. *Energy Policy, 38,* 4520–4527.

Ye Lim, S., Ghazali, M., & Mun Ho, C. (2011). Export and economic growth in Southeast Asia current Newly Industrialized Countries: Evidence from nonparametric approach. *Economics Bulletin, 31*(3), 2683–2693.

Chapter 8
Analyzing the Tourism Development and Ecological Footprint Nexus: Evidence From the Countries With Fastest-Growing Rate of Tourism GDP

Ilyas Okumus and Sinan Erdogan

Abstract Sustainable development is a holistic approach that aims to do future-friendly planning with its economic, environmental and social dimensions by establishing a balance between the needs of human life and the sustainability of natural resources. Tourism, contributing to the economic development of both developed and developing countries, includes environmental, social and economic dimensions of sustainable development with this structure. Therefore, analyzing tourism development and environmental quality nexus is a crucial issue for policymakers to design effective policies for a sustainable life. The main purpose of this study is to investigate the impacts of tourism investments on the ecological footprint in the six countries (Ecuador, Egypt, Turkey, Uzbekistan, Tunisia and Sri Lanka), which are in the top 20 countries, fastest-growing in terms of tourism GDP over the period of 1995–2014. In addition to tourism investments, our quadratic EKC model includes economic growth, energy use and individual internet use as independent variables. Empirical findings reveal that tourism investments and internet uses have negative impacts on ecological footprint. On the other hand, energy consumption increases environmental degradation. Also, the existence of the EKC is confirmed in these countries.

Keywords Tourism investments · Ecological footprint · Environmental Kuznets curve hypothesis

I. Okumus (✉)
Department of Public Finance, Hatay Mustafa Kemal University, Antakya, Turkey
e-mail: ilyasokumus@mku.edu.tr

S. Erdogan
Department of Economics, Hatay Mustafa Kemal University, Antakya, Turkey
e-mail: phderdogan@gmail.com

D. Balsalobre-Lorente et al. (eds.), *Strategies in Sustainable Tourism, Economic Growth and Clean Energy*, https://doi.org/10.1007/978-3-030-59675-0_8

8.1 Introduction

Sustainable development is one of the most crucial pathways to combat global climate crisis issues. According to the IPCC (2019), the rise in global temperatures in the period 1951–2010 was precisely due to human activities. This increasing temperature in recent years has caused extreme weather events. The extreme weather events show its effects all over the world, from drought and record temperatures in northern Europe to forest fires in the US, heatwaves in China and drought to the extraordinarily powerful monsoon that devastated vast areas of South India. If the human activities that cause about 1 °C global warming compared to the pre-industrial period continue, it will cause the temperature to exceed the 1.5 °C limit between 2030 and 2050. The 1.5 °C limit is critical to sustainable development and poverty prevention. Limiting global warming to 1.5 °C means avoiding many permanent effects on ecological systems and habitats. To limit 1.5 °C of global warming, it has become imperative to harmonize human activities with sustainable development goals (SDGs) in every field or sector such as industry, energy, tourism, transportation and agriculture. These severe environmental disasters that we experience reveal the necessity of analyzing the environmental results of the activities and investments made in each sector, but the tourism sector becomes prominent because of several reasons. Tourism is a sector that not only contributes to the tourism revenues, but it also contributes to other sectors such as transportation, accommodation, construction, food, entertainment and agriculture (Katircioğlu 2014). Due to their contribution to economic performance and other sectors' activities, tourism development and tourism investments are encouraged by every country (Adamou and Clerides 2009). According to the annual data of the World Travel and Tourism Council (WTTC) measuring global economic impacts of the tourism industry, the tourism industry in 2018, corresponds to 10.4% of global GDP and 10% of global employment creating approximately 319 million jobs.

Moreover, the tourism industry is expected to attract approximately $ 941 billion of capital investment in 2018. This is expected to increase by 4.4% in 2019 and increase by 4.2% over the next decade to $1.489 billion in 2029 (WTTC 2019). Travel and tourism investments are expected to represent approximately 4.4% of total national investments in 2019 (WTTC 2019). Besides, the tourism sector is one of the leading sectors of development for some countries and regions. Therefore, conducting research on the sustainability of the tourism sector is essential for a more livable world.

Sustainable tourism provides economic and social benefits while creating an opportunity to minimize negative impacts on environmental and cultural heritage. According to WTTC (2015), tourism investments including the development of accommodation and maintenance of new buildings, the maintenance of furniture and equipment to renewing current touristic places, tourist transport such as airplane, cruise ships and buses; investments for restorations of famous attractions and tourist sites, information and communication technology (ICT) projects related to tourism, arecrucial factors for sustainable tourism development. On the one hand, tourism investments have some environmental benefits such as improving energy efficiency, proper protection of water and waste and conservation of ecosystems and biodiversity. On the other hand, unplanned tourism investments can have serious pressures on the environment and natural resources. In the existing literature, researchers have claimed that tourism actions, especially transportation and accommodation, are mentioned among the major responsible for energy consumption. They also argued that this energy consumption also largely met by fossil fuels (Paramati et al. 2018).

Even though greenhouse gases resulting from the use of fossil fuels are major responsibilities of climate crisis problems, the unplanned and environmentally insensitive tourism investments and activities can damage water resources, fishing areas, forest areas, agricultural areas, grazing areas and building areas. If there is no comprehensive collection and disposal system in regions where tourism activities are intense, wastes accumulating over time may pose a serious problem (Akadiri et al. 2019). When the current literature is reviewed, it is seen that CO_2 emission is used as an indicator of environmental pollution in the majority of the studies analyzing tourism industry and environmental pollution nexus. Considering the potential effects of the tourism sector on the environmental components, it can be said that a holistic approach is required on investigating the ecological effects of tourism. Therefore, the use of a more comprehensive indicator as an environmental indicator will give more consistent results when analyzing the relationship between tourism and the environment. Ecological footprint index is a comprehensive indicator of environmental degradation. The ecological footprint is a method developed to calculate the ecosystem balances deteriorated by human activities and to determine the amount to be regained to the ecosystem. In other words, it calculates the "number of worlds" that will be required for a sustainable future, both against the resources people demand from nature and disrupting the natural balance (Wackernagel and Rees 2004). The ecological footprint is tracked by the sum of six different surface area components that show productivity, including farmland, grazing land, fishing area, area for carbon demand, building area and forest demand areas. The ecological footprint is a crucial indicator for the environment as it helps governments and local leaders to understand and improve the impacts of public project investments on the planet (National Footprint Accounts 2016). Therefore, it may be more rational to utilize the ecological footprint as an environmental indicator to analyze the effect of the tourism industry on the environment.

Despite significant rises in tourism investments, the number of studies investigating the impacts of tourism investments on environmental quality is limited. Therefore, the main goal of this study is to analyze the effects of tourism investments on environmental degradation in the six[1] countries (Ecuador, Egypt, Turkey, Uzbekistan, Tunisia and Sri Lanka), which are in the top 20 countries, fastest-growing in terms of tourism GDP. Tourism investments in Ecuadorincreased from about 341 million dollars to $1.18 billion between 1995 and 2014. Tourism investments in Egypt ascended by 2.76 times in the period covering 1995–2014. Similarly, tourism investments in Turkey, Uzbekistan, Tunisia and Sri Lanka scaled up roundly by 2.8,

[1]Due to data constraints we investigated environment-tourism nexus in six countries, and these countries are Ecuador, Egypt, Turkey, Uzbekistan, Tunisia and Sri Lanka.

2.1, 2 and 2.7 times, respectively (WTTC 2020). It could be said that there is a need to harmonize the effects of tourism investments and environmental quality. Conducting a research on the effects of tourism investments on the environment could contribute to establishing eco-friendly tourism investments composition, and it may contribute to achieving ecological aims of SDGs (SDGs 6, 13, 14 and 15). Furthermore, due to limited proofs, efficient policy design may be a challenge for decision-makers. Therefore, this study may provide evidence to create key policy proposals. Investigation of interactions between tourism investments and environmental pollution may extend the existing literature and give an idea for researchers.

8.2 Literature Review

Researchers widely investigated the determinants of environmental quality by using the EKC hypothesis since the seminal paper of Grossman and Krueger (1991), had released. The earliest studies focused on the role of macroeconomic, demographic and resource-based indicators such as economic growth, population, energy-based indicators. Afterwards, researchers focused on the role of economic activities, as well as traditional variables, such as trade, investments, tourism, industrialization and so forth. Some of the studies on the determinants of environmental quality were reported in Table 8.1.

The common features of the existing literature can be briefly summed up as follows. First, some of the studies examined the determinants of environmental quality within the context of EKC, while other parts did not use the EKC hypothesis. It can be said that there is no convincing evidence of the validity of the EKC hypothesis. Second, if one examines the existing literature on tourism-environmental quality nexus, he/she can easily find out that most of the researchers are focused on the role of either tourism arrivals or tourism receipts. However, there is a limited number of studies conducted to unveil tourism investments and environment interactions, it could be inferred that there is a research gap on this topic. It can be said there is no consensus on the effect of tourism on environmental quality. Third, on the methodological aspects, panel data methods were extensively preferred. This study mainly focuses on the effects of tourism investments in six countries with the fastest growth rate in terms of tourism GDP for the period from 1995 to 2014, by employing panel data methods and aims to fill the research gap in the literature.

Table 8.1 Former literature review

Author(s)	Countries-Period	Empirical Methodology	Empirical Results	EKC
Katircioglu et al. (2014)	Cyprus, 1970–2009	ARDL	International Tourist Arrivals (+) EC (+)	N
Saenz-de-Miera and Rossello (2014)	Mallorca, 2003M1–2007M12	Generalized Additive Model	Number of Tourist (+)	N
Katircioğlu (2014)	Singapore, 1971–2010	DOLS	Tourist Arrivals (−) EC (+)	Valid
De Vita (2015)	Turkey, 1960–2009	DOLS	Tourist Arrivals (+) EC (+)	Valid
Zaman et al. (2016)	Developed and Developing Countries, 2005–2013	2SLS	Tourism Development (M) EC (+) Health Expenditures (M) Domestic Investment (+)	Valid
Ozturk et al. (2016)	144 Countries, 1988–2008	GMM and System GMM	GDP growth from Tourism (M) EC (M) TR (M) UR (M)	M
Paramati et al. (2018)	Developed and Developing Countries,	FMOLS	International Tourism (+) Receipts (+) Population (+) GDP (+) Energy Efficiency (+) Service Sector (+) Industrialization (+)	Valid
Zhang and Gao (2016)	Regions of China, 1995–2011	FMOLS	International Tourist Arrivals (−) EC (M)	M
Salahuddin et al. (2016)	Australia, 1985–2012	ARDL	Internet Use (*) GDP (−) Financial Development (+)	N
Dogan et al. (2017)	OECD Countries,	DOLS	Tourism (+) EC (+) TR (−)	Invalid
Narad da Gamage et al. (2017)	Sri Lanka, 1974–2013	DOLS	Tourism Receipts (+) EC (+)	Invalid

(continued)

Table 8.1 (continued)

Author(s)	Countries-Period	Empirical Methodology	Empirical Results	EKC
Dogan and Aslan (2017)	EU and candidate countries, 1995–2011	OLS-FE, FMOLS, DOLS, Mean Group Estimator	Tourism (−) GDP (−) EC (+)	N
Paramati et al. (2017)	European Union Countries, 1991–2013	FMOLS	GDP (+) FDI (M) TR (M)	M
Raza et al. (2017)	USA, 1996(1)–2015(3)	Wavelet-Based Anaylsis	Tourism Development (−)	N
Park et al. (2018)	2001–2014	Pooled Mean Group Estimator	Internet Use (−) Financial Development (−) GDP (−) TR (−) EC (+)	N
Ozcan and Apergis (2018)	20 Emerging Economices, 1990–2015	FMOLS	Internet Use (−) GDP(+) Financial Development (*) TR (+) EC(+)	N
Azam et al. (2018)	3 ASEAN Countries, 1990–2014	FMOLS	Tourist Arrivals (M) EC (+)	M
Bella (2018)	France, 1995–2014	Vector Error Correction Model	Tourist Arrivals (−)	Valid
Paramati et al. (2018)	28 EU Countries, 1990–2013	Panel ARDL	Tourism Investment (−) Population (+) GDP (+) Technology (−) TR (−)	N
Saint Akadiri et al. (2019)	Turkey, 1970–2014	ARDL Bound Testing Approach	GDP (+) Tourist Arrival (+) Globalization (*)	N
Koçak et al. (2020)	Top 10 Most Visited Countries, 1995–2014	Continuously Updated Fully Modified (CUP-FM) Continuously Updated Bias-Corrected (CUP-BC)	Tourism Arrivals (+) Tourism Receipts (−) GDP (+) UR (+) Energy Intensity (−)	N

M: Mixed Results, TR: Trade Openness, UR: Urbanization, EC: Energy Consumption: *: No Statistically Significant Effect, FE: Fixed Effects, GMM: Generalized Method of Moment, FMOLS: Fully Modified Ordinary Least Squares, DOLS: Dynamic OLS, 2SLS: Two-Stage Least Squares, FDI: Foreign Direct Investment, N: Not Tested

8.3 Model, Data, Methodology and Empirical Results

8.3.1 Model and Data

To investigate the effects of tourism investments on environmental degradation within the context of the EKC hypothesis in the six countries, which are in the top 20 countries, fastest-growing in terms of tourism GDP, countries for the period from 1995 to 2014. We utilized a logarithmic model, which has shown in Eq. 8.1.

$$\ln EF_{it} = \beta_{0i} + \beta_1 \ln Y_{it} + \beta_2 \ln Y_{it}^2 + \beta_3 \ln TO_{it} + \beta_4 \ln EC_{it} + \beta_5 \ln INT_{it} + \varepsilon_{it} \quad (8.1)$$

where *EF* is ecological footprint per capita, *Y* is GDP per capita (constant, 2010 US $), Y^2 is GDP per capita square, *TO* is tourism investment per capita (constant, 2010 US $), *EC* is energy consumption per capita (kg of oil equivalent per capita), *INT* is individuals using the internet (% of the population). We obtained data for *EF* from The Global Footprint Network (2019), while we retrieved data for *Y*, *EC* and *INT* from World Bank (2020). Last, we obtained *TO* data from World Travel & Tourism Council (2019).

8.3.2 Methodology

The cross-section dependence frequently occurs in panel data estimations and could cause biased estimations and policy inferences (Erdogan et al. 2020; Sarafidis et al. 2009). Therefore, we implemented a cross-section dependence test proposed by Pesaran (2004), as a diagnostic test. Pesaran utilizes the following test statistic, which is based on the average of pairwise correlation coefficients, to test the existence of cross-section dependence

$$CD = \sqrt{2T / N(N-1)} \sum_{i=1}^{N-1} \sum_{j=i+1}^{N} \left(T\widehat{\rho}_{ij} - 1 \right) \quad (8.2)$$

where $\widehat{\rho}_{ij}$ is the estimate of the pairwise correlation of the error terms. CD approach utilizes "cross-section independence" in the null, while "cross-section dependence" in the alternative. Moreover, we tested the slope homogeneity by using the Delta approach proposed by Pesaran and Yamagata (2008). Pesaran and Yamagata (2008) utilize "slope homogeneity ($H_0 : \beta_i = \beta$ for all (i)" in the null, while "slope heterogeneity ($H_1 : \beta_i \neq \beta_j$)" in the alternative. Equation (8.3) is used for a relatively large sample, while the small sample performance of delta test could be improved by using the specification shown in Eq. (8.3) (Acaravci and Erdogan 2017)

$$\tilde{\Delta} = \sqrt{N}\left(N^{-1}\tilde{S} - k\Big/\sqrt{2k}\right) \quad (8.3)$$

$$\tilde{\Delta}_{adj} = \sqrt{N}\left(N^{-1}\tilde{S} - E(\tilde{z}_{iT})\Big/\sqrt{Var(\tilde{z}_{iT})}\right) \quad (8.4)$$

In order to diagnose the integrational level of the variables, we implemented a panel unit root test proposed by Breitung (2000). This approach uses the following data generating process to test "unit root ($H_0 : \rho_i = 0\ for\ all\ i$) "in the null against the alternative hypothesis of "no unit root ($H_1 : \rho_i < 0$)".

$$y_{it} = \alpha_i + \beta_i t + x_{it} \quad (8.5)$$

where $x_{it} = \sum_{k=1}^{p+1} \alpha_{ik} x_{i,t-k} + \varepsilon_{it}$. To test the existence of cointegration in the employed model, we utilized Westerlund's (2005) method, which is based on the variance ratio. Westerlund (2005) suggests two statistics computed under the assumption of homogeneity and heterogeneity, respectively. Due to the existence of slope heterogeneity, we employed group-mean variance ratio statistics and it can be estimated by the following specification:

$$VR_G = \sum_{i=1}^{N} \sum_{t=1}^{T} \widehat{E}_{it}^{2} \widehat{R}_i^{-1} \quad (8.6)$$

where $\widehat{E}_{it} \equiv \sum_{j=1}^{t} \hat{e}_{ij}$, $\widehat{R}_{it} \equiv \sum_{t=1}^{T} \hat{e}_{it}^{2}$ and $\hat{e}_{it}$ is the residual term, obtained by the data generating process of Westerlund, and the null of "no cointegration" is tested against the alternative of "some panels are cointegrated" through this specification. Moreover, Westerlund states that the variance ratio-based cointegration test has a satisfactory small sample performance. Finally, we used the FMOLS estimation procedure proposed by Pedroni (2000), to estimate long-run parameters. The panel FMOLS strategy can be applied as $\widehat{\beta}_{GFMOLS} = N^{-1} \sum_{i=1}^{N} \beta_{FMOLSi}$, were β_{FMOLSi} estimated by utilizing country-specific FMOLS estimation of Eq. (8.1), and related t-ratio can be obtained as $t_{\beta_{GFMOLS}} = N^{-1/2} \sum_{i=1}^{N} t_{\beta_{FMOLSi}}$ (Erdogan 2020).

8.3.3 Empirical Results

We started our analysis by implementing CD and Delta tests as diagnostic tests and reported the results in Table 8.2. CD results show that the null hypothesis of "cross-section independence" is accepted for variables and model. Therefore, cross-section dependence does not exist in the variable and model, and first-generation panel data estimation techniques can be implemented. Moreover, Delta tests show that the null hypothesis of "slope homogeneity" is rejected at different significance levels. If it is considered that the adjusted version of the delta test has better small sample properties, it could be more rational to consider it. Therefore, the null hypothesis of "slope

Table 8.2 CD and Delta Tests Results

Test	lnEF	lnY	lnY^2	lnTO	lnEC	lnINT	MODEL
CD	4.43 (0.425)	16.49 (0.952)	16.49 (0.952)	7.79 (0.459)	4.58 (0.752)	16.85 (0.973)	1.15 (0.250)
$\tilde{\Delta}$							1.451 (0.073)
$\tilde{\Delta}_{adj}$							1.777 (0.038)

Note Probability values were reported in parenthesis

homogeneity" is rejected at a 5% significance level. Therefore, slope coefficients are heterogeneous.

After determining the non-existence of cross-section dependence and slope coefficients are homogenous, we investigated the integrational level of the variables by using the methodology of Breitung (2000), and the results reported in Table 8.3. According to the results, all variables exhibit unit root processes, whereas all of them are stationary at first difference.

Having a similar integrational level of variables as *I(1),* we investigated cointegration in the model by implementing a cointegration method based on the variance ratio approach. The findings show that the null hypothesis of "no cointegration" was rejected at a 5% significance level. It can be inferred that there is a long-run relationship between variables (Table 8.4).

After confirming the existence of cointegration between variables, we estimated the long-run coefficients by implementing the FMOLS procedure. According to the findings reported in Table 8.5, real GDP per capita *(Y)* and real GDP per capita square (Y^2) has a positive and negative coefficient, respectively, and these coefficients are statistically significant. There is an inverted-U shaped nexus between economic

Table 8.3 Unit Root Test Results

	Level	1st Difference
	Trend + Constant	Trend + Constant
EF	0.793 (0.786)	−3.020 (0.001)
Y	0.718 (0.763)	−2.563 (0.005)
Y^2	0.695 (0.756)	−2.740 (0.003)
TO	−0.213 (0.415)	−2.774 (0.002)
EC	−0.627 (0.265)	−2.851 (0.002)
INT	2.441 (0.992)	−4.460 (0.000)

Note Maximum lag-length was determined as k = 2. Probability values were shown in parenthesis

Table 8.4 Cointegration test results

Test	Statistics	Prob.
VR_G	−1.890	0.029

Table 8.5 FMOLS estimation results

Variables	FMOLS results	t-Stat	Prob.
Y	13.564	6.529	0.000
Y^2	−0.782	−6.215	0.000
TO	−0.024	−2.134	0.035
EC	0.470	9.090	0.000
INT	−0.020	−7.898	0.000

Note We estimated the long-run covariance by utilizing the Bartlett kernel and Andrews bandwidth approach

growth and environmental degradation; hence, EKC is valid. This may be a result of the scale, composition and technique effects of EKC. On the one hand; once the economies are at the upward slope portion of the EKC, the increase of resource use and economic growth may lead to deterioration of environmental conditions by increasing emissions, reducing the regenerative capacity of nature.

On the other hand, when the economies are at the downward slope portion of EKC, composition and technique effects occur and promote extensions of clean economic activities and more use of eco-friendly production technologies (Erdogan and Acaravci 2019). Moreover, the reducing effect of economic growth may be related to an increase of clean environment demand by the rise of the economic development level (Acaravci and Akalin 2017; Lopez and Islam 2008). These findings are consistent with the results of Katircioglu et al. (2014), De Vita et al. (2015), Zaman et al. (2016), Paramati et al. (2017), Bella (2018), whereas in contrast with findings of Dogan et al. (2017), Gamage et al. (2017). Furthermore, we approximately calculated the turning point[2] of income level as 5,837 US$. Based on this finding, it can be said that Ecuador (5,412.1 US$), Egypt (2, 648.2 US$), Uzbekistan (2,025.3 US$), Tunisia (4302.4 US$) and Sri Lanka (3505.5 US$) were at the upward slope portion of the EKC, whereas Turkey (13,277.7 US$) was at the downward slope portion of the EKC.

Moreover, the increase in tourism investments (TO) has a negative and statistically significant effect on ecological footprint. Hence, an 1% increase in tourism investment reduces environmental degradation by 0.024%. This may be because of the direct and indirect effects of tourism investments. First, the direct decreasing effect of tourism investment on environmental degradation can occur through the composition of tourism investments. For instance, eco-friendly tourism investments may help to substitute tourism facilities with old and pollutant technologies with eco-friendly ones. Therefore, negative externalities of tourism facilities could be internalized, and polluting effects of tourism investments on ecology may be hampered. Second, Khadaroo and Seetanah (2008), asserted that the indirect decreasing effect of tourism investments occur by transport-related tourism investments. For instance, investments on new roads such as highways could help to reduce fuel consumption, which, in turn, helps to reduce air pollution. This evidence is consistent with Raza et al. (2017) and Paramati et al. (2018).

[2]The turning point of income level was estimated by utilizing following specification: $\beta_1 / -2\beta_2$.

The energy consumption *(EC)* has a positive and statistically significant effect on environmental degradation; this result is consistent with theoretical expectations. Hence, an 1% increase in energy consumption increases environmental degradation by 0.470%. This may be a result of the increasing effect of energy consumption on greenhouse gas and its detrimental effect on ecological sustainability. This result confirms the results of Katircioglu et al. (2014), Katircioğlu (2014), De Vita (2015), Dogan et al. (2017), Gamage et al. (2017), Dogan and Aslan (2017), Ozcan and Apergis (2018), Azam et al. (2018), while partly in contrast with Ozturk et al. (2016), Zhang and Gao (2016). Furthermore, internet use (INT) has a negative and statistically significant effect on environmental pollution. Therefore, an 1% increase in internet use reduces environmental pollution by 0.020%. In the countries, where information and communication technologies are widely used, individuals can easily collect information about the current environmental actions. Moreover, they can quickly become organized on online platforms to raise awareness about environmental issues and put pressure on their governments to take necessary actions. This finding confirms the results of Park et al. (2018) and Ozcan and Apergis (2018), whereas in contradiction with Salahuddin et al. (2016).

8.4 Conclusion

Tourism-environment nexus has been widely discussed recently, and most of the studies were focused on either the role of tourism arrivals or tourism receipts on environmental quality, yet there is a limited number of studies that focused on the role of tourism investments on environmental quality. The primary purpose of this study is to unveil tourism investments-environmental quality interaction in the six countries, which are in the top 20 countries, fastest-growing in terms of tourism GDP, countries for the period of 1995–2014. Based on empirical estimations, there is no cross-section dependence in variables and model, and slope coefficients are heterogeneous. There is a cointegration nexus among variables, and long-run coefficient estimation results confirm that the EKC hypothesis holds, tourism investments and internet use have a negative and statistically significant effect on environmental degradation, whereas energy consumption has a positive and statistically significant effect.

These empirical findings imply that EKC is valid. Therefore, the countries at the upward slope portion of the EKC should increase their economic growth to pass over the turning point (Danish et al. 2020), while the countries at downward slope portion of the EKC should adopt eco-friendly production technologies to internalize negative externalities of economic growth. Tourism investments improve environmental quality in these countries; therefore, increasing tourism investment can help achieve environmental aims of SDGs in those countries. In this context, policymakers should consider making tax reductions or subsidy in eco-friendly tourism investments. Policymakers should consider creating new legislations on the renovation of old facilities and new tourism facility investments. Moreover, the obligation of the installation of renewable energy plants for tourism facilities may be another eco-friendly tourism investment option.

The use of non-renewable energy has a positive effect on environmental degradation. Non-renewable energy is one of the main inputs of production, and it has a direct effect on economic growth. Therefore, policymakers should not dramatically change the production composition to consider maintaining the economic performance of the country and the welfare of individuals. Instead, policymakers should focus on internalizing the negative externalities of non-renewable energy use. Increasing the efficiency of non-renewable energy inputs could be an efficient policy. Policymakers should encourage research and development activities on energy efficiency.

Furthermore, filter inserts on factories, ban on the use of the vehicles with old technologies, and uncompromising implementation of current environmental legislations are significant. Therefore, regulatory authorities should take an active role in monitoring the execution of these policies. Internet usage hurts environmental degradation. Therefore, policymakers should consider investing more on information and communication technologies that may create positive externalities on economic growth. Moreover, reducing taxes on information and communication may increase of internet use; hence, individuals may easily get exact information on environmental quality and organize on online platforms to show actions on environmental issues.

References

Acaravci, A., & Akalin, G. (2017). Environment–economic growth nexus: A comparative analysis of developed and developing countries. *International Journal of Energy Economics and Policy, 7*(5), 34–43.

Acaravci, A., & Erdogan, S. (2017). The relationship between institutional structure and economic growth: A comparative analysis for selected countries. *International Journal of Economics and Financial Issues, 7*(6), 141.

Adamou, A., & Clerides, S. (2009). Prospects and limits of tourism-led growth: The international evidence. *The Rimini Centre for Economic Analysis WP*, 41–09.

Akadiri, S. S., Akadiri, A. C., & Alola, U. V. (2019). Is there growth impact of tourism? Evidence from selected small island states. *Current Issues in Tourism, 22*(12), 1480–1498.

Azam, M., Alam, M. M., & Hafeez, M. H. (2018). Effect of tourism on environmental pollution: Further evidence from Malaysia, Singapore and Thailand. *Journal of cleaner production, 190*, 330–338.

Bella, G. (2018). Estimating the tourism induced environmental Kuznets curve in France. *Journal of Sustainable Tourism, 26*(12), 2043–2052.

Breitung, J. (2000). The local power of some unit root tests for panel data. Retrieved from https://edoc.hu-berlin.de/bitstream/handle/18452/3936/69.pdf?sequence=1.

Danish, D., Ulucak, R., & Khan, S. U.-D. (2020). Determinants of the ecological footprint: Role of renewable energy, natural resources, and urbanization. *Sustainable Cities and Society, 54*, 101996.

De Vita, G., Katircioglu, S., Altinay, L., Fethi, S., & Mercan, M. (2015). Revisiting the environmental Kuznets curve hypothesis in a tourism development context. *Environmental Science and Pollution Research, 22*(21), 16652–16663.

Dogan, E., & Aslan, A. (2017). Exploring the relationship among CO_2 emissions, real GDP, energy consumption and tourism in the EU and candidate countries: Evidence from panel models robust to heterogeneity and cross-sectional dependence. *Renewable and Sustainable Energy Reviews, 77,* 239–245.

Dogan, E., Seker, F., & Bulbul, S. (2017). Investigating the impacts of energy consumption, real GDP, tourism and trade on CO_2 emissions by accounting for cross-sectional dependence: A panel study of OECD countries. *Current Issues in Tourism, 20*(16), 1701–1719.

Erdogan, S. (2020). Analyzing the Environmental Kuznets Curve Hypothesis: The Role of Disaggregated Transport Infrastructure Investments. *Sustainable Cities and Society*, 102338. DOI: https://doi.org/10.1016/j.scs.2020.102338.

Erdogan, S., & Acaravci, A. (2019). Revisiting the convergence of carbon emission phenomenon in OECD countries: new evidence from Fourier panel KPSS test. *Environmental Science and Pollution Research*, 1–14.

Erdogan, S., Akalin, G., & Oypan, O. (2020). Are shocks to disaggregated energy consumption transitory or permanent in Turkey? New evidence from fourier panel KPSS test. Energy, 117–174.

Grossman, G. M., & Krueger, A. B. (1991). Environmental impacts of a North American free trade agreement (No. w3914). National Bureau of economic research.

Katircioğlu, S. T. (2014). Testing the tourism-induced EKC hypothesis: The case of Singapore. *Economic Modelling, 41,* 383–391.

Katircioglu, S. T., Feridun, M., & Kilinc, C. (2014). Estimating tourism-induced energy consumption and CO_2 emissions: The case of Cyprus. *Renewable and Sustainable Energy Reviews, 29,* 634–640.

Khadaroo, J., & Seetanah, B. (2008). The role of transport infrastructure in international tourism development: A gravity model approach. *Tourism management, 29*(5), 831–840.

Koçak, E., Ulucak, R., & Ulucak, Z. Ş. (2020). The impact of tourism developments on CO_2 emissions: An advanced panel data estimation. *Tourism Management Perspectives, 33,* 100611.

Lopez, R. E., & Islam, A. M. (2008). *Trade and the Environment.*

Naradda Gamage, S. K., Hewa Kuruppuge, R., & Haq, I. U. (2017). Energy consumption, tourism development, and environmental degradation in Sri Lanka. *Energy Sources, Part B: Economics, Planning, and Policy, 12*(10), 910–916.

National Footprint Accounts. (2016). Working Guidebook to the National Footprint Accounts. Retrieved July 7, 2020, from http://www.footprintnetwork.org/content/documents/National_Footprint_Accounts_2016_Guidebook.pdf2016.

Ozcan, B., & Apergis, N. (2018). The impact of internet use on air pollution: Evidence from emerging countries. *Environmental Science and Pollution Research, 25*(5), 4174–4189.

Ozturk, I., Al-Mulali, U., & Saboori, B. (2016). Investigating the environmental Kuznets curve hypothesis: the role of tourism and ecological footprint. *Environmental Science and Pollution Research, 23*(2), 1916–1928.

Paramati, S. R., Alam, M. S., & Chen, C.-F. (2017a). The effects of tourism on economic growth and CO_2 emissions: A comparison between developed and developing economies. *Journal of Travel Research, 56*(6), 712–724.

Paramati, S. R., Shahbaz, M., & Alam, M. S. (2017b). Does tourism degrade environmental quality? A comparative study of Eastern and Western European Union. *Transportation Research Part D: Transport and Environment, 50,* 1–13.

Paramati, S. R., Alam, M. S., & Lau, C. K. M. (2018). The effect of tourism investment on tourism development and CO_2 emissions: Empirical evidence from the EU nations. *Journal of Sustainable Tourism, 26*(9), 1587–1607.

Park, Y., Meng, F., & Baloch, M. A. (2018). The effect of ICT, financial development, growth, and trade openness on CO_2 emissions: An empirical analysis. *Environmental Science and Pollution Research, 25*(30), 30708–30719.

Pedroni, P. (2000). Fully modified OLS for heterogeneous cointegrated panels. *Advances in econometrics, 15,* 93–130.

Pesaran, M. H. (2004). General diagnostic tests for cross section dependence in panels. Retrieved from https://papers.ssrn.com/sol3/papers.cfm?abstract_id=572504.

Pesaran, M. H., & Yamagata, T. (2008). Testing slope homogeneity in large panels. *Journal of Econometrics, 142*(1), 50–93.

Raza, S. A., Sharif, A., Wong, W. K., & Karim, M. Z. A. (2017). Tourism development and environmental degradation in the United States: Evidence from wavelet-based analysis. *Current Issues in Tourism, 20*(16), 1768–1790.

Saenz-de-Miera, O., & Rosselló, J. (2014). Modeling tourism impacts on air pollution: The case study of PM10 in Mallorca. *Tourism Management, 40,* 273–281.

Saint Akadiri, S., Alola, A. A., & Akadiri, A. C. (2019). The role of globalization, real income, tourism in environmental sustainability target. Evidence from Turkey. *Science of the Total Environment, 687,* 423–432.

Salahuddin, M., Alam, K., & Ozturk, I. (2016). Is rapid growth in Internet usage environmentally sustainable for Australia? An empirical investigation. *Environmental Science and Pollution Research, 23*(5), 4700–4713.

Sarafidis, V., Yamagata, T., & Robertson, D. (2009). A test of cross section dependence for a linear dynamic panel model with regressors. *Journal of econometrics, 148*(2), 149–161.

The Global Footprint Network. (2019). Ecological Footprint. Retrieved from https://www.footprintnetwork.org/our-work/ecological-footprint/.

The Intergovernmental Panel on Climate Change. (2018). Global warming of 1.5 °C. Retrieved from July 7, 2020, from https://report.ipcc.ch/sr15/pdf/sr15_spm_final.pdf.

Wackernagel, M., & Rees, W. (2004). What is an ecological footprint?. *The sustainable urban development reader*, 211–219.

Westerlund, J. (2005). New simple tests for panel cointegration. *Econometric Reviews, 24*(3), 297–316.

World Bank. (2020). World Development Indicators. Retrieved from https://databank.worldbank.org/source/world-development-indicators.

World Travel & Tourism Council. (2019). Travel & Tourism: Economic Impact 2019. Retrieved from https://www.wttc.org/-/media/files/reports/economic-impact-research/regions-2019/world2019.pdf.

WTTC. (2015). *European travel & tourism: Where are the greatest current and future investment needs?*. London, United Kingdom: WTTC.

WTTC. (2020). *Travel & Tourism: Economic impact 2019 World*. London, UK: WTTC.

Zaman, K., Shahbaz, M., Loganathan, N., & Raza, S. A. (2016). Tourism development, energy consumption and Environmental Kuznets Curve: Trivariate analysis in the panel of developed and developing countries. *Tourism Management, 54,* 275–283.

Zhang, L., & Gao, J. (2016). Exploring the effects of international tourism on China's economic growth, energy consumption and environmental pollution: Evidence from a regional panel analysis. *Renewable and Sustainable Energy Reviews, 53,* 225–234.

Chapter 9
Investigating the Tourism Originating CO_2 Emissions in Top 10 Tourism-Induced Countries: Evidence from Tourism Index

Asli Ozpolat, Ferda Nakipoglu Ozsoy, and Mehmet Akif Destek

Abstract Because of the structural change in the economy, it is observed that the emphasis on the service sector brings along some problems. The link between the tourism sector and environmental pollution has started to draw attention to the issues in recent years, and it reveals the need for policy-makers to take measures to prevent environmental pollution. For this purpose, the impact of economic growth, tourism index, urbanization, and energy intensity on environmental pollution has been investigated for selected 10 countries with the international tourism revenue during 1995–2014 using first-generation panel data (MG) estimator, second-generation panel data (CCE-MG) estimator, and heterogeneous panel causality test. According to the empirical analysis, individual results indicate that the impact of tourism on environmental pollution differs in each country. While tourism increases environmental pollution in Germany, France, and Italy, where the tourism sector is the most developed, there is not any environmental pollution increasing effect in the US and Australia. In addition, according to the CCE-MG panel results, tourism, energy intensity, and per capita income increase environmental pollution.

Keywords Tourism index · Environmental pollution · Energy consumption · Panel data

A. Ozpolat
Department of Management and Organization, Gaziantep University, Gaziantep, Turkey
e-mail: ozpolat@gantep.edu.tr

F. N. Ozsoy
Department of International Relations, Gaziantep University, Gaziantep, Turkey
e-mail: nakipoglu@gantep.edu.tr

M. A. Destek (✉)
Department of Economics, Gaziantep University, Gaziantep, Turkey
e-mail: adestek@gantep.edu.tr

D. Balsalobre-Lorente et al. (eds.), *Strategies in Sustainable Tourism, Economic Growth and Clean Energy*, https://doi.org/10.1007/978-3-030-59675-0_9

9.1 Introduction

The continuous increase in the negative effects of environmental pollution obliges governments worldwide to take measures that are more stringent on the environment. The fact that environmental destruction is an inevitable consequence of economic growth draws attention to the necessity to examine the relationship between environmental pollution and economic growth, especially within the scope of scale, technique, and composition effects (Grossman and Kruger 1991). The scale effect leads to the use of more natural resources to meet the increase in demand with increasing production levels, and it causes environmental destruction with increasing use of fossil fuels, which are cheaper than renewable energy sources in production (Grossman and Kruger 1991; Owusu and Asumadu 2016). Technique effect represents that economic growth has a positive effect on environmental pollution. There is an increase in funds allocated for R&D, mainly in high-income countries, while R&D investments form the basis of the development of environmentally friendly technologies and trigger an increase in energy efficiency. Therefore, environmental quality starts to increase with increasing efficiency and clean technology (Reppelin-Hill 1998: 283–284). The composition effect states that there is no clear result about the effect of economic growth on the quality of the environment, and that the quality of the environment changes depending on the structural change in the economy. Because of the increase in income level and structural transformation, it is known that the transformation from the agricultural sector to industry sector causes more intensive energy use and thus environmental destruction increases. On the other hand, it is stated that the factors causing environmental pollution decrease with the transition from pollution-intensive industries to service-oriented industries in high-income post-industrialization economies (Akbostancı et al. 2009: 862; Sarkodie and Strezov 2018). Despite it is expressed that the transition from pollution-intensive industries to the service sector will decrease the environmental damage due to the composition effect, when the researches in this field are examined, it is seen that countries generally put environmental quality and environmental awareness in the second plan. Therefore, a wrong perception arises that the service sector is an environmentally friendly sector. There is more than one sector in the service sector, and some of them are known to have serious environmental impacts (Alcántara and Padilla 2006). Based on this, in recent studies, it is seen that the effects of the sectors on the environment are tried to be brought to the fore in order to take measures on a sectoral basis in reducing environmental pollution. Accordingly, while it is stated that environmental pollutants occur in the infrastructure construction and expansion activities in the transportation sector and natural resources are affected negatively, it is stated that multimodal transportation also causes environmental destruction (Rondinelli and Berry 2000). On the other hand, studies examining the effects of different sectors on carbon emissions are also emphasized in the important role played by service activities on environmental pollution in many sectors, from manufacturing to transportation (Alcántara and Padilla 2006; Tarancón and del Río 2007). As a result of the relationship between ICT and environmental pollution for Sub-Saharan Africa countries, there are also studies that show that Information Communication Technologies increase carbon emissions per capita (Asongu et al. 2018; Avom et al. 2020). There are also studies that found that the relationship between the education sector and environmental pollution causes higher greenhouse gas emissions in cities where there is a lower level of education in the Czech Republic (Branis and Linhartova 2012). On the other hand, the tourism sector, which contributes the most to the national economies and has the highest share in the service sector in the transition of the country's economies from industrialization to the service sector, also has environmental destructions. While the tourism sector is expressed as one of the sectors that causes the greenhouse gas emissions the most, it is stated that the increase in energy demand in tourism-related activities such as air transportation or accommodation increases carbon emission emissions and environmental pollution (Tovar and Lockwood 2008). It is observed that tourist travels cause environmental destruction, so the opinions that argue that the pressure on environmental pollution can be reduced with less travel or environmentally friendly work preference (Simpson et al. 2008; Dickinson et al. 2011), but they also advocate otherwise (Gao et al. 2019; Akadiri et al. 2020).

Each country emphasizes the importance of sustainable development in order to raise the current standard of living and implement various effective policy recommendations by ensuring efficient use of its resources. At this point, the possible contribution of tourism, which is one of the sectors with increasing economic and socio-cultural importance in the world, to economic growth is not ignored. On the other hand, the developments in the tourism sector, the increase in the number of international tourists or incomes not only have an impact on economic growth but also raise the increase in energy demand. In this context, it is seen in the literature that the relationship between economic growth and tourism is focused and this relationship is explained through different hypotheses such as "Growth-oriented hypothesis", "Conservation hypothesis", "Neutrality hypothesis", and "Feedback hypothesis" (Marin 1992; Oh 2005; Tugcu 2014; Sokhanvar et al. 2018). There are studies focusing on the relationship between tourism and environmental pollution, which is relatively less in the literature (Katircioglu 2014b). Some of these studies examine the relationship between tourist arrivals or tourism revenues and carbon emissions (Lee and Brahmasrene 2013; Solarin 2014; Durbarry and Seetanah 2015; Shi et al. 2019); others have tested the relationship between tourism and energy consumption (Dogan et al. 2017). As a result of the findings obtained from the studies, it has been observed that the development of tourism around the world generally harms the nature. Negative developments are encountered such as pollution caused by mass tourism, plundered natural resources for tourism, opened natural areas for rent. Therefore, the close relationship between tourism and environmental pollution has been exposed.

The fact that more people travel for tourism purposes every year, the development of the economic structure of the countries and the increase in per capita income are the factors that stimulate tourism demand. Increasing tourism activities, utilizing services such as transportation or accommodation increase the demand for energy, and increasing energy demand causes energy consumption and environmental pollution. It is known that tourism revenues and tourist spending in many countries contribute significantly to the GDP of countries. The report of the Top 10 (France, Spain, USA, China, Italy, Turkey, Mexico, Germany, Thailand, and UK) destinations by international tourism receipts in 2018 shows that these countries receive 40% of worldwide arrivals and earners account for almost 50% of total tourism receipts (UNWTO 2019). In 2019, the impact of Travel & Tourism's direct on world's GDP is US$8.9 trillion and on global GDP is 10.3% (WTTC 2019). These figures show how important the tourism industry is in the global economy. In addition, carbon emissions in these countries are remarkable. According to the International Energy Agency (2020) data, carbon emission levels in China, Korea, US, Germany, Italy, Canada, France, Australia, and UK are constantly increasing between 1995 and 2017. In addition, the carbon emission data of countries are still at high levels compared to the world average. CO_2 emission per capita in the world in 2017 is 4.4 t CO_2/capita. Accordingly, CO_2 emissions are quite high in Australia with 15.63 t CO_2/capita in 2017; Canada with 14.99 t CO_2/capita and US with 14.61 t CO_2/capita. In addition, Fig. 9.1 shows the CO_2 share of international top 10 countries in the world.

Considering Fig. 9.1, the CO_2 emission of these countries in 1995 constituted 59% of the total CO_2 emission, while this rate decreased to 57% in 2017. The total share has decreased by 2% over the years, but the share of these 10 countries in total CO_2 emissions is quite high.

In the light of these data, the importance of tourism for the countries and environmental quality, the development, and sustainability of tourism by protecting the nature is possible only with realistic and balanced policies and practices. Sustainable Tourism is a concept created after the United Nations Rio Sustainable Development Summit in 1992. Sustainable Development in Tourism, in keeping with protecting

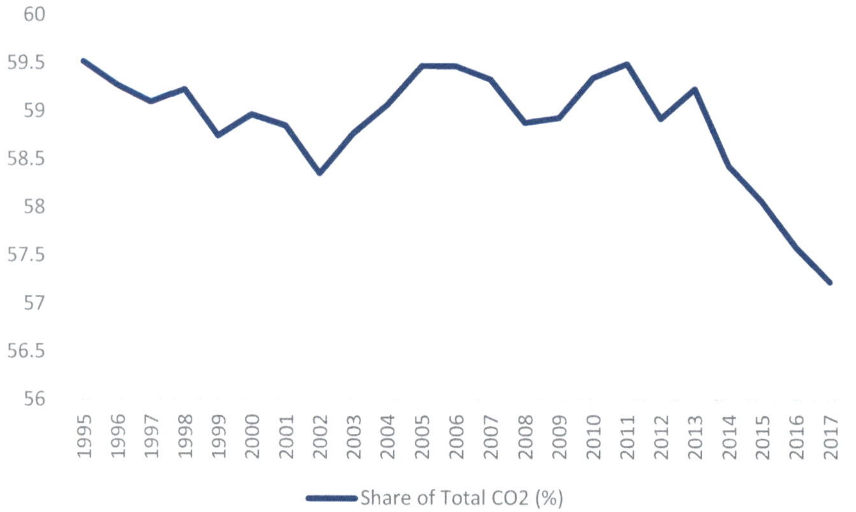

Fig. 9.1 CO_2 emissions of international tourism top 10 in total world CO_2 emissions (IEA 2020)

and developing future opportunities, adopts the principle of meeting the needs of today's tourists and host regions. In this way, continuity of cultural integrity, compulsory ecological processes, and biological diversity is ensured, while economic, social, and aesthetic requirements are also emphasized (UN 1999). While the tourism sector in the world is a sector that is open to development in the medium and long terms, which can offer a wide variety of tourism alternatives and has the potential of rapid growth.

In this study, the impact of economic growth, tourism index, urbanization, and energy consumption on environmental pollution has been investigated for selected 10 countries with the international tourism revenues in the period of 1995–2014. The possible contributions of the study to the literature are as follows: (i) to the best of our knowledge, this is the first study to examine the relationship between tourism and carbon emissions in selected 10 countries with the most international tourism revenues; (ii) to clearly observe the impact of tourism on environment, a tourism index is constructed with principal component analysis; (iii) to avoid a possible omitted variable bias, economic growth, urbanization, and energy density are also used as explanatory variables; and (iv) second-generation panel data methodologies are employed to consider the cross-sectional dependence among observed countries.

9.2 Literature Review

In the environmental economic literature, it is possible to come across many studies examining the effects of economic growth, urbanization, and energy on environmental pollution. In studies that examine the relationship between environmental pollution and urbanization, it is seen that a common idea cannot be reached on the impact of urbanization on environmental quality. In some studies, it is stated that there is an increase in greenhouse gas emissions released to the environment due to increased urbanization causing more energy use (Jones 1991; Parikh and Shukla 1995), while other studies show that urbanization increases efficiency in public infrastructure and therefore environmental pollution decreases (Sharma 2011). The relationship between economic growth and environmental pollution has been mostly examined within the framework of EKC hypothesis (Murthy and Gambhir 2018; Naz et al. 2019; Do and Dinh 2020), and different results are obtained as a result of different methods, observation range, or variables. In studies examining the relationship between energy and environmental pollution, the effects on environmental pollution with variables such as energy consumption, renewable energy, or non-renewable energy consumption are examined (Liu et al. 2017; Chen and Lei 2018; Bekun et al. 2019; Destek and Aslan 2020) but wondering what the effects of the sectors in the economy will be on environmental pollution has started to attract attention recently. Tourism sector is one of the leading sectors in the economy. For these reasons, in this study, it aims to examine the possible effects of economic growth, urbanization, energy density, and tourism on environmental pollution. In line with this purpose, firstly, studies examining the effects of economic growth, energy density, and urbanization on environmental pollution are included, and then studies that investigate the relationship between tourism and environmental pollution are presented.

9.2.1 Economic Growth, Urbanization, Energy Density, and Environmental Pollution

The focus of the studies in the literature is based on the examination of the relationship between economic growth and carbon emissions. According to this relationship examined by the Environmental Kuznets Hypothesis, as per capita income increases, environmental destruction firstly increases. However, even if income per capita increases after a certain income level, environmental degradation improves. Many researchers closely followed this relationship (Grossman and Kruger 1991; Jha 1997; He and Richard 2010; Tiwari et al. 2013; Omri et al. 2015; Destek et al. 2018; Destek and Sarkodie 2019; Rahman et al. 2019; Destek and Sinha 2020). In the following process, the development of different econometric methods and obtaining more up-to-date data sets have allowed the study of the effects of different factors on environmental pollution, and the validity of the EKC hypothesis has been moved to different dimensions with using different variables such as urbanization and energy. In the literature, no consensus has yet been reached on the effects of urbanization on the environment. Some studies state that urbanization increases energy demand and more emissions are generated, and thus environmental quality is negatively affected (Parikh and Shukla 1995; Cole and Neumayer 2004; York 2007; Destek 2019). On the other hand, some studies claim that urbanization and urbanization intensity reduce energy use and emissions emitted and increase the effective use of public infrastructure (Newman and Kenworthy 1989; Chen et al. 2008).

Parikh and Shukla (1995) tested the relationship between economic development, urbanization, energy use, and greenhouse gas effects for developed countries and developing countries, and the results concluded that urbanization increased energy consumption. In addition, three effects of urbanization on energy use are mentioned with the realization of urban growth and changing lifestyles. The first effect is the transformation in energy use. With technological development, production has shifted mostly to products where high energy is used intensively. The second effect is an increase in energy consumption as the trend toward demand for goods and services increases. The last effect is the effect on the environment through direct transportation and household consumption. The shift of the workers in the agricultural sector to the industry has led the population largely to the cities. This mobility brought the importance attached to rural-urban freight transport, resulting in higher cost and energy use. All these developments affect local pollution and global warming directly or indirectly.

Jones (1991) examined the relationship between urbanization, population, and energy consumption for 59 developing countries using 1980 data and reported that an increase in the population also increases per capita energy consumption. He concluded that a 10% increase in the population increased the per capita energy consumption by 4.5–4.8%. On the other hand, the increase in urbanization rate increases energy consumption. The increasing population and urbanization cause environmental destruction. York et al. (2003) tested the impact of urbanization on national energy footprints and emissions. According to the results of the STIRPAT model applied for 142 countries having 97% of the world population and economic output, it has been determined that urbanization positively affects national energy footprints and emissions. Similarly, Cole and Neumayer (2004) found the positive relationship between carbon emissions and urbanization. According to the results of the study, 10% growth in urbanization increased carbon emissions by 7% in 86 countries. Alam et al. (2007) for Pakistan between 1971 and 2005 and Adusah-Poku (2016) covering 45 Sub-Saharan African countries in years 1990 and 2010 determined the existence of the positive relationship between urbanization and carbon emission. Moreover, Poumanyvong and Kaneko (2010) analyzed the effects of urbanization on carbon emissions and energy use for 99 countries during the period from 1975 to 2005 and the results reflected that urbanization increases the energy use of the middle- and high-income groups, while decreasing the energy use in the low-income group. In addition, although the effect of urbanization on carbon emissions is felt more in the middle-income group, this effect was positive for all income groups. Zhang and Lin (2012) investigated the urbanization, energy consumption, and carbon emissions in the case of China during 1995–2010. Their results showed that although the effects of urbanization on energy use differ between regions, urbanization generally increases energy consumption and carbon emissions. Rafiq et al. (2016) claimed that urbanization increases energy intensity, but there is an insignificant increase in carbon emissions. In addition, Dogan and Turkekul (2016) found a positive relationship between urbanization, energy consumption, and carbon emissions over the period 1960–2010. Moreover, EKC hypothesis is invalid between real GDP end carbon emissions in the USA. Shahbaz et al. (2016) found U-shaped relationship between carbon emissions and urbanization in Malaysia during 1970Q1-2011Q4. Ozatac et al. (2017) reported that EKC hypothesis is valid between real GDP and carbon emissions in Turkey, and the findings supported that energy consumption and urbanization have positive effect on carbon emissions. Liu and Bae (2018) tested the nexus between energy intensity, real GDP, urbanization, and carbon emissions for China in 1970–2015 and found that all variables increase carbon emissions.

Ehrhardt-Martinez et al. (2002) examined the relationship between urbanization and deforestation rate in developing countries using the EKC model in 1980–1995, and as a result of the findings they obtained that the rate of deforestation increased in the early stages of urbanization but decreased with the advancement of urbanization. Martínez-Zarzoso and Maruotti (2011) analyzed the effect of urbanization on carbon emissions for developing countries covering the time span 1975–2003 and found that there is an inverted U-shaped between variables. Similarly, Lin and Zhu (2017) found that urbanization first increases the energy and carbon intensity and then decreases it. In short, there is an inverted U-shaped relationship in China's 30 provinces in 2000–2015. Sharma (2011) focused on the real GDP, energy consumption, urbanization, and carbon emissions' relationship in 1985–2005 for 69 countries. The findings showed that the real GDP and energy consumption affect carbon emissions positively and urbanization has a negative effect on carbon emissions. Moreover, Xu and Lin (2015) also reported an inverted U-shaped between urbanization and carbon emissions in China from 1990 to 2011. Li and Lin (2015) reported different results according to the income groups for 73 countries. They found negative relationship between urbanization and energy consumption but positive nexus between carbon emissions and urbanization in the low-income group. In addition, urbanization increases energy consumption and carbon emissions in the middle- and high-income groups. Wang et al. (2015) also reported an inverse U-shaped relationship between urbanization and carbon emissions in OECD countries from 1960 to 2010. Bekhet and Othman (2017) tested the relationship between real GDP, carbon emissions, urbanization, and energy consumption and found that EKC hypothesis is valid between carbon emissions and urbanization in the long run, and there is a bidirectional causality between energy consumption, real GDP, and carbon emissions for Malaysia. Saidi and Mbarek (2017) also claimed that there is a positive monotonic relationship between real GDP and carbon emissions, and there is a negative relationship between urbanization and carbon emissions. On the other hand, Shaheen et al. (2020) investigated the nexus between urbanization, energy consumption, and carbon emissions in Pakistan 1972–2014 years. According to the results of the study, there was an insignificant relationship between urbanization and carbon emissions.

9.2.2 Tourism and Environmental Pollution

In the literature, earlier studies generally take into account the relationship between economic growth and tourism (Hall 1998; Balaguer and Cantavella-Jorda 2002; Durbarry 2004; Bernini 2009; Arslanturk et al. 2011; Balcilar et al. 2014; Katircioglu 2014a; Bilen et al. 2015; Brida et al. 2016; Eyuboglu and Eyuboglu 2020). However, although these studies contribute to the literature, they do not provide any information about the effects of tourism on environmental quality. Recently, researchers have focused on examining the effect of tourism on environmental pollution. While most of these researchers have found that tourism positively affects environmental pollution, some express opposite results. Therefore, there is no consensus between tourism and environmental pollution. Katircioglu (2014b) researched the nexus between tourism and environmental pollution in Turkey for the period from 1960 to 2010, and the results depicted that tourism has a positive effect on carbon emissions in both short and long runs. Solarin (2014) examined the nexus between tourism arrivals and carbon emissions and expressed that tourist arrivals cause environmental pollution in Malaysia covering the time span 1972–2010. Moreover, there was a one-way causality from tourist arrivals to carbon emissions. Similarly, Durbarry and Seetanah (2015) implied that tourist arrivals caused environmental pollution by increasing carbon emissions in Mauritius 1978–2011 years. Dogan et al. (2017) investigated the validity of EKC hypothesis between economic growth and carbon emissions and tested the relationship between energy consumption, tourism, economic growth, and carbon emissions for OECD countries during 1995–2010. The DOLS method results indicated that tourism and energy consumption increase environmental pollution, and there is any evidence for validity of EKC hypothesis. De Vita et al. (2015) also reported that the tourist arrivals and economic growth have positive effect on environmental pollution in Turkey. Sghaier et al. (2018) investigated the interaction between tourism and environmental pollution for Tunisia, Egypt, and Morocco in 1980–2014. According to the results of the study, there is no statistically significant relationship between tourist arrivals and environmental pollution in Morocco. On the other hand, while the tourist arrival in Egypt improves environmental pollution, it causes environmental destruction in Tunisia. Shi et al. (2019) researched the effect of tourist inflows and tourism expenditures on carbon emissions for 1995–2015. Their results expressed that tourism expenditures create environmental pollution in low-income countries. On the other hand, there was a positive relationship between tourism inflows and carbon emissions in both high and low-income countries. In addition, Eyuboglu and Uzar (2019) examined the effect of tourist arrival, economic growth, and energy consumption on carbon emissions over the period from 1960 to 2014 in Turkey, and the results indicated that economic growth, energy consumption, and tourism increase the carbon emissions in both short and long runs.

On the other hand, some studies claimed that there is a negative relationship between tourism and environmental pollution. Lee and Brahmasrene (2013) investigated the nexus between tourism and carbon emissions in EU countries for the period from 1988 to 2009. Their results showed that tourism has a negative effect on carbon emissions as a result of the effective management of tourism in the EU countries. Moreover, a 1% increase in tourism revenues decreases carbon emissions by 0.105%. Katircioglu (2014a) also supported negative relationship between tourism and carbon emissions in his study. This study is stated that the implementation of successful energy-saving policies in Singapore plays an important role in tourism development and environmental quality. Therefore, tourism increases the environmental quality. Zhang and Gao (2016) analyzed the nexus between international tourism and carbon emissions in China over the period 1995–2011 and reported that tourism increases environmental quality in the eastern region. Naradda Gamage et al. (2017) investigated the interaction between tourism receipts and carbon emissions in 1974–2013 years in Sri Lanka and reported that tourism receipts improve environmental degradation. Gao et al. (2019) examined the nexus between tourism development and carbon emissions for 18 Mediterranean countries during 1995–2010 and reported that tourism affects carbon emissions negatively. 1% increase in tourism improves environmental quality by 0.019%. Akadiri et al. (2020) found that tourism has a negative effect on carbon emissions in 16 small islands in 1995–2014 years. In addition, Koçak et al. (2020) examined the tourism and environmental pollution nexus for 10 most visited countries during 1995–2014 and reported that tourist arrivals cause environmental pollution, while tourism receipts improve environmental quality.

9.3 Empirical Strategy

In order to investigate link between tourism and CO_2 emissions and compare relative relationship between GDP per capita, urbanization, and energy intensity, the annual data has been collected for the period of 1995–2017 for top 10 countries with the International Tourism Expenditure (Australia, Canada, China, France, Germany, Korea, Italy, Russia, United State, and United Kingdom). The panel model is constituted according to Koçak vd., 2020. However, the main difference among papers is that tourism index was computed by using international tourism variables in this paper. The panel model is as follows:

$$\mathrm{InCO}_{i,t} = \alpha_0 + \alpha_1 \mathrm{InY}_{i,t} + \alpha_2 \mathrm{InTIND}_{i,t} + \alpha_3 \mathrm{INURB}_{i,t} + \alpha_4 \mathrm{InEN}_{i,t} + \varepsilon_{i,t} \quad (9.1)$$

In the model, t, i, and $\varepsilon_{i,t}$ indicate that time, cross-section, and residual term, respectively. Moreover, $\mathrm{InCO}_{i,t}$ is natural log of CO_2 emissions (metric tons per capita), $\mathrm{InTIND}_{i,t}$ is the tourism index, $\mathrm{InEN}_{i,t}$ is the natural log of energy intensity level of primary energy (MJ/$2011 PPP GDP), $\mathrm{InY}_{i,t}$ is the natural log of GDP per capita (constant 2010 US $), and $\mathrm{InURB}_{i,t}$ is the natural log of urban population. Tourism index includes three tourism data, which are international tourism, expenditures (% of total imports), international tourism, number of arrivals and international tourism, and receipts (% of total exports). The tourism index was computed using principal component analysis (PCA). The data of GDP per capita, urban population, and international tourism data were sourced from the World Development Indicators 2015 (World Bank); CO_2 emissions and energy intensity were obtained from international energy agency. Since there are a limited number of empirical studies using tourism index in the literature, the effect of the variables (α_2) in the model on the CO_2 emissions is not clear. In addition, in general, when the literature is evaluated, the coefficient in the model, α_3, is also not certain. For case in point, York et al. (2003), Poumanyvong and Kaneko (2010) find that urbanization increases environmental pollution, while Sharma (2011) indicates a negative relationship among variables and Shaheen et al. (2020) achieved an insignificant relationship. Therefore, although the relationship between the variables varies by country, it also depends on the stages of the urbanization process. However, α_1 and α_4 coefficients are expected to be positive and significant in generally. Sharma (2011) and Liu and Bae (2018) stated that per capita income and energy consumption increase environmental pollution.

In the analysis, it has been analyzed the impact of tourism on CO_2 emissions. For this purpose, firstly, the cross-sectional dependency in panel data analysis was investigated by CD test. This commitment is estimated by the CD test developed by Pesaran (2004). The CD test is calculated as follows:

$$CD = \sqrt{\frac{2T}{N(N-1)}} \left(\sum_{i=1}^{N-1} \sum_{j=i+1}^{N} \hat{\rho}_{ij} \right) \quad (9.2)$$

In the model, T represents the time dimension of the panel; N is the cross-sectional dimension of the panel, and $\rho_{\hat{ij}}$ is the binary OLS correlation sample estimate of the remains (Pesaran 2004: 1–7). After determining the cross-sectional dependence and homogeneity, the stationary of the variables was investigated by using CIPS unit root analysis developed by Pesaran (2007). The CIPS unit root analysis is derived from CADF statistics in Eq. 9.3.

$$\Delta y_{i,t} = a_i + \rho_i y_{it-1} + \beta_i \bar{y}_{t-1} + \sum_{j=0}^{k} \tau_{ij} \Delta \bar{y}_{it-1} + \sum_{j=0}^{k} \delta_{ij} y_{it-1} + \varepsilon_{i,t} \tag{9.3}$$

a_i The term deterministic in the equation refers to the number of lags k, $\bar{y}_t$ the cross-sectional average of time. The CIPS unit root model created accordingly is as follows:

$$\text{CIPS} = \left(\frac{1}{N}\right) \sum_{i=1}^{N} t_i(N, T) \tag{9.4}$$

After this stage, Westerlund Durbin-Hausman Test (2008) was estimated to determine the long-term relationship between the variables in the study. In Westerlund Durbin-Hausman Cointegration (2008) test, the horizontal cross-sectional dependence is taken into account and the variables need not be equally stable. In this method, two separate test statistics (DH_g, DH_p) are calculated as groups and panels. While panel statistic is (DH_p), (DH_g) is expressed by group statistics. The statistics are as follows:

$$DH_g = \sum_{i=1}^{n} \hat{S}_i \left(\tilde{\phi}_i - \hat{\phi}_i\right)^2 \sum_{t=2}^{T} \hat{e}_{it-1}^2 \quad DH_p = \hat{S}_n \left(\tilde{\phi} - \hat{\phi}\right)^2 \sum_{i=1}^{n} \sum_{t=2}^{T} \hat{e}_{it-1}^2 \tag{9.5}$$

After specifying the cointegration, Common Factor Correlated Effect (CCE) panel estimator has been tested to predict cointegration estimator. The test can be estimated in the presence of heterogeneity and cross-sectional dependence among countries. Pesaran (2006) developed CCE panel estimator computed by OLS estimator $\hat{B}_{i,cce}$. Therefore, the individual CCE estimator is as follows:

$$\hat{B}_{i,cce} = \sum_{i=1}^{N} \hat{B}_{i,cce} \tag{9.6}$$

Finally, the causality relationship between variables has been estimated by Heterogeneous Granger Causality Test (2012).

The test is written as follows:

$$y_{i,t} = \alpha_1 + \sum_{k-1}^{K} \gamma_i^{(k)} y_{i,t-k} + \varepsilon_{i,t} \tag{9.7}$$

According to Eq. (9.7), x and y are the stationary variables for N individuals on T periods. K is denoted by lag order and autoregressive parameters $\gamma_i^{(k)}$ and regression coefficients slopes $\beta_i^{(k)}$ refer to different groups. According to the equation, the null hypothesis is not a homogeneous causality in the panel. The test is written as follows.

9.4 Empirical Results and Discussions

In the first stage of the study, the horizontal cross-sectional dependency and homogeneity between the variables have been investigated. If there is a horizontal cross-sectional dependency among the variables, using tests that take into account the dependency will make the results more significant. In this context, the Pesaran (2004) Cross-Sectional Dependence (CD) test is estimated. The test results are shown in Table 9.1.

According to the results in Table 9.1, there is a cross-sectional dependency and heterogeneity among the variables. Therefore, the cointegration relationship between variables has been calculated through tests that take into account cross-sectional dependency. At this stage, CIPS test has been applied. According to the CIPS test results, all variables except the urbanization have been obtained as stationary in the first degree, that is, all series except urbanization are integrated at $I(1)$.

In addition, the long-term relationship between CO_2 emission, tourism index, GDP per capita, energy intensity, and urbanization has been investigated for all models. The existence of long-term relationship was estimated by Westerlund DH cointegration analysis. The results are given in Table 9.2.

Table 9.1 Cross-sectional dependency and homogeneity test result

	LM	CD_{LM}	CD	LM_{adj}
InCO	971.881 (0.000)	97.702 (0.000)	31.159 (0.000)	10.156 (0.000)
InY	152.514(0.000)	11.333 (0.000)	7.723 (0.000)	10.256 (0.000)
InURB	178.675 (0.000)	14.091 (0.000)	7.388 (0.000)	10.370 (0.000)
InEN	235.531 (0.000)	20.084 (0.000)	9.673 (0.000)	17.410 (0.000)
InTIND	479.853 (0.000)	45.838 (0.000)	18.460 (0.000)	6.997 (0.000)
	Statistics	p-values		
$\tilde{\Delta}$	6.161	0.000		
$\tilde{\Delta}_{adj}$	6.607	0.000		

Table 9.2 Panel unit root and cointegration test result

Panel unit root	CIPS-stat (level)	CIPS-stat (first differences)
InCO	−0.551	−4.177***
InTIND	−2.018	−4.114***
InY	−2.040	−3.030***
InURB	−0.807	−1.237
InEN	−2.295	−4.857***
Panel cointegration	Statistics	***P***-value
Westerlund-DH-g	1.395	0.081
Westerlund-DH-p	1.488	0.068

Critical values for the CIPS are 2.76, −2.94, and −3.3 at 10, 5, and 1% level; *** indicate statistical significance at 10, 5, and 1% level, respectively

Westerlund DH cointegration analysis is a test that can also be used if the variables are not equally stable. According to the results obtained, there is a long-term relationship between the variables. After determining the cointegration relationship, parameter estimates have been made for the panel model. Panel group results have been shared primarily in parameter estimation. Thus, mean group results and CCE results can be examined comparatively. The main reason for comparing the two models is that the mean group estimator does not take into account the cross-sectional dependency. The first-generation panel tests, which ignored the cross-sectional dependency, may cause the results to be statistically insignificant.

When the results in Table 9.3 are analyzed, it is seen that panel results are significant in both test groups. According to the MG estimator results, while tourism and urbanization decrease CO_2 emissions, GDP per capita and energy consumption increase CO_2 emissions. However, some of the signs of coefficients are obtained opposite from CCE-MG estimation. This difference is possibly sourced from that CCE-MG estimation allows cross-sectional dependence among countries, while MG estimation does not. Therefore, the findings from CCE-MG can be accepted as more robust. According to the results of the CCE-MG estimation, increasing tourism increases pollution. In addition, GDP per capita and energy consumption increase carbon dioxide emissions. On the other hand, urbanization reduces CO_2 emissions. In addition to panel results, individual parameter estimates are given in Table 9.4. The CCE estimator considers the cross-sectional dependency, and so test can give

Table 9.3 Estimator results of cointegration for panel

	Mean Group (MG)				CCE-MG			
	InTIND	InY	InURB	InEN	InTIND	InY	InURB	InEN
Panel	−0.036**	2.511**	−2.494***	0.733***	0.010**	2.498***	−0.909*	0.974***

Table 9.4 Estimator results of cointegration for individual countries

	lnTIND	lnY	lnURB	lnEN
China	0.010***	2.535***	0.522	1.444***
US	−0.048*	2.640**	0.680	1.034*
Germany	0.029*	1.849*	−1.992***	0.912**
UK	0.010	0.888*	−0.303	0.816*
France	0.008***	3.382**	3.340**	0.479*
Australia	−0.029*	1.705**	4.134***	0.307***
Russia	0.005	2.263***	1.800*	0.976***
Canada	−0.024	2.635**	1.769	0.609**
Korea	0.016	4.020***	−6.406***	1.157*
Italy	0.037*	3.858**	0.842	0.894**

Note *, **, and *** indicate statistical significance at 10, 5, and 1% levels, respectively

results that are more robust. CCE test, in which long-term coefficients are estimated, has been applied for the individual test results.

When the results are evaluated, the tourism index in China, Germany, France, and Italy has a positive effect on CO_2 emissions, while negative link has been obtained among the variables in the US and Australia. On the contrary, in UK, Russia, Canada, and Korea, the relationship between variables is insignificant. When the literature is investigated, it can be seen that there is no consensus on the relationship between tourism and environmental pollution. Therefore, tourism policies implemented by countries are determinant on the impact of tourism on pollution. In addition, there are studies showing that the environmental pollution of tourism revenues in general decreases, but pollution increases as the number of tourists increases (see, for example, Lee and Brahmasrene 2013; Eyuboglu and Uzar 2019; Koçak et al. 2020; Khan et al. 2020; Akadiri et al. 2020). According to effect of GDP on CO_2 emissions, the link between per capita income and CO_2 emissions is positive and significant in all countries. That is, as the per capita income increases, carbon dioxide emissions increase. It is known that increase in national income due to industrialization and increase in production can increase the emission of CO_2. Considering that the variables in the analysis are developed and industrialized countries, it is natural to expect a positive relationship between the variables. According to this approach, a positive relationship can be expected between urbanization and CO_2 emissions. However, there is not any consensus in the literature between urbanization and CO_2. The dimensions of the link between the variables may change in the first stages of urbanization or depending on the urbanization policies implemented. When the results obtained accordingly are evaluated, while urbanization is positive and significant in France and Australia, it is negative and significant in Germany and Korea and urbanization is insignificant in Italy, Canada, UK, US, and China. As stated before, urbanization has three different effects on the environment. According to the results of the study, it can be stated that increasing energy consumption and transportation increase CO_2 emission with urbanization in France and Australia. In Germany and Korea, urbanization process policies are thought to have a positive impact on the environment. Finally, the relationship between energy use and CO_2 emission is positive and significant in all countries. The fact that energy is currently the most important factor for production, growth, industrialization, and other factors of growth shows the sustainability of energy demand. Therefore, depending on the type of energy used, energy density and energy consumption have a negative impact on environmental pollution. In the last stage of the analyses, heterogeneous panel causality test was estimated to investigate the causality relationship between the variables.

According to the results presented in Table 9.5, there is a bidirectional causality between tourism index, urbanization and CO_2 emissions, and a unidirectional causality from GDP per capita to CO_2 emissions. While there is a one-way causality relationship between energy density and urbanization, a two-way causality relationship is obtained between per capita income and tourism index. A one-way causality relationship has been found between the urbanization and tourism index from the tourism index to urbanization. Finally, while there is a bidirectional causality between per capita income and urbanization, there is a bidirectional causality from energy

Table 9.5 Heterogeneous panel causality test results

Null hypothesis:	W-Stat.	Prob.
lnEN does not homogeneously cause lnCO	6.00824	0.0000
lnCO does not homogeneously cause lnEN	1.66493	0.3175
lnURB does not homogeneously cause lnCO	8.68346	0.0000
lnCO does not homogeneously cause lnURB	10.3633	0.0000
lnY does not homogeneously cause lnCO	8.45266	0.0000
lnCO does not homogeneously cause lnY	1.89560	0.1554
lnTIND does not homogeneously cause lnCO	5.79230	0.0000
lnCO does not homogeneously cause lnTIND	2.45916	0.0143
lnURB does not homogeneously cause lnEN	5.59752	2.E−16
lnEN does not homogeneously cause lnURB	12.1178	0.0000
lnY does not homogeneously cause lnEN	3.44830	2.E−05
lnEN does not homogeneously cause lnY	1.39951	0.6067
lnTIND does not homogeneously cause lnEN	4.02214	1.E−07
lnEN does not homogeneously cause lnTIND	1.03382	0.8783
lnY does not homogeneously cause lnURB	7.9631	0.0000
lnURB does not homogeneously cause lnY	3.23588	0.0001
lnTIND does not homogeneously cause lnURB	3.60866	5.E−06
lnURB does not homogeneously cause lnTIND	1.73822	0.2570
lnTIND does not homogeneously cause lnY	2.77032	0.0025
lnY does not homogeneously cause lnTIND	1.13323	0.9773

intensity to CO_2 emissions. In addition, the result is supported by the growth hypothesis. Therefore, it can be stated that the decisions taken bilaterally in tourism have a positive effect on economic growth. Therefore, policy-makers subsidize tourism and provide an increase in economic growth. Panel results show that tourism has a direct and indirect effect on environmental pollution.

9.5 Conclusions

The current study aims to search the effect of tourism on CO_2 emissions and to compare the relationship between energy density, GDP per capita, and urbanization for the period from 1995 to 2017 in international top 10 countries. For this purpose, the relationship among variables examines with first-generation panel data (MG) estimator and second-generation panel data (CCE-MG) estimator and heterogeneous panel causality test. In general, when individual results are evaluated, the impact of tourism on environmental pollution varies by country. In countries such as Germany, France, and Italy where tourism sector is the most developed, tourism increases environmental pollution. This effect is thought to depend on the number of tourist and tourism expenditures. In US and Australia, it can be stated that tourism, which has a pollution-reducing effect, is related to the policies implemented. According to CCE-MG panel results, tourism, energy density, and per capita national income increase environmental pollution.

Overall, the findings show that the policies implemented to increase the number of tourists and tourism revenues based on economic concerns cause environmental destruction. However, the destructive effects of tourism on the environment can be eliminated through the effectiveness of resource management, policies that ensure the sustainability of cultural integrity, mandatory ecological processes, biodiversity, and life support systems. Moreover, as a result of increasing tourism activities, the implementation of effective environmentally sensitive tourism policies and successful energy-saving practices will lead to both the development of tourism and the increase of environmental quality.

References

Adusah-Poku, F. (2016). Carbon dioxide emissions, urbanization and population: Empirical evidence from SubSaharan Africa. *Energy Economics Letters, 3*(1), 1–16.

Akadiri, S. S., Lasisi, T. T., Uzuner, G., & Akadiri, A. C. (2020). Examining the causal impacts of tourism, globalization, economic growth and carbon emissions in tourism island territories: Bootstrap panel Granger causality analysis. *Current Issues in Tourism, 23*(4), 470–484.

Akbostancı, E., Türüt, S., & Tunç, İ. (2009). The relationship between income and environment in Turkey: Is there an environmental Kuznets Curve? *Energy Policy, 37,* 862–863.

Alam, S., Fatima, A., & Butt, M. S. (2007). Sustainable development in Pakistan in the context of energy consumption demand and environmental degradation. *Journal of Asian Economics, 18,* 825–837.

Alcántara, V., & Padilla, E. (2006). *An input–output analysis of the "key" sectors in CO_2 emissions from a production perspective: An application to the Spanish economy.* Working Paper, 06.01. Department of Applied Economics, Autonomous University of Barcelona.

Arslanturk, Y., Balcilar, M., & Ozdemir, Z. A. (2011). Time-varying linkages between tourism receipts and economic growth in a small open economy. *Economic Modelling, 28*(1), 664–671.

Asongu, S. A., Le Roux, S., & Biekpe, N. (2018). Enhancing ICT for environmental sustainability in sub-Saharan Africa. *Technological Forecasting and Social Change, 127,* 209–216.

Avom, D., Nkengfack, H., Fotio, H. K., & Totouom, A. (2020). ICT and environmental quality in Sub-Saharan Africa: Effects and transmission channels. *Technological Forecasting and Social Change, 155,* 120028.

Balaguer, J., & Cantavella-Jorda, M. (2002). Tourism as a long-run economic growth factor: The Spanish case. *Applied Economics, 34*(7), 877–884.

Balcilar, M., Van Eyden, R., Inglesi-Lotz, R., & Gupta, R. (2014). Time-varying linkages between tourism receipts and economic growth in South Africa. *Applied Economics, 46*(36), 4381–4398.

Bekhet, H. A., & Othman, N. S. (2017). Impact of urbanization growth on Malaysia CO_2 emissions: Evidence from the dynamic relationship. *Journal of Cleaner Production, 154,* 374–388.

Bekun, F. V., Alola, A. A., & Sarkodie, S. A. (2019). Toward a sustainable environment: Nexus between CO_2 emissions, resource rent, renewable and nonrenewable energy in 16-EU countries. *Science of the Total Environment, 657,* 1023–1029.

Bernini, C. (2009). Convention industry and destination clusters: Evidence from Italy. *Tourism Management, 30*(6), 878–889.

Bilen, M., Yilanci, V., & Eryüzlü, H. (2015). Tourism development and economic growth: A panel Granger causality analysis in the frequency domain. *Current Issues in Tourism,* 1–6.

Branis, M., & Linhartova, M. (2012). Association between unemployment, income, education level, population size and air pollution in Czech cities: Evidence for environmental inequality? A pilot national scale analysis. *Health & Place, 18*(5), 1110–1114.

Brida, J. G., Cortes-Jimenez, I., & Pulina, M. (2016). Has the tourism-led growth hypothesis been validated? A literature review. *Current Issues in Tourism, 19*(5), 394–430.

Chen, H., Jia, B., & Lau, S. S. Y. (2008). Sustainable urban form for Chinese compact cities: Challenges of a rapid urbanized economy. *Habitat International, 32,* 28–40.

Chen, W., & Lei, Y. (2018). The impacts of renewable energy and technological innovation on environment-energy-growth nexus: New evidence from a panel quantile regression. *Renewable Energy, 123,* 1–14.

Cole, M. A., & Neumayer, E. (2004). Examining the impact of demographic factors on air pollution. *Population and Environment, 26*(1), 5–21.

De Vita, G., Katircioglu, S., Altinay, L., Fethi, S., & Mercan, M. (2015). Revisiting the environmental Kuznets curve hypothesis in a tourism development context. *Environmental Science and Pollution Research, 22*(21), 16652–16663.

Destek, M. A., Ulucak, R., & Dogan, E. (2018). Analyzing the environmental Kuznets curve for the EU countries: The role of ecological footprint. *Environmental Science and Pollution Research, 25*(29), 29387–29396.

Destek, M. A. (2019). Financial development and environmental degradation in emerging economies. In *Energy and environmental strategies in the era of globalization* (pp. 115–132). Cham: Springer.

Destek, M. A., & Sarkodie, S. A. (2019). Investigation of environmental Kuznets curve for ecological footprint: The role of energy and financial development. *Science of the Total Environment, 650,* 2483–2489.

Destek, M. A., & Aslan, A. (2020). Disaggregated renewable energy consumption and environmental pollution nexus in G-7 countries. *Renewable Energy, 151,* 1298–1306.

Destek, M. A., & Sinha, A. (2020). Renewable, non-renewable energy consumption, economic growth, trade openness and ecological footprint: Evidence from organisation for economic co-operation and development countries. *Journal of Cleaner Production, 242,* 118537.

Dickinson, J. E., Lumsdon, L. M., & Robbins, D. (2011). Slow travel: Issues for tourism and climate change. *Journal of Sustainable Tourism, 19,* 281–300.

Do, T., & Dinh, H. (2020). Short-and long-term effects of GDP, energy consumption, FDI, and trade openness on CO_2 emissions. *Accounting, 6*(3), 365–372.

Dogan, E., & Turkekul, B. (2016). CO_2 emissions, real output, energy consumption, trade, urbanization and financial development: Testing the EKC hypothesis for the USA. *Environmental Science and Pollution Research, 23*(2), 1203–1213.

Dogan, E., Seker, F., & Bulbul, S. (2017). Investigating the impacts of energy consumption, real GDP, tourism and trade on CO_2 emissions by accounting for cross-sectional dependence: A panel study of OECD countries. *Current Issues in Tourism, 20*(16), 1701–1719.

Dumitrescu, E. I., & Hurlin, C. (2012). Testing for Granger non-causality in heterogeneous panels. *Economic Modelling, 29,* 1450–1460.

Durbarry, R. (2004). Tourism and economic growth: The case of Mauritius. *Tourism Economics, 10*(4), 389–401.

Durbarry, R., & Seetanah, B. (2015). Assessing the impact of tourism and travel on climate change. *Journal of Hospitality Marketing & Management, 24*(4), 401–410.

Ehrhardt-Martinez, K., Crenshaw, E. M., & Jenkins, J. C. (2002). Deforestation and the environmental Kuznets Curve: A cross-national investigation of intervening mechanism. *Social Science Quarterly, 83*(1), 226–243.

Eyuboglu, K., & Uzar, U. (2019). The impact of tourism on CO_2 emission in Turkey. *Current Issues in Tourism,* 1–15.

Eyuboglu, S., & Eyuboglu, K. (2020). Tourism development and economic growth: An asymmetric panel causality test. *Current Issues in Tourism, 23*(6), 659–665.

Gao, J., Xu, W., & Zhang, L. (2019). Tourism, economic growth, and tourism-induced EKC hypothesis: Evidence from the Mediterranean region. *Empirical Economics,* 1–23.

Grossman, G. M., & Kruger, A. B. (1991). *Environmental impacts of the North American free trade agreement.* NBER Working Paper 3914, 1–39.

Hall, D. R. (1998). Tourism development and sustainability issues in Central and Southeastern Europe. *Tourism Management, 19*(5), 423–443.

He, J., & Richard, P. (2010). Environmental Kuznets Curve for CO_2 in Canada. *Ecological Economics, 69*(5), 1083–1093.

International Energy Agency. (2020). https://www.iea.org/.16.06.2020.

Jha, S. K. (1997). The Kuznets curve: A reassessment. *World Development, 24,* 773–780.

Jones, D. W. (1991). How urbanization affects energy-use in developing countries. *Energy Policy, 19*(7), 621–630.

Katircioglu, S. T. (2014a). Testing the tourism-induced EKC hypothesis: The case of Singapore. *Economic Modelling, 41,* 383–391.

Katircioglu, S. T. (2014b). International tourism, energy consumption, and environmental pollution: The case of Turkey. *Renewable and Sustainable Energy Reviews, 36,* 180–187.

Khan, A., Chenggang, Y., Hussain, J., Bano, S., & Nawaz, A. (2020). Natural resources, tourism development, and energy-growth-CO_2 emission nexus: A simultaneity modeling analysis of BRI countries. *Resources Policy, 68,* 101751.

Koçak, E., Ulucak, R., & Ulucak, Z. Ş. (2020). The impact of tourism developments on CO_2 emissions: An advanced panel data estimation. *Tourism Management Perspectives, 33,* 100611.

Lee, J. W., & Brahmasrene, T. (2013). Investigating the influence of tourism on economic growth and carbon emissions: Evidence from panel analysis of the European Union. *Tourism Management, 38,* 69–76.

Li, K., & Lin, B. (2015). Impacts of urbanization and industrialization on energy consumption/CO_2 emissions: Does the level of development matter? *Renewable and Sustainable Energy Reviews, 52,* 1107–1122.

Lin, B., & Zhu, J. (2017). Energy and carbon intensity in China during the urbanization and industrialization process: A panel VAR approach. *Journal of Cleaner Production, 168,* 780–790.
Liu, X., & Bae, J. (2018). Urbanization and industrialization impact of CO_2 emissions in China. *Journal of Cleaner Production, 172,* 178–186.
Liu, X., Zhang, S., & Bae, J. (2017). The nexus of renewable energy-agriculture-environment in BRICS. *Applied Energy, 204,* 489–496.
Marin, D. (1992). Is the export-led hypothesis valid for industrialized countries? *Review of Economics and Statistics, 74,* 678–688.
Murthy, K. V., & Gambhır, S. (2018). Analyzing environmental Kuznets Curve and pollution haven hypothesis in India in the context of domestic and global policy change. *Australasian Accounting, Business and Finance Journal, 12*(2), 134–156.
Martínez-Zarzoso, I., & Maruotti, A. (2011). The impact of urbanization on CO_2 emissions: Evidence from developing countries. *Ecological Economics, 70*(7), 1344–1353.
Naradda Gamage, S. K., Hewa Kuruppuge, R., & Haq, I. (2017). Energy consumption, tourism development, and environmental degradation in Sri Lanka. *Energy Sources, Part B: Economics, Planning and Policy, 12*(10), 910–916.
Naz, S., Sultan, R., Zaman, K., Aldakhil, A. M., Nassani, A. A., & Abro, M. M. Q. (2019). Moderating and mediating role of renewable energy consumption, FDI inflows, and economic growth on carbon dioxide emissions: Evidence from robust least square estimator. *Environmental Science and Pollution Research, 26*(3), 2806–2819.
Newman, P. W. G., & Kenworthy, J. R. (1989). *Cities and automobile dependence: An international sourcebook.* Aldershot: Gower Technical.
Oh, C.-O. (2005). The contribution of tourism development to economic growth in the Korean economy. *Tourism Management, 26*(1), 39–44.
Omri, A., Daly, S., Rault, C., & Chaibi, A. (2015). Financial development, environmental quality, trade and economic growth: What causes what in MENA countries. *Energy Economics, 48,* 242–252.
Owusu, P., Asumadu, & S.S., 2016. A review of renewable energy sources, sustainability issues and climate change mitigation. *Cogent Engineering, 3,* 1167990. https://doi.org/10.1080/23311916.2016.1167990.
Ozatac, N., Gokmenoglu, K. K., & Taspinar, N. (2017). Testing the EKC hypothesis by considering trade openness, urbanization, and financial development: The case of Turkey. *Environmental Science and Pollution Research, 24*(20), 16690–16701.
Parikh, J., & Shukla, V. (1995). Urbanization, energy use and greenhouse effects in economic development. *Global Environmental Change, 5*(2), 87–103.
Pesaran, M. H. (2004). *General diagnostic tests for cross section dependence in panels.* IZA Discussion Paper, 1240.
Pesaran, M. H. (2006). Estimation and inference in large heterogeneous panels with a multifactor error structure. *Econometrica, 74*(4), 967–1012.
Pesaran, M. H. (2007). A simple panel unit root test in the presence of cross-section dependence. *Journal of Applied Economics, 22*(2), 265–312.
Poumanyvong, P., & Kaneko, S. (2010). Does urbanization lead to less energy use and lower CO_2 emissions? A cross-country analysis. *Ecological Economics, 70*(2), 434–444.
Rafiq, S., Salim, R., & Nielsen, I. (2016). Urbanization, openness, emissions, and energy intensity: A study of increasingly urbanized emerging economies. *Energy Economics, 56,* 20–28.
Rahman, Z. U., Cai, H., & Khattak, S. I. (2019). Energy production-income-carbon emissions nexus in the perspective of NAFTA and BRIC nations: A dynamic panel data approach. *Economic Rersearch, 32,* 3378–3391.
Reppelin-Hill, V. (1998). Trade and environment: An empirical analysis of the technology effect in the steel industry. *Journal of Environmental Economics and Management, 38,* 283–301.
Rondinelli, D., & Berry, M. (2000). Multimodal transportation, logistics, and the environment: Managing interactions in a global economy. *European Management Journal, 18*(4), 398–410.

Saidi, K., & Mbarek, M. B. (2017). The impact of income, trade, urbanization, and financial development on CO_2 emissions in 19 emerging economies. *Environmental Science and Pollution Research, 24*(14), 12748–12757.

Sarkodie, S. A., & Strezov, V. (2018). Assessment of contribution of Australia's energy production to CO_2 emissions and environmental degradation using statistical dynamic, 144. Sarkodie, S. A., & Strezov, V. (2019). *Science of the Total Environment, 649*, 128–145 approach. *Science of the Total Environment,* 639, 888–899.

Sghaier, A., Guizani, A., Ben Jabeur, S., & Nurunnabi, M. (2018). Tourism development, energy consumption and environmental quality in Tunisia, Egypt and Morocco: A trivariate analysis. *GeoJournal, Wttc, 2017,* 1–17.

Shahbaz, M., Loganathan, N., Muzaffar, A. T., Ahmed, K., & Jabran, M. A. (2016). How urbanization affects CO_2 emissions in Malaysia? The application of STIRPAT model. *Renewable and Sustainable Energy Reviews, 57,* 83–93.

Shaheen, A., Sheng, J., Arshad, S., Salam, S., & Hafeez, M. (2020). The dynamic linkage between income, energy consumption, urbanization and carbon emissions in Pakistan. *Polish Journal of Environmental Studies, 29*(1).

Sharma, S. S. (2011). Determinants of carbon dioxide emissions: Empirical evidence from 69 countries. *Applied Energy, 88,* 376–382.

Shi, H., Li, X., Zhang, H., Liu, X., Li, T., & Zhong, Z. (2019). Global difference in the relationships between tourism, economic growth, CO_2 emissions, and primary energy consumption. *Current Issues in Tourism*, 1–16.

Simpson, M. C., Gössling, S., Scott, D., Hall, C. M., & Gladin, E. (2008). Climate change adaptation and mitigation in the tourism sector: Frameworks, tools and practices. In: *Climate change adaptation and mitigation in the tourism sector: Frameworks, tools and practices.*

Sokhanvar, A., Çiftçioğlu, S., & Javid, E. (2018). Another look at tourism-economic development nexus. *Tourism Management Perspectives, 26,* 97–106.

Solarin, S. A. (2014). Tourist arrivals and macroeconomic determinants of CO_2 emissions in Malaysia. *Anatolia, 25*(2), 228–241.

Tarancón, M. A., & del Río, P. (2007). A combined input–output and sensitivity analysis approach to analyse sector linkages and CO_2 emissions. *Energy Economics, 29,* 578–597.

Tiwari, A. K., Shahbaz, M., & Hye, Q. M. A. (2013). The environmental Kuznets Curve and the role of coal consumption in India: Cointegration and causality analysis in an open economy. *Renewable Sustainable Energy Review, 18,* 519–527.

Tovar, C., & Lockwood, M. (2008). Social impacts of tourism: An Australian regional case study. *International Journal of Tourism Research, 10*(4), 365–378.

Tugcu, C. T. (2014). Tourism and economic growth nexus revisited: A panel causality analysis for the case of the Mediterranean Region. *Tourism Management, 42,* 207–212.

United Nations Department of Economic and Social Affairs. (1999). *Tourism and sustainable development: The global importance of tourism background paper.* https://sustainabledevelopment.un.org/content/documents/398wttc.pdf, June 14, 2020.

UNWTO. (2019). World Tourism Organization. *International Tourism Highlights*, 2019 Edition https://www.e-unwto.org/doi/pdf/10.18111/9789284421152, May 24, 2020.

Wang, Y., Zhang, X., Kubota, J., Zhu, X. & Lu, G. (2015). A semi-parametric panel data analysis on the urbanization-carbon emissions nexus for OECD countries. *Renewable and Sustainable Energy Reviews,* 704–709.

Westerlund, J. (2008). Panel cointegration tests of the fisher effect. *Journal of Applied Econometrics, 23,* 193–233.

WTTC. (2019). *Travel & tourism regional performance.* https://wttc.org/Research/Economic-Impact, May 24, 2020.

Xu, B., & Lin, B. (2015). How industrialization and urbanization process impacts on CO_2 emissions in China: Evidence from nonparametric additive regression models. *Energy Economics, 48,* 188–202.

York, R. (2007). Demographic trends and energy consumption in European Union Nations, 1960–2025. *Social Science Research, 36*(3), 855–872.

York, R., Rosa, E. A., & Dietz, T. (2003). Footprints on the earth: The environmental consequences of modernity. *American Sociological Review*, 279–300.

Zhang, C., & Lin, Y. (2012). Panel estimation for urbanization, energy consumption and CO_2 emissions: A regional analysis in China. *Energy Policy, 49,* 488–498.

Zhang, L., & Gao, J. (2016). Exploring the effects of international tourism on China's economic growth, energy consumption and environmental pollution: Evidence from a regional panel analysis. *Renewable and Sustainable Energy Reviews, 53,* 225–234.

Chapter 10
Sustainable Tourism Production and Consumption as Constituents of Sustainable Tourism GDP: Lessons from a Typical Index of Sustainable Economic Welfare (ISEW)

Angeliki N. Menegaki

Abstract Based on the New Economics trend, only sustainable income should be regarded as a genuine income. The conventionally measured income through GDP is neither sustainable nor genuine. The Index of Sustainable Economic Welfare (ISEW) incorporates all aspects of income generation and or income destruction in a triple-level consideration: economy, environment and society. In this chapter, I propose transferring this logic to the measurement of tourism income, as part of national GDP. Many countries boast high percentages of tourism GDP, with subsequent direct, indirect and induced effects. However, there is a question of how much of that income is sustainable and genuine and how much cost that income incurs during the process of its generation and consumption. This chapter attempts transferring the paradigm of the ISEW as a proxy for sustainable GDP into a tourism ISEW as a proxy for sustainable tourism GDP.

Keywords Defensive tourism expenses · Genuine tourism generated welfare · Sustainable income · Sustainable tourism · Tourism ISEW

10.1 Introduction

This chapter aims to conceptualize tourism ISEW. Before presenting that, it is useful to provide some background knowledge on how we got from GDP to the ISEW in mainstream economics. GDP was invented after the Great Depression, and up to date has been widely used as a smart policy tool by economists and politicians. However, it does not distinguish welfare improving activity from welfare reducing activity (Talbreth et al. 2007). To explain this phrase, I provide an example: Fast food chains sales increase GDP, but they create obesity which reduces our quality of life.

A. N. Menegaki (✉)
Department of Economics and Management of Tourist Units, Agricultural University of Athens, 75 Iera Odos st, Athens, Greece
e-mail: amenegaki@aua.gr

Open University of Cyprus, Latsia, Nicosia, Cyprus

D. Balsalobre-Lorente et al. (eds.), *Strategies in Sustainable Tourism, Economic Growth and Clean Energy*, https://doi.org/10.1007/978-3-030-59675-0_10

Moreover, GDP does not include, for example, the positive value created from "Do-it-yourself" activities. It can include only official financial or marketed transactions. With the way GDP has been used up to date, it has confounded growth with development (Costanza et al. 2009), or prosperity with growth (Jackson 2012). Bergh (2009) evaluates the reasons why researchers acknowledge the criticism against GDP but do not accept its relevance, and therefore, continue to use in their analyses. This is a paradox described in his paper. Moreover, green accounting has been first applied by Costanza et al. (1997). These authors stressed the importance of incorporation of ecosystem values in conventional economic accounting. Nevertheless, sustainability concepts had been expressed and defined in much earlier bibliography. For example, Kuznet (1934), had first objected to the welfare of a nation being decided only by its GDP, because the latter measures together assets and consumer goods, using values that are based on the existing distribution of income, while also failing to include intangibles such as negative or positive externalities. Referring to the same value of intangibles, Nordhaus and Tobin (1972), had reported the existence of activities beyond market transaction that also affected human and economic welfare and developed the Measure of Economic Welfare (MEW) as the forerunner of later measures of sustainable GDP, namely the ISEW, the GPI or others. Daly and Cobb (1989) had warned that national accounting treated the planet as a business in liquidation, to stress the fact that countries and people did not care about building new values, but only liquidating the existing resources. Daly and Cobb were also the fathers of the Index of sustainable economic welfare index (ISEW), whose tourism counterpart I will present in this chapter.

Therefore, the need to replace or complement GDP (with environmental and social indicators) was born. Using different measures of economic progress can have a profound impact on any policy. This need is particularly felt during turbulent economic times, such as the current ones due to the COVID-19 pandemic outbreak. Goossens et al. (2007) divide the GDP indicators into three groups: those replacing, those supplementing and those adjusting GDP. Also, Boyd (2007), suggests a green GDP which will account only for the nonmarket benefits of nature excluded the services and goods measured in common GDP. Next, Fig. 10.1 shows the difference between current GDP from other more comprehensive progress measures which are discussed next.

(i) GDP comprises (a) personal consumption (durables, non-durables and services), (b) investment (business investments, construction, changes in business inventories), government spending (without transfer payments) and net exports of goods and services.
(ii) GPI—another measure of economic well-being that accounts for both the benefits and costs of economic production- is an improved version of ISEW, that started appearing interchangeably with ISEW, mostly in the 1990s (Lawn 2013). Besides all the economic constituents described in GDP, it also includes social and environmental constituents. Social constituents for the ISEW or GPI are divided to those that add positive value (like assets in a balance sheet) and those which add negative value (like liabilities in a balance sheet). In the

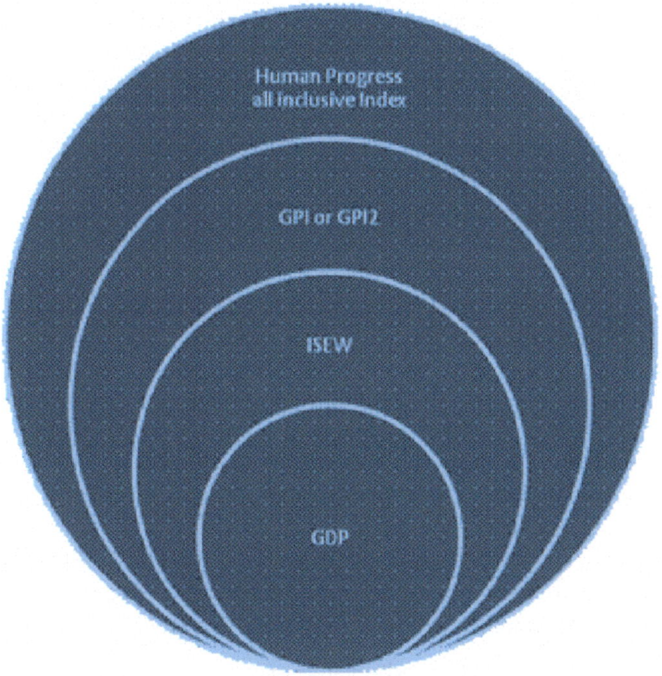

Fig. 10.1 Schematic representation of a Human Progress All Inclusive Index (HPAII). *Source* Authors' configuration

first group belong household labour or volunteer labour, while in the second group, namely with a negative nuance, belong vehicle crashes, commuting, lost leisure, underemployment, crime and family breakdown. Eleven years after Lawn's work (2003), who had provided a list of all the items used to calculate the GPI for the USA, Bagstad et al. (2014), suggest that an improvement of GPI should be made, and this is what is called GPI.2 by them. In GPI.2 they suggest the inclusion of underemployment with a negative sign.

(iii) Environmental constituents for the ISEW or GPI on the other hand, encompass wetland ecosystem services, forest services and farmland services, all with a positive sign. With a negative sign, the following parameters are suggested: pollution abatement, water pollution, air pollution, noise pollution, climate change, ozone depletion and non-renewable resource use. For GPI2 Bagstad et al. (2014), also suggest the inclusion of the cost of extreme events, cost of replacing conventional energies with renewable energies, cost of water scarcity, depletion of mining materials and groundwater, importation of hazardous materials, as well as with a negative sign within the environmental constituents. They also suggest eliminating the ozone depletion parameter and provide for both quality and quantity dimensions in services from wetlands, forests and farmland.

(iv) Last, the economic parameters in the ISEW and the GPI also include income inequalities, while the inventors of GPI2 also suggest some transfer payments with particular care though to avoid double counting. Needless to mention that there are dozens of additional economic, social and environmental parameters and variations of them that can form different variations of the ISEW type indexes. Suppose we could include all of them in one Index that could form the Human Progress all-inclusive Index (HPAII).

(v) The most significant hindrance for the implementation of the above-presented progress indexes is that the data required to build them remain at a theoretic level for the majority of countries. Namely, most national statistical agencies have not compiled such datasets yet. In addition to this, different countries and regions, depending on the particular conditions in them, estimate modified social and environmental parameters depending on local issues of relevance. Bagstad et al. (2014) recognize that a more formal process is required through which to incorporate the Genuine Progress Index (GPI) into mainstream economic policy, and this will not affect comparability across countries. Also, most GDP accounting will stop relying on ad hoc measures. Last but not least, Lawn (2013), beware that both the ISEW and GPI are destined to measure welfare but not sustainability. To achieve the latter, we need to include biophysical indicators.

Tourism causes many positive and negative externalities that should be taken into account when calculating the income and welfare generated by tourism activities. The most important aspects pertinent to this situation are the allocation of benefits and income generated from tourism that is not equal and fair for all social groups. The damages of the natural environment are most of the times irreversible. The damages of the social environment may sometimes take significant dimensions, and they may cause a perpetual alienation of local societies. All these aspects may require defensive expenditures which should be derived from the generated income. Particularly the environmental damages require allowances that compensate for them. Net investment should be taken into consideration because tourism infrastructure becomes old and needs to be renewed. Household work and volunteer work should not be underestimated particularly when many tourism businesses are family ones and its members may not be paid with money and proper transactions.

The rest of this chapter is structured as follows: After this introduction which motivates the paper and explains the conceptualization of the ISEW and other relevant indexes, I continue with part 2 which adds to the background material that is useful for a better understanding of the concept and its history, as well as other supplementary material, that can be used in ISEW accounting and sustainability accounting. Part 3 presents the detailed ISEW methodology and Part 4 explains the parallels to tourism GDP. Last, Part 5 offers some concluding remarks.

10.2 Some Stylized Facts for High-Income Countries

10.2.1 More on Sustainable Wealth Indexes

This section adds to the supplementary background material which strengthens the understanding of readers about the ISEW, the reasons why it was born and some basic sustainability accounting tools and concepts that go together. The section also provides a historical overview of all relevant indexes together with a summary of values for statistical life which, strictly speaking, should be perused in sustainability accounting and cases of cost-benefit analysis. To start with, there is a misunderstanding that rich countries are rich in all aspects. There is also a common belief that in rich countries the consumption needs of all or most people are covered and everybody is leading a happy life. The facts, however, show a different reality: For example, rich Scandinavian countries have the highest suicide rates (Szalavitz 2011). Alexander (2011) attributes the high suicide rates in the USA to the materialism which reigns in American society. Young people are not taught the principles of hard work and ethics. They have been accustomed to model lives shown on TV (which for most Americans would mean to live beyond their means) and they wish to have wealthy lives with little personal effort. As far as divorce rates are concerned, Sweden is the 3rd country with the highest divorce rates worldwide, Belgium is 7th, Finland is 8th, the UK is 10th and the USA is 12th (Divorce.com 2014). Income inequality is an aspect that is usually overseen when one is using GDP accounting and following statements about the wealth of a country's citizens. Inequality hurts health and consequently, on economic growth. A plausible explanation for that is that it increases people's stress and anxiety to reach a certain status in society (Rowlingson 2011). Overall, income inequality is dysfunctional because it slows economic growth; it results in both health and social problems, generates political instability and leads to severe inequalities, particularly among children (Ortiz and Cummins 2011). Some of the richest countries host some of the poorest people. USA was reported with the highest poverty rate in households with children, Italy 3rd and UK 4th followed by Canada, Germany and Belgium (Smeeding 2006). However, given the orientation of the so-called "New Economics" on genuine progress and sustainable economic welfare, we suggest that the concept of tourism GDP is myopic because it does not tell us what the genuine effect and contribution of tourism on sustainable economic welfare is. To further elaborate this crucial rationale, we mean to suggest that: The GDP of each economy has a different structure and has been produced with ways that may bear different repercussions on human well-being. For example, a highly industrialized country may have produced too much pollution, may have induced extreme urban sprawling with low quality of life, family disintegration caused by the increased working hours of the labour force and the list of the negative consequences is endless. On the other hand, a less developed country, probably with a lower GDP per capita may enjoy a cleaner environment, tighter human bonds, less family breakdown and overall may consist of happier people. Moreover, an industrialized country produces more harm to the environment than a country that produces services. Construction, petrochemical activities or agriculture are typically the most polluting activities in an economy. Correspondence to the above can also be made about tourism GDP. In some countries, tourism is of all-inclusive type and in some others, it is of alternative form, ecological, slow or it generally adopts forms which are more environmental and socially friendly. This kind of tourism attracts high-quality clients with more respectful attitudes towards local societies and cultures. Life satisfaction trend is well below GDP/capita trend in major European countries such as France, Ireland, Greece and Portugal. While until the 90s, both in Greece and Portugal, life satisfaction exceeded the GDP/capita trend, after the 90 s, life satisfaction has fallen dramatically (Heinz-Herbert 2006). According to a research by Nolan and Whelan (2009), household deprivation was divided into three types: consumption, housing facilities and neighbourhood environment. While in consumption the most deprived countries seemed to be Hungary, Lithuania, Poland, Latvia and Slovakia, the top housing deprived countries were Estonia, Lithuania and Latvia, while most surprisingly, the top deprived of a neighbourhood environment point of view were Latvia, Cyprus and Germany followed by UK, Portugal, Netherlands, Italy, Spain, Belgium and Estonia. The least deprived on this parameter appeared to be Sweden. To further stress the importance of psychological, non-material wealth, UNICEF (2007), in a study that assesses children's well-being, ranks Sweden and Finland among the highest total scores. However, the score consists of several parameters: material well-being, health and safety, educational well-being, family and peer relationships, behaviours and risks and subjective well-being. As regards family and peer relationships, Sweden and Finland achieve the lowest ranks, while Italy (with a middle total rank) and Portugal (with one of the lowest total ranks) achieve the highest scores, thus revealing that well-being is not a single-dimensional thing, but rather a multi-faceted situation where a country should aim to achieve the highest score in all dimensions. Life expectancy was found to be unrelated to spending on health care in rich countries. Homicides and longer working hours are more common in countries with higher inequalities. Countries with less inequality are more innovative and recycle their waste more (Wilkinson and Pickett 2006).

Posner and Costanza (2011) provide a summary-review of studies till 2008, that use the ISEW and GPI measure and the interested reader should turn to that for an overview. The following table shows the latest studies, namely from 2009 onwards.

Based on Table 10.1, we observe that there are several studies available for specific years or data-spans. For example, with an interest in Portugal and the USA, Beça and Santos (2014a), compare both the GDP and ISEW measures of economic welfare and show that the ISEW is more enlightening when it comes to aspects such as resource use intensity and decoupling. Their results were not insensitive to the measure used each time. Another example could be provided by Kubiszewski et al. (2013), who find GPI to be a far better approximation of economic welfare than GDP, although GPI itself is not the perfect economic welfare indicator. Particularly they find that while global GDP has tripled since 1950, GPI has decreased since 1978. Also, Li and Fang (2014), peruse an integrated method with geographic information systems and a comprehensive dataset and create a synthetic global and national green GDP maps. For reasons of space consideration in this chapter, we will not describe in more

Table 10.1 Studies from 2009 onwards with countries for which ISEW, GPI or other measures of sustainable GDP have been estimated

	Study	Countries	Method	Years
1	Greasley et al. (2014)	UK	Genuine savings	1760
2	Condon and Tsigaris (2003)	Canada	Genuine wealth growth rates	1997–2009
3	Mota and Domingos (2013)	Portugal	Genuine savings and green net national income	1990–2005
4	Kubiszewski et al. (2013)	17 countries (meta-analysis)	GPI	1950–2003
5	Bagstad and Shammin (2012)	Northeast Ohio	GPI	1990–2005
6	Packard and Chapman (2012)	Wellington region, N. Zealand	GPI	2001–2008
7	Danilishin and Veklich (2010)	Ukraine	GPI	2000–2007
8	Beça and Santos (2014)	Portugal & USA	ISEW	1960–2010
9	Gigliarano et al. (2014)	Regional Italy	ISEW	1999–2009
10	Bagstad et al. (2014)	USA state level	GPI	Various
11	Bleys (2013)	Flanders, Belgium	ISEW	1990–2009
12	Pulselli et al. (2012)	Tuscany, Italy	ISEW	1971–2006
13	Li and Fang (2014)	Global	Green GDP	2009
14	Ferreira and Moro (201)	Ireland	Genuine savings	1995–2005
15	Xu et al. (2010)	Wuyishau, China	Green GDP	2005

Source Author's compilation

detail what each study does. The interested reader can get a good idea of the content of the study if he or she refers to the original papers, but Table 10.1, suffices to know what they are about.

10.2.2 The Value of Statistical Life

When lives are lost in a country, due to an unsuccessful structure of an economy, this should also be taken into account for the calculation of its sustainable wealth. As we will see next, there are many components in the social part of the ISEW which take into account deaths and regard that as a negative aspect of the economy which is causing them. Below, I give a summarized overview of how the value of life is handled and accounted for in economic literature (Table 10.1). This should also be taken into account when we come to the calculation of the sustainable GDP for tourism.

There are many methods used to calculate the value of statistical life (Boiteux and Baumstark 2001) by which we mean the amount a representative sample of the population is willing to pay for a policy to save human life and reduce the annual risk of dying from 3 in 10,000 to 2 in 10,000 (World Health Organization 2014). Miller (2000) in a meta-analysis of about 68 studies on 13 countries, uses a model to calculate the value of statistical life for many countries among which 23 European ones for the years 1995 and 1998. He concludes that the values of statistical life are about 120 times higher than the GDP/capita corresponding to each country. At the year of this publication, Miller had found out that there were only 13 countries in which this type of WTP studies had been performed. The European default value for the statistical life is 2,487 m (for the WHO European region), 3,387 m (EU-27 countries) or 3,371 m for EU-countries plus Croatia (World Health Organization 2014).

There are mainly two groups of countries for which we do not have the value of statistical life (Table 10.2 and 10.3). One group is the Balkan countries (Bosnia, Bulgaria, Kosovo, FYROM, Romania, Serbia, Slovakia and Slovenia). The other group is former Soviet Union countries (Estonia, Latvia, Lithuania, Ukraine and Croatia). For those countries, we could make some ad hoc assumptions. For example, If we divide the $GDP/capita in 1997, for each country with the calculated VSL, then we get about 0.14. I will multiply this amount with the 1997 GDP/capita of all the countries with missing VSL, to get an ad hoc estimation of that.

10.3 The Methodology of the Tourism ISEW Based on the Conventional ISEW

This section describes the three parts of the construction of the conventional ISEW and draws the parallels to the compilation of the tourism ISEW.

Welfare is an ambiguous and multi-faceted concept. Therefore, a composite indicator is needed to reflect it. Understandably, some of the welfare dimensions are tangible, and some are intangible. Intangible ones are mainly psychological parameters that eventually contribute to happiness and well-being. Intangible ones are more difficult to calculate than tangible ones. There are means to calculate intangibles, such as revealed or stated preference techniques. But even if they have been calculated in one country, there is not any institutional framework to oblige or enable other countries to calculate them too. Hence, unless commonly accepted calculation means are established, cross country comparisons cannot be made, neither in the conventional ISEW nor in the tourism ISEW. This makes difficult the calculation of a complete ISEW which can host all possible parameters affecting the well-being. However, depending on the degree of institutional progress of the country, some countries have had more progress in sophisticated statistical data keeping, while others have not. The first ISEW was produced by Daly and Cobb in 1989, for the US and then was improved in 1994. The ISEW conception has many supporters and many opponents.

Table 10.2 The values of statistical life in Europe

Country	$1997 GDP/capita	Best estimate of VSL (000 $)	Values from new studies (2000 onwards, €)
EU mean	20,714	2,730	
Austria	24,418	2,680	3,021,948 (Leiter and Pruckner 2009)
Belgium	22,824	3,000	
Czech Republic	4,839	680	
Denmark	30,834	3,990	2,651,682 (Desaigues et al. 2007)
Finland	22,340	2,930	
France	22,795	2,990	
Germany	24,406	3,190	
Greece	10,950	1,490	
Hungary	4,275	610	
Ireland	19,194	2,540	
Italy	19,081	2,520	3,598,485 (Alberini and Chiabai 2007)
NRL	22,307	2,930	
Norway	33,360	4,300	
Poland	3,362	480	795,082 (Giergiczny 2008)
Portugal	9,758	1,330	
Russia	2,556	370	
Spain	12,965	1,750	
Sweden	24,670	3,230	7,693,884 (Svensson 2009)
Switzerland	34,397	4,430	4,362,827 (Rheinberger 2009)
UK	28,206	3,670	

Source Adapted from Miller (2000) and OECD (2012)

The Index has been criticized for the fact that it measures welfare and sustainability together (Neumayer 2000), and for the methodological treatment of flows and stocks (Beça and Santos 2010). Responses to the former criticism support that the ISEW indicator is an aggregate indicator for both current and future well-being. Future well-being is an aspect of utility for the current generation, because of the satisfaction they receive from knowing they will not damage the utility of their offspring (Cobb and Cobb 1994). This is further elaborated by Lawn (2003), who draws principles from Irving Fisher's "net psychic income" and thus supports why each item in the ISEW contributes to the psychic income. Regardless of the hesitations posed by the ISEW opposers, the current ISEW is better than nothing (Lawn and Clarke 2008), in the sense that it does a good job but not a perfect one. Or as reported in Posner and Costanza (2011), it is better to be approximately right than perfectly wrong.

Table 10.3 Countries with missing VSL

Country	GDP/capita $ current (1)	GDP/capita 1997$ (2)	Best estimate of VSL (000 $)	VSL $
Bosnia	1038 (*0.93)	965 (*0.14)	135,1	135,100
Bulgaria	1210	1125	157,5	157,500
Kosovo	–	–	–	–
Macedonia FYROM	1875	1743	244,02	244,020
Romania	1565	1455	203,7	203,700
Serbia	2738	2546	356,44	356,440
Slovakia	5023	4671	653,94	653,940
Slovenia	10282	9562	1342	1,342,000
Cyprus	13277	12347	1728,58	1,728,580
Croatia	5140	4780	669,2	669,200
Estonia	3609	3356	469,84	469,840
Hungary	4522	4205	588,7	588,700
Iceland	27378	25461	3564,54	3,564,540
Latvia	2521	2344	328,16	328,160
Lithuania	2833	2634	368,76	368,760
Ukraine	991	9216	1290,24	1,290,240
Malta	9683	9005	1260,7	1,260,700
LXB	44140	41050	5747	5,747,000

Source Authors calculation based on Miller (2000)

The GPI and the ISEW have minor differences between then and this is the reason they are often used interchangeably (Fig. 10.2). Figure 10.1 shows the components of the GPI with a sign in front of them. The Index is divided into three major parts (economic, environmental and social) similarly to the ISEW. Then, Table 10.4, shows the components of the ISEW.

10.3.1 A Simplified ISEW Version Upon Data Availability

Given the fact that the ISEW as suggested theoretically, consists of many components (Table 10.3), that may not be directly available from statistical agencies across the different countries, a simplified version of the ISEW has been recommended in literature and has been used in many applications (Menegaki and Tugcu 2017; Menegaki and Tiwari 2016; Menegaki et al. 2017; Menegaki and Tsagarakis 2015; Menegaki 2018). The formal expression of the ISEW is described in Eq. 10.1.

$$ISEW = C_w + G_{eh} + K_n + S - N - C_s \quad (10.1)$$

Fig. 10.2 The constituent parts of the GPI. *Source* Berik and Gaddis (2011), Bagstad et al. (2014)

where Cw is the weighted consumption, Geh stands for non-defensive public expenditure, Kn is the net capital growth, S is the unpaid work benefit, N is the depletion of the natural environment and Cs is the cost from social problems, which has not been measured in the current calculations due to lack of data. We understand that environmental or ecological degradation involves many more problems, for which, few data are available which are not comparable across countries. For example, the cost of water pollution or the cost of the loss of land and wetlands is not available in the publicly available official databases such as Eurostat, OECD and World Bank. The same applies to the lack of social data. The inability to include costs from social problems leads to a simplification of Eq. 10.1 into Eq. 10.2

$$\text{ISEW} = C_w + G_{eh} + K_n + S - N \tag{10.2}$$

The approach in Eqs. 10.1 and 10.2, is also suggested in Pulselli et al. (2012), Gigliarano et al. (2014), Menegaki and Tsagarakis (2015) and other literature as aforementioned (Tables 10.5 and 10.6).

Tourism is one of the sectors that build the GDP and equivalently one of the sectors that build the ISEW. One part of the ISEW that is attributed solely to tourism,

Table 10.4 Summary of the ISEW methodology

Item and Impact	Description/ Brief methodology
+Adjusted private consumption expenditures	Private consumption adjusted to take account of income inequality (Gini) and poverty (headcount ratio). Formula: (Private consumption expenditures) × (adjustment index)
+Services from domestic labor	Value of unpaid domestic work. Formula: (domestic labor hours) × (categories of population in labor age) × (household labor estimated wage)
+Services from durable consumer goods	Value of the services provided annually by the durable goods, net of the expenditure for their purchase. The service is estimated from total stock of durables, annual depreciation and real interest rates.
+Public expenditures on health and education	Part of public expenditure useful to increase well-being (health, education) and not to restore a deteriorated situation (defensive expenditure)
−Costs of commuting	Cost of time-use and transport expenses for repeated travel for work. Direct cost of transport and opportunity cost of time spent
−Costs of car accidents	Material costs, moral costs and costs due to the loss of production caused by car accidents
−Costs of water pollution	Costs caused by human pressure on the water. Depuration for each equivalent inhabitant
−Costs of air pollution	Costs arising from the emission of each pollutant. Formula: (economic value per pollutant unit) × (annual pollutant emission)
−Costs of noise pollution	Costs due to noise pollution (willingness to pay for reduction of noise)
−Loss of natural and agricultural land	Costs due to the loss of natural areas. Formula: (economic value per area unit) × (annual pollutant emission)
−Depletion of non-renewable resources	Depletion of non-renewable resource. Formula: (fossil fuel consumption) × (substitution cost)
Long-term environmental damage	Costs arising from environmental damage with long-term consequences. Formula: (greenhouse gas emissions) × (marginal abatement cist)
+Net capital growth	Change in the stock of capital net of the budget needed for the new workers
+Net balance of payments	Export-Import Balance
=ISEW	

Source Adapted from Gigliarano et al. (2014)

Table 10.5 The construction of a simplified ISEW for Greece based upon data availability

Type of component	Variables	Computation
Benefits	(+) Adjusted private consumption	Personal consumption adjusted for income inequality. It is also adjusted for the value of durable products
	(+) Public expenditure on education and health	Since some of this expenditure is defensive, we follow Jackson and Stymne (1996) and include only half of this amount
	(+) Services from unpaid family workers	We multiply the percentage of unpaid employment with total employment and then with the basic annual wages (Eurostat 2014)
Benefits/Costs	(±) Net capital growth	We use gross capital minus the consumption of fixed capital and calculate its growth rate
Costs (environmental)	(−) Mineral depletion	Is the ratio of the stock of mineral resources to the remaining reserve lifetime (capped at 25 years)
	(−) Energy depletion	Is the ratio of the stock of mineral resources to the remaining reserve lifetime (capped at 25 years)
	(−) Long-term environmental damage from carbon emissions	The number of tons of emitted carbon is multiplied with $20 per ton (This is the unit damage in 1995 US$)
	(−) Cost of local pollution	It is estimated as WTP to avoid mortality attributable to particulate emissions Pandey (2014)
Costs (social)	(−) Cost of divorces	We multiply the divorce number per 10,000 inhabitants and then by the price of consensual divorce, i.e. 800 € (Kathimerini 2014). I have also used the British annual cost of family dissolution (http://www.centreforsocialjustice.org.uk/policy/pathways-to-poverty/family-breakdown) adapted for Greece and the final consumption expenditure as done in (2008)
	(−) Cost of road fatalities	We multiply the number of road fatalities per million inhabitants with the value of statistical life (Rackwitz 2006; Giannopoulos (2010)
	(−) Cost of suicides	We multiply the number of suicides per 100,000 population with the value of statistical life (Rackwitz 2006; Giannopoulos 2010)

(continued)

Table 10.5 (continued)

Type of component	Variables	Computation
	(−) Cost of noise pollution	We multiply the percentage of total population who state that suffer from noise from neighbors or from streets with the population and I multiply the suffering population with 137.2 € per person Caulfield and O'Mahony (2007) adapted by the final consumption expenditure increase, as done in Noury (2008)

Source Menegaki and Tsagarakis (2015)
Note All data series come from WDI (2014), except for noise pollution: codeilc_mddw01 (Eurostat 2014), suicides and road fatalities (OECD 2013). *Note* Social costs that have not been estimated specifically for this application for Greece in Table 5. For example, the cost of family dissolution has been adapted from other countries where this cost has been estimated. The same was done for the price of statistical life where the price was calculated by Giannopoulos (2010) and adapted for all the time span of our analysis as indicated in Rackwitz (2006) and Giannopoulos (2010)

Table 10.6 The suggested construction of a basic tourism ISEW

Type of component	Variables
Benefits	(+) Adjusted private consumption
	(+) Public expenditure on education and health
	(+) Services from unpaid family workers
Benefits/Costs	(±) Net capital growth
Costs (environmental)	(−) Mineral depletion
	(−) Energy depletion
	(−) Long-term environmental damage from carbon emissions
	(−) Cost of local pollution
Costs (social)	(−) Cost of divorces
	(−) Cost of worker burn-out*
	(−) Cost of road fatalities
	(−) Cost of cultural alienation*
	(−) Cost of suicides
	(−) Cost of commuting and traffic congestion**
	(−) Cost of noise pollution

Source Author's compilation. *Notes* One asterisk denotes components that do not exist in the conventional ISEW. Two asterisks denote that this magnitude exists in the conventional ISEW but in a slightly different form could be regarded as the tourism ISEW. The total economic contribution of tourism to GDP can be identified through the tourism satellite account and with the aid of input-output analysis. Analogous calculations can be made for the calculation of the economic component of the ISEW to reach the amount that corresponds solely to tourism contributions. The environmental and social parts of the tourism ISEW will be the most difficult to calculate for the same reasons they are difficult to calculate for the conventional ISEW. Environmental data are difficult to get, but social data are the hardest. Therefore, when a magnitude has been already calculated for the total conventional ISEW, it is convenient (based on sound assumptions) to end up to the contributing share of tourism and transfer that amount to the tourism ISEW. However, if these magnitudes are not used in the calculation of the total conventional ISEW, or they have never before been calculated for the total conventional ISEW, then they should be directly and exclusively sought for the tourism ISEW.

The cost of workers' burn down is a situation that is due to the seasonality that characterizes the tourism sector. Also, it is since the tourism sector is one of labour intensity and low productivity rate. The fact that wages in tourism follow the general pattern of the ones in the rest of the economy, combined with the lower productivity though leads to the Baumol disease or Baumol effect. Workers in the tourism sector must work hard during the tourism season and this leads to their exhaustion (burn-out). This has a serious effect on workers' health and future capability to work. Thus, it may have an impact on the labour capacity of the sector and the incurred medical costs which are covered by taxpayers.

As far as the cultural alienation is concerned, tourists bring with them habits and behaviours which may not be beneficial for a local community. There are destinations popular among young tourists who become drunk and adopt violent or criminal behaviours. Local people may also adopt these behaviours and become used to types of entertainment and pastimes which were previously unknown. These are not beneficial for the destination and the latter gradually loses its reputation for the rest of the tourist groups. The valuation of cultural alienation is not an easy task and one of the ways it can take place is through stated or revealed preference techniques.

Regarding commuting time, this is included in the ISEW but not in the form of congestion time. These are the two different situations, but commuting time sometimes become longer due to congestion. However, congestion does not presuppose commuting. Congestion can occur in touristic areas and can worsen the lives of local people.

10.4 Conclusion

Nowadays when humanity has reached a critical point of the earth's sustainability, it is high time we considered measures of sustainable income for the measurement and comparison of our well-being with others. The ISEW is a big step forwards, albeit not free of theoretical and more of applied nature problems. The allocation of the total conventional ISEW into the various sectors that constitute the economy, is a major challenge. Tourism is one of these sectors.

Currently ISEW studies, except for the fact that they expand as national applications of acceptable measures of sustainable welfare for different countries, they aim to enrich the ISEW aggregate indicator with more sophisticated variables and measurements. The challenge lies in agreeing upon common measurements across nations even on the most objectively measured components of the ISEW, such as the social variables. Tourism causes many positive and negative externalities that should be taken into account when calculating the income and welfare generated by tourism activities. The most important aspects pertinent to this situation is the allocation of benefits and income generated from tourism is not equal and fair for all social groups. The damages of the natural environment are most of the times irreversible. The damages of the social environment may sometimes take significant dimensions and they may cause a perpetual alienation of local societies. All these aspects may require defensive expenditures which should be derived from the generated income. Particularly the environmental damages require allowances that compensate for them. Net investment should be taken into consideration because tourism infrastructure becomes old and needs to be renewed. Household work and volunteer work should not be underestimated particularly when many tourism businesses are family ones, and its members may not be paid with money and proper transactions. Therefore, different sectors may have to address particular caveats in specific unique components.

We need to place great care for any measure of welfare or sustainability. We must not add together measurements components that are underpinned one by weak and one by strong sustainability. For example, we cannot assume that the same value amount of human capital depreciation equals natural capital depreciation, because the value of human life is different from the value of natural resources. This field of the construction and continuous improvement of the ISEW, as well as its sectoral disaggregation, promises much additional future research.

References

Alberini, A., & Chiabai, A. (2007). Urban environmental health and sensitive populations: How much are the Italians willing to pay to reduce their risks? *Regional Science and Urban Economics, 37*, 239–258.

Alexander, R. (2011). Why are suicides so high in the wealthiest country in the world?, Townhall 27/10/2011. Retrieved 8 Feb 2014, from http://townhall.Com/columnists/rachelalexander/2011/10/27/why_are_suicides_so_high_in_the_wealthiest_country_in_the_world/page/full.

Bagstad, K. J., Berik, G., & Gaddis, E. J. B. (2014). Methodological developments in US state-level genuine progress indicators: Toward GPI 2.0. *Ecological Indicators, 45*, 474–485.

Bagstad, K. J., & Shammin, M. R. (2012). Can the genuine progress indicator better inform sustainable regional progress?—A case study for Northeast Ohio. *Ecological Indicators, 18*, 330–341.

Beça, P., & Santos, R. (2014). A comparison between GDP and ISEW in decoupling analysis. *Ecological Indicators, 46*, 167–176.

Beça, P., & Santos, R. (2010). Measuring sustainable welfare: A new approach to the ISEW. *Ecological Economics, 69,* 810–819.

Bergh, Jeroen C. J. M. van den. (2009). The GDP paradox. *Journal of Economic Psychology* 30, 117–135.

Berik, G., Gaddis, E. (2011). The Utah Genuine Progress Indicator (GPI), 1990–2007: A Report to the People of Utah, Executive Summary. Retrieves 13, June 2020, from https://utahpopulation.org/wp-content/uploads/2014/11/Utah_GPI_Exec_Summary.pdf.

Bleys, B. (2013). The regional Index of sustainable economic welfare for Flanders, Belgium. *Sustainability (Switzerland), 5,* 496–523.

Boiteux, M., & Baumstark, L. (2001). Transports: choix de investissements et cout des nuisances. Mimeo, Commissariat General du Plan.

Boyd, J. (2007). Nonmarket benefits of nature: What should be counted in green GDP? *Ecological Economics, 61,* 716–723.

Caulfield, B., & O'Mahony, M. M. (2007). An examination of the public transit information requirements of users. *Intelligent Transportation Systems. IEEE Transactions, 8*(1), 21–30.

Cobb, C. W., & Cobb, B. (1994). *The Green National Product: a Proposed Index of Sustainable Economic Welfare.* Lanham: University Press of America.

Condon, A. O., & Tsigaris, P. (2013). Genuine wealth growth rates across Canada from 1997 to 2008. *International Journal of Sustainability Policy and Practice, 9,* 34–45.

Costanza, R., D'Arge, R., De Groot, R., Farber, S., Grasso, M., Hannon, B., et al. (1997). The value of the world's ecosystem services and natural capital. *Nature, 387,* 253–260.

Costanza, R., Hart, M., Posner, S., Talberth, J., (2009). Beyond GDP: The need for new measures of progress, the pardee papers, no.4, January. Retrieved 26, July 2014, from http://www.Bu.Edu/pardee/files/documents/pp-004-gdp.Pdf.

Daly, H. E., & Cobb, J. B. J. (1989). *For the common good: Redirecting the economy toward community, the environment and a sustainable future.* Boston: Beacon Press.

Danilishin, B. M., & Veklich, O. A. (2010). Genuine progress indicator as an adequate macroeconomic indicator of public welfare. *Studies on Russian Economic Development, 21,* 644–650.

Desaigues, B., Rabl, A., Ami, D., Kene, B. M., Masson, S., Salomon, M. A., et al. (2007). Monetary value of a life expectancy gain due to reduced air pollution: Lessons from a contingent valuation in France. *Revue d'Economie Politique, 117,* 674–698.

Divorce.com. (2014). Worldwide divorce statistics. Retrieved 2, Aug 2014 http://divorce.Com/worldwide-divorce-statistics/.

Eurostat. (2014). Eurostat database. Retrieved 30, July 2014, from http://epp.eurostat.ec.europa.eu/portal/page/portal/statistics/search_database.

Ferreira, S., & Moro, M. (2011). Constructing genuine savings indicators for Ireland, 1995-2005. *Journal of Environmental Management, 92,* 542–553.

Giannopoulos, G. (2010). Εκτίμηση και αξιολόγηση κοινωνικού ρίσκου στην ναυτιλία (in Greek). Dissertation, National Metsovio Technical University, School of marine architect engineering, Section of ship studies and transport. Retrieved 1, Aug 2014, from http://dspace.lib.ntua.gr/bitstream/123456789/3657/3/giannopoulosi_riskassessment.pdf.

Giergiczny, M. (2008). Value of a statistical life—the case of Poland. *Environmental & Resource Economics, 41,* 209–221.

Gigliarano, C., Balducci, F., Ciommi, M., & Chelli, F. (2014). Going regional: An index of sustainable economic welfare for Italy, Computers. *Environment and Urban Systems, 45,* 63–77.

Goossens, Y., Makipaa, A., Schepelmann, P., van de Sand, I., Kuhndtand, M., Herrndorf, M., (2007). Alternative progress indicators to gross domestic progress (GDP) as a means towards sustainable development. Ip/a/envi/st/2007-10.Brussel, Belgium: Policy department, economic and scientific policy (European Parliament). Retrieved 26, July 2014, from http://www.Pedz.Uni-mannheim.De/daten/edz-ma/ep/07/est19990.Pdf.

Greasley, D., Hanley, N., Kunnas, J., McLaughlin, E., Oxley, L., & Warde, P. (2014). Testing genuine savings as a forward-looking indicator of future wellbeing over the (very) long-run. *Journal of Environmental Economics and Management, 67,* 171–188.

Heinz-Herbert, N., (2006). European survey data: Rich countries for quality of life research, Zuma centre for survey research and methodology, Mannheim, 7th ISQOLS conference "prospects for quality of life in the new millennium" Rhodes University, Crahamstown, 17–20 Juli 2006. Retrieved 2, Aug 2014, from http://www.Gesis.Org/fileadmin/upload/institut/wiss_arbeitsbereiche/soz_indikatoren/publikationen/noll-european-qualityoflife-survey-data.Pdf.

Jackson, T. (2012). *Prosperity without growth: Economics for a finite planet*. London: Routledge.

Jackson, T., & Stymne, S. (1996). Sustainable economic welfare in Sweden-A pilot index 1950–1990. Retrieved 25, July 2014, from http://www.sei-international.org/mediamanager/documents/Publications/Policy-institutions/sustainable_economic_welfare_sweden.pdf.

Kathimerini. (2014). Διαζύγια λόγω διαθεσιμότητας (in Greek). Retrieved 11, Aug 2014, from from:[https://www.kathimerini.gr/49069/article/epikairothta/ellada/diazygia-logw-dia8esimothtas.

Kubiszewski, I., Costanza, C. R., Franco, P., Lawn, J., Talberth, T., & Jackson, Aylmer, C. (2013), Beyond GDP: Measuring and achieving global genuine progress. *Ecological Economics* 93, 57–68.

Kuznet, S., (1934). National income 1929–1932. A report to the US Senate, 73rd congress, 2nd session. Washington DC. US government printing office.

Lawn, P. A. (2003). A theoretical foundation to support the Index of Sustainable Economic Welfare (ISEW), Genuine Progress Indicator (GPI), and other related indexes. *Ecological Economics, 44*(1), 105–118. https://doi.org/10.1016/S0921-8009(02)00258-6.

Lawn, P., & Clarke, M. (2008). (Eds.). *Sustainable Welfare in the Asia-Pacific: Studies Using the Genuine Progress Indicator*. Edward Elgar, Cheltenham, UK.

Lawn, P. A. (2013). The failure of the ISEW and GPI to fully account for changes in human-health capital—A methodological shortcoming not a theoretical weakness. *Ecological Economics, 88,* 167–177.

Leiter, A. M., & Pruckner, G. J. (2009). Proportionality of willingness to pay to small changes in risk: The impact of attitudinal factors in scope tests. *Environmental & Resource Economics, 42,* 169–186.

Li, G., & Fang, C. (2014). Global mapping and estimation of ecosystem services values and gross domestic product: A spatially explicit integration of national 'green GDP' accounting. *Ecological Indicators, 46,* 293–314.

Menegaki, A. N. (2018). The basic, the solid and the full Index of sustainable economic welfare (ISEW) for Turkey. *Environments, 6*(24), 1–18. https://doi.org/10.3390/economies6020024.

Menegaki, A. N., & Tiwari, A. K. (2016). An Index of Sustainable economic welfare in the energy-growth nexus for American countries. *Ecological Indicators, 72*(1), 494–509.

Menegaki, A. N., Marques, A. C., & Fuinhas, J. A. (2017). Redefining the energy-growth nexus with an index for Sustainable Economic Welfare in Europe. *Energy, 141,* 1254–1268.

Menegaki, A. N., & Tsagarakis, K. P. (2015). More indebted than we know? Informing fiscal policy with an ISEW index for Greece. *Ecological Indicators, 57,* 159–163.

Menegaki, A. N., & Tugcu, C. T. (2017). Energy consumption and sustainable economic welfare in G7 countries; A comparison with the conventional nexus. *Renewable and Sustainable Energy Reviews, 69,* 892–901.

Miller, T. R. (2000). Variations between countries in values of statistical life. *Journal of Transport Economics and Policy, 34*(2), 169–188.

Mota, R. P., & Domingos, T. (2013). Assessment of the theory of comprehensive national accounting with data for Portugal. *Ecological Economics, 95,* 188–196.

Neumayer, E. (2000). On the methodology of ISEW, GPI and related measures: Some constructive suggestions and some doubt on the 'threshold' hypothesis. *Ecological Economics, 34,* 347–361.

Nolan, B., Whelan, C.T. (2009). Using non-monetary deprivation indicators to analyze poverty and social exclusion in rich countries: Lessons from Europe?, UCD school of applied social science, wp09/11. Retrieved 2, Aug 2014, from http://www.Ucd.Ie/t4cms/wp11%2009%20noaln%20whelan.Pdf.

Nordhaus, W., & Tobin, J. (1972). Is growth obsolete? In: Economic growth, national bureau of economic research series, no 96e, N. York: Columbia, University press.

Nourry, M. (2008). Measuring sustainable development: Some empirical evidence for France from eight alternative indicators. *Ecological Economics, 67*(3), 441–456.

OECD (2013). OECD Factbook. Retrieved 1, Aug 2014, from http://www.oecd-ilibrary.org/sites/factbook-2013-en/06/02/03/index.html?contentType=&itemId=/content/chapter/factbook-2013-50-en&containerItemId=/content/serial/18147364&accessItemIds=&mimeType=text/html.

OECD. (2012). The value of statistical life: A meta-analysis, env/epoc/wpnep(2010)9/final. Retrieved 21, Aug 2014, from http://www.OECD.Org/officialdocuments/publicdisplaydocument pdf/?Cote=env/epoc/wpnep(2010)9/final&doclanguage=en.

Ortiz, I., & Cummins, M. (2011). Global inequality: Beyond the bottom billion: A rapid review of income distribution in 141 countries, INICEF. Retrieved 2, Aug 2014, from [http://www.Unicef.Org/socialpolicy/files/global_inequality.Pdf.

Packard, A., & Chapman, R. (2012). An evaluation of the wellington regional genuine progress index. *New Zealand Geographer, 68,* 1–13.

Pandey, K. D. (2014). The Human costs of air pollution: New estimates for developing countries, Working Paper, OECD.

Posner, S. M., & Costanza, R. (2011). A summary of ISEW and GPI studies at multiple scales and new estimates for Baltimore city, Baltimore County, and the state of Maryland. *Ecological Economics, 70,* 1972–1980.

Pulselli, F. M., Bravi, M., & Tiezzi, E. (2012). Application and use of the ISEW for assessing the sustainability of a regional system: A case study in Italy. *Journal of Economic Behavior & Organization, 81,* 766–778.

Rackwitz, R. (2006). The effect of discounting, different mortality reduction schemes and predictive cohort life tables on risk acceptability criteria. *Reliability Engineering and System Safety, 91*(4), 469–484.

Rheinberger, C. M. (2009). Paying for safety: Preferences for mortality risk reductions on alpine roads, Fondazione Eni Enrico Mattei working papers, working paper 338. Retrieved 21, Aug 2014, from http://ageconsearch.Umn.Edu/bitstream/54350/2/77-09.Pdf.

Rowlingson, K. (2011). Does income inequality cause health and social problems?. Retrieved 2, Aug 2014, from http://www.Jrf.Org.Uk/sites/files/jrf/inequality-income-social-problems-full.Pdf.

Smeeding, T. (2006). Poor people in rich nations: The united states in comparative perspective. *Journal of Economic Perspectives, 20,* 69–90.

Svensson, M. (2009). Precautionary behavior and willingness to pay for a mortality risk reduction: Searching for the expected relationship. *Journal of Risk and Uncertainty, 39,* 65–85.

Szalavitz, M. (2011). Why the happiest states have the highest suicide rates, time April 25. Retrieved 2, Aug 2014, from http://healthland.Time.Com/2011/04/25/why-the-happiest-states-have-the-highest-suicide-rates/.

Talberth, J., Cobb, C., & Slattery, N., (2007). The genuine progress indicator 2006: Redefining progress, Oakland CA.

UNICEF, (2007). An overview of child wellbeing in rich countries. Retrieved 2, Aug 2014, from http://www.Unicef.Org/media/files/childpovertyreport.pdf.

WDI (2014). World Development Indicators. Retrieved July 2014, from http://data.worldbank.org/products/wdi.

Wilkinson, R., & Pickett, K. (2006). The spirit level, why greater equality makes societies stronger. Retrieved 2, Aug 2014, from http://www.Tantor.Com/extras/b0505_spiritlevel/b0505_spiritlevel_pdf_1.Pdf.

World Health Organization. (2014). Heat, health assessment tool. Retrieved 21, Aug 2014, from http://heatwalkingcycling.Org/index.Php?Pg=requirements&act=vsl.

Xu, L., Yu, B., & Yue, W. (2010). A method of green GDP accounting based on eco-service and a case study of Wuyishan, China. *Procedia Environmental Sciences, 2,* 1865–1872.

Chapter 11
Developments and Challenges in the Greek Hospitality Sector for Economic Tourism Growth: The Case of Boutique Hotels

Aimilia Vlami

Abstract The hospitality industry is a rapidly evolving market, where the goal now is "differentiation of experience". This article explores the growth of boutique hotels in Greece, as it is one of the dominant trends in the country's hotel industry. Specifically, it is attempted to determine the structural and functional characteristics of boutique hotels operating in Greece. In that framework, adopting the triangulated method (Secondary and Primary research qualitative and quantitative research) strengthen the complementary nature of the findings and enrich our understanding of the multidimensional and multifaceted status of boutique hotel. The findings of this effort concern the main characteristics of Greek boutique hotels, mainly from a supply perspective. Greek boutique hotels are characterized by low capacity and their location, which is an indispensable "ingredient" of the product, they stand out for their unique design concept, aim and work towards upgraded amenities & tailored services, and usually, invest in providing technology.

Keywords Boutique hotel · Economic sustainability · Structural characteristics · Amenities and services

11.1 Introduction

During the 1970s and 1980s, the hotel product experienced an intense era of homogenization and standardization (Freund de Klumbis and Munsters 2005; Teo and Chang 2009). It was the era during which the global hospitality industry grew following the Ford production model, whereby hotel units, in the framework of a "McDonaldisation" production process (Aliukeviciute 2012), aimed at satisfying large numbers of tourists, with small profit margins. The development of differentiated hotel products, which can be placed in the mid-1980s, is directly connected to the gradual

The original version of this chapter was revised: The author name has been amended. The correction to this chapter is available at https://doi.org/10.1007/978-3-030-59675-0_15

A. Vlami (✉)
Affiliate Lecturer of the Hellenic Open University, Athens, Greece
e-mail: vlami.aimilia@ac.eap.gr

D. Balsalobre-Lorente et al. (eds.), *Strategies in Sustainable Tourism, Economic Growth and Clean Energy*, https://doi.org/10.1007/978-3-030-59675-0_11

transition of the hotel industry from the Ford model to a more flexible production model, where providing tailored accommodation services, is the added value of the accommodation (Judd 2006; Rogerson 2010). This transition in the hotel sector was directly connected to the changes in the consumer behaviour of tourists, who were increasingly seeking unique accommodation experiences (Freund de Klumbis and Munsters 2005; Aggett 2007; Timothy and Teye 2009).

11.2 The Historical Evolution of Boutique Hotels

Blakes Hotel is considered the precursor of boutique hotels. It was founded by actress Anouska Hempel in 1978, in London. The hotel had a capacity of 50 rooms, each with its design, decoration and amenities. The hotel in question had a small bar and restaurant, adapted to the funky chic aesthetic that dominated the interior design of the era (Callan and Fearon 1997). However, the term boutique hotel was first used in the USA, in 1984, when Ian Schrager founded Morgan hotel while trying to describe a new accommodation product, which aimed at reproducing the atmosphere and style of small historical European hotels, and which focused on specific target markets with interests such as rock & roll, wine tourism, avant-garde architecture, etc. (Mintel 2011 in Jones and Quadri-Felitti 2013). The Small Luxury Hotel (SLH) was founded in 1991; it was the first collective trademark for boutique hotels. The international marketing chain in question, after 30 years of operation, has more than 500 hotel units as members in 80 countries around the world. Necessarily, the term boutique hotel was broadened and used to promote small hotel accommodations with unique design and high-quality services. These hotels enriched their product by developing differentiated accommodation, entertainment and recreation services, which resulted in many of them specializing as "boutique spa hotels", "boutique gourmet hotels", etc. (Callan and Fearon 1997; Anhar 2001; Lea 2002; Adner 2003; Lim and Endean 2009). During the 2000s, boutique hotels attracted the intense interest of the press, construction companies, and consumers. International hotel chains, following the "if you can't beat them, join them" adage, started creating their boutique brands (such as Starwood with W brand and Hyatt with Andaz hotels), as they observed that boutique hotels were taking a significant portion of their clientele. Multinational brands, capitalizing on economies of scale and their excellent management, increasingly focused on the development of lifestyle hotels, with a capacity of 100–200 rooms.

In 2009, the independent Boutique & Lifestyle Association (BLLA) was created as a response to the demands of a fragmented industry for a collective voice (https://www.blla.org/about). At present, the Association has more than 750 members and is a forum for networking, promotion, communication, cooperation and know-how exchange among boutique hotels, hotel sector suppliers, the travel industry and naturally, tourists—consumers.

According to the findings of the annual Summit on boutique hotels (Boutique & Lifestyle Hotel Summit, London, May 23–24, 2016), it was ascertained that the term boutique hotel is experiencing a transition from small luxury hotels to budget boutique hotels, which are addressing middle-income visitors, offering them unique accommodation experiences.

11.3 The Definition and Characteristics of a Boutique Hotel

The complicated nature of boutique hotels has led to the creation of several definitions, none of which, however, have been broadly accepted, (Callan and Fearon 1997; Teo et al. 1998; Albazzaz et al. 2003; Forsgren and Franchetti 2004; Freund de Klumbis 2004; Horner and Swarbrooke 2005; Victorino et al. 2005; McIntosh and Siggs 2005; Drewer 2005; Van Hartesvelt 2006; Erkutlu and Chafra 2006; Aggett 2007; Lim and Endean 2009; McNeill 2009; Rogerson 2010; Aliukeviciute 2012; Jones et al. 2013; Goh 2014; Arifin et al. 2014) as it is a hotel product that stands out for its diversity as regards its design, services and even customer service. The BLLA, in the framework of the initiative it undertook in 2012, to express a generally accepted definition for boutique hotels and to identify their main similarities and differentiation with lifestyle hotels, concluded that: "boutique hotels are small accommodations which aim at offering an authentic cultural experience to visitors, by providing high-level services" (Jones et al. 2013). Specifically, the main findings of this research effort are the following: They are small hotel units which stand out for their unique architecture and the decoration of their interior spaces, and they are housed either in recently constructed buildings with modern decor or transformed historical and listed buildings. Transformations that correspond to modern lifestyles, respecting, however, the architectural characteristics and identities of the buildings, the goal being for guests to experience a "uniquely authentic" experience. Units that offer tailored services, aiming at producing the feelings of "intimacy" and "warm hospitality" in tourists—consumers. These results, essentially, connect a series of keywords with what boutique hotels are or are not (see Table 11.1).

According to the international experience (Callan and Fearon 1997; Prentice 1997; Teo 1998; Schmitt 1999; Prentice and Anderson 2000; Albrecht and Johnson 2002; Lea 2002; Albazzaz et al. 2003; Horner and Swarbrooke 2004; McIntosh and Siggs 2005; Caterer 2005; Providence Hospitality 2006; RAC 2006; Van Hartesvelt 2006; Aggett 2007; Lim and Endean 2009; Olga 2009; Timothy and Teye 2009, Rogerson 2010; Sarheim 2010; Aliukeviciute 2012; Schrager 2015; Fuentes-Moraleda et al. 2019; McKenney 2015) the main characteristics of boutique hotels are: (a) their small size, (b) their location, (c) their unique design concept, (d) their upgraded amenities & tailored services and (d) their focus on specific target markets. In summary, it is ascertained that the term boutique hotel refers to small hotels, at which the authenticity of the experience is the main concern so that it registers in consumers' consciousness as a "unique accommodation experience", which can only be repeated by repeating their stay. The main goal is to create a unique identity which is difficult to copy, by capitalizing on its location, taking advantage of technology and promoting local cultural heritage; by combining comfort with personal services, so

Table 11.1 What a boutique hotel is or is not

What a boutique hotel is	What a boutique hotel is not
A small hotel	A medium—large hotel
Independent hotel units or members of small Boutique Hotel chains	Classic hotel chains
Unique architecture and decoration	
Creation of unique experience	Standard hotel design
Personal service, warm hospitality	Providing a standardised product
Hi-Tech	Standard service
Inspired	Low-tech
Exclusive	Conventional
Hip & cool hotel	All-inclusive

Source BLLA (2012)

that the tourist—consumer feels "like they are home", namely, that they are not a "faceless number" behind a close the door on a long hallway; that they are, instead, a visitor in a hospitable environment.

11.4 Methodology

The goal of this article is to define the structural and functional characteristics of the boutique hotels operating in Greece. The question that arises spontaneously and without pressure, is the following: "What is, at present, the capacity of boutique hotels (expressed in units and beds) operating in Greece?" A question that is not easily answered, due to the inability to institute boutique hotels as a distinct type of major hotel accommodation (Law 4276/2014), resulting in the accommodations in question receiving permits and being recorded on the Tourism Business Register of the Ministry of Tourism as hotels, and they're being certified by the Hellenic Chamber of Hotels (HCH) into star categories, with no mention of the term "boutique". Thus, a significant lack of necessary and reliable statistical data for quantitative and qualitative analysis and interpretation of the structural evolution of boutique hotels in Greece was identified. An effort was made by the HCH in the following years to cover—to a certain extent—this shortcoming. In 2015, the HCH moved to the copyrighting of the "Boutique hotel" collective trademark with the General Secretariat of Trade and Consumer Protection of the Ministry of Economy & Development, and at present, a recognition and accreditation system is being implemented for a boutique hotel in Greece which has been documented to be offering a differentiated hotel accommodation product. The result of this recognition is that incorporated hotels may add the term "boutique" to their trade name. Specifically, for the needs of this effort and to cover the aforementioned lack of information, we have applied a mixed-method, adopting the matter of triangulating quantitative and qualitative research (Flick 2006). Triangulation is broadly defined by Denzin (1978: 291), "as the combination of methodologies in the study of the same phenomenon". It is a complicated process during which each method affects the other, to maximize the credibility of research ventures (Denzin 1978). Adopting the triangulated method in this effort will strengthen the complementary nature of the findings and enrich our understanding of the multidimensional and multifaceted status of Boutique Hotel (BH) (Fig. 11.1).

Thus, the methodology framework is structured as follows: *Secondary research*: (1) researching international bibliography and experience to define the term boutique hotel and its main characteristics. (2) Exploring Greek hotels to record boutique hotels which: are either members of international boutique hotels trademarks, or have received awards as "boutique hotels" by TripAdvisor, or claim to have developed this accommodation product on their official website. *Primary research*: (1) planning qualitative research through content analysis of the official websites of the approximately 300 self-proclaimed boutique hotels, which were identified and recorded during the secondary research. Codification and development of a database with their main characteristics (a type of accommodation, location, star category, size, operation period, sectors, services provided, etc.). From this phase, 190 hotels were selected to be analyzed based on their location so that all regions in Greece could be represented and mainly based on the accommodation product offered and comparing it to the findings of the international experience study. (2) planning the quantitative research through an online questionnaire (with the use of Google forms), on a sample of 190 hotels, the main goals of which are to define: the concept of design and operation of boutique hotels in Greece (goal and main aims). Their structural characteristics (the type of accommodation, star categories, operation period, average size, etc.). This effort contributed to the production of a typology of boutique hotels based on their location. Their main operational characteristics, and especially their main amenities and services provided to guests. The main characteristics demanded of boutique

Fig. 11.1 The methodology framework

hotels (mix of guests, age groups, etc.). The questionnaire designed consisted of 29 closed questions and seven open questions, which were part of eleven thematic units related to the spaces, sectors and services of the hotel, according to international bibliography, the term BH refers to the vision that is applied throughout all structures and operations of an accommodation. In that framework, the thematic units are structured as follows: Building, Reception—Common use areas, Rooms/Apartments, Bathroom, Catering, Entertainment—Sports—Recreation, Other Services, Special Certifications, Staff, Various, Demand. The research was carried out from April to May 2019 (its pilot application was carried out on 15–30 May 2019). The specific period was selected: a. so that hotel managers would have time to complete the questionnaire, as it is low season for Greece's hotels and b. for hotel units which operate seasonally (and are usually starting to operate in this period) to participate in the research. In the framework of this research, telephone communication was also carried out in three consecutive phases: (a) finding the e-mail of the owner and/or manager of the accommodation and sending them a message motivating them to participate in the research, (b) personal communication with the head of the hotel to inform them regarding the specific research being carried out and (c) a reminder, ten days after the questionnaire had been sent, regarding the completion of the questionnaire, to achieve the most satisfactory representativeness in the sample possible. The high percentage of response to the research, as more than 75% of the sample (144 of the hotel units) responded to the questionnaire, led to drawing relatively credible conclusions. At this point, however, it should be pointed out that this article is not attempting to analyze all the issues of supply and demand related to the development of boutique hotels in Greece. Specifically, this effort is focusing on defining and analyzing the main characteristics of Greek boutique hotels from a supply perspective. Future research could focus on defining and analyzing the target markets of Greek boutique hotels.

11.5 Boutique Hotels Concepts in Greece

According to the results of this approach, the essential characteristics of an average boutique hotel in Greece, are its small size, the owners'/managers' vision, which is reflected in the overall design concept of the indoor and outdoor spaces of the hotel, and which runs through all aspect of the hotel: accommodation, catering, recreation, etc. the unique design of its indoor spaces which must inspire "homely warmth", while still creating the feeling of "private" space and comfort, its upgraded infrastructure and amenities in the rooms, its provision of high-level personal services. They are hotels that meet the specifications of their quality category (namely, they are the excellent hotels in their category) and are characterized by an overall design concept (a narrative) that runs through all parts and aspects of the hotel, and which makes the accommodation "unique" in relation to its competition. Accurately, the boutique hotels in the research appear to have the following common characteristics as regards their architecture and decoration: the overall design concept of the hotel is often recorded and signed by a specialist (architect/interior designer/decorator) and the owner/manager of the hotel. Its thematic differentiation from its competitors is the main goal of its architectural approach, both as regards its external appearance and as regards its interior layout and decoration. Its design themes are applied consistently in all aspects of the hotel's operation: rooms, catering, recreation, staff, stationery, brochures, etc. The specially designed outdoor spaces and the determination of specific functions for the indoor common use spaces, which aim at creating "warm" spaces. The special attention paid to the furniture, decorative elements and the use of works of art contribute to the creation of an atmospheric, overall consistent space. The appropriate lighting design (artificial and/or natural), as well as the appropriate selection of colours, is combined with the overall design theme. The aesthetic result of each room may be completely different, following the overall design concept of the accommodation.

11.6 Boutique Hotels Structural Characteristics

- **Accommodation Type**. The boutique hotels operating in Greece mainly concern classic hotel accommodation and self-catering hotel accommodation (75 and 25% correspondingly). Furthermore, a significant number of the hotel housed in traditional buildings, offering differentiated accommodation services was recorded. However, this form of boutique hotels, historic boutique hotels, are not the subject of study of this article, as they present significant particularities and for that reason will be the subject of a later study to be carried out.
- **Size**. Boutique hotels are characterized by low capacity (a limited number of rooms). Ian Schrager supported that small size is the main characteristic of boutique hotels, as low capacity allows visitors to quickly become accustomed to the environment and to enjoy the expected warm hospitality. However, there is no determined limit on the size of these accommodations, the scales are different from country to country, according to the numbers of the domestic hotel industry. The average size of the hotels that have been members of the SLH over the last 29 years has been decreasing, which has resulted in the current average being 48 rooms (in contrast to the average accommodation size of 1991, which was 62 rooms, SLH, 2019). Boutique hotels in Greece are small and very small in size. They are usually small independent units, whose owner or manager (who are often the same person) is the main party in charge of the design, organization and management of the accommodation, and they are not part of a National or International chain. Specifically, the average size of hotel units is 38 rooms, which breaks down to 21, 43 and 56 rooms for 3*, 4* and 5* hotel units correspondingly. The reason that the number of rooms is determined essentially concerns the capability of the accommodation to provide tailored services, as well as to secure the integrity of communication between guests and staff. However, there is a significant parameter in defining the size of a BH, as it depends on the market scale, namely, in Las Vegas it may have several hundred rooms, while in other

Table 11.2 Open for business and operating period

	Open for business (%)	Operating period					
Star categories	1950–1979	1980–1989	1990–1999	2000–2009	2010–2019	Continuous (%)	Seasonal (%)
5*	11,5	19,2	19,2	26,1	23,1	38,5	61,5
4*	27,6	3,4	6,9	27,6	34,5	64	36
3*	14,3	12,2	8,2	26,5	38,8	73,5	26,5

markets it may be less than 100 rooms. The findings of the research agree with the majority of the bibliography that finds that BH has less than 100 rooms (Aggett 2007; Callan and Fearon 1997; Teo 1998; Freund de Klumbis 2005; Lim and Endean 2009), which is one of their defining characteristics.

- **Star Categories**. There is a feeling that due to the uniqueness of boutique hotels, it would not be possible to grade them or place them in categories (Callan and Fearon 1997). Certain researchers support that boutique hotel should be categorized at least as four-star accommodations, while others note that boutique hotels may be three-, four-, or five-star accommodations (AA 2006; RAC 2006; Teo et al. 1998; Van Hartesvelt 2006; Aggett 2007; Lim and Endean 2009: 42; Rogerson 2010; Aliukeviciute 2012: 1), although some may have no stars. Thus, based on the research results, 49% of Greek BH are 3-star units, 26% are 4-star units and 24% are 5-star units.
- **Open for Business and Operating Period**. To identify the opening and operation period of boutique hotels, data were drawn and compared from the listed members of the HCH register, where the main data of the entire hotel sector of Greece are listed, along with their possible changes (such as hotel name change, relocation, initial operation date, operation cessation for a while, operating period, etc.).

Thus, as can be seen in Table 11.2, the following became evident: (a) 5* hotel units started being founded mainly since the 1980s, with an increasing trend from 2000 onwards, and their majority (61.5%) operates seasonally, (b) a similar picture was ascertained regarding 4* and 3* hotel units, as the majority of these units started being created in the early twenty-first century, and have presented increased activity over the last 8 years. The majority of these accommodations operate throughout the year.

- **Boutique Hotels Typology**. The boutique hotels identified, taking into account their location, can be grouped into two main categories: i. city boutique hotels (40%) and ii. resort boutique hotels (60%). Specifically, city boutique hotels are located: (a) in the urban centres of Athens, Thessaloniki, Patras, Volos, and especially in important commercial regions, (b) in city suburbs, on the outskirts of cities, and especially in areas with high per capita income (Kifissia, Glyfada, Voula, Ekali, etc.) and (c) in the urban sectors of tourist areas (Nafplio, Kastoria, Heraklion, Rethymnon, Argostoli, city of Rhodes, etc.). Furthermore, it was ascertained that units located in the centre or the suburbs of major cities are mainly housed in neoclassical buildings and blocks of flats, examples of Greek urban architecture of the twentieth century. The units located in smaller cities (such as Nafplio, Agrinio, Kastoria, etc.) are usually housed in traditional buildings, examples of local architectural tradition. Boutique hotels in major urban centres feature the existence of a hip restaurant, a lounge and bar, small conference halls, while there is an intense hi-tech element in the rooms. On the other hand, boutique hotels resorts are mainly located in distant areas instead of tourist centres, as well as in locations of particular natural beauty. They are small units where there is a combination of traditional architecture with modern aesthetic elements, the room's technological infrastructure is "hidden", so as not to adulterate the aesthetic of the room and the broader environment. The majority of these hotels have a restaurant, pool (common use and private), balneotherapy services—spa, spaces for events such as weddings, etc.

– **Boutique Hotels Upgraded Amenities and Services**. Greek boutique hotels focus on differentiating the hotel experience. When one mentions the aim of providing a differentiated accommodation experience at present, they do not simply mean providing all the services that will cover the basic needs of sleep, food and hygiene. They mainly mean the cultural experience they expect to take away from their stay. Therefore, the main requirement of a boutique hotel is not so much the sale of goods and services, as much as the experiences that will move and carry away tourists - consumers, providing them with emotions they do not feel in their daily lives. Based on the results of the research the main upgraded amenities and services usually provided by boutique hotels are listed (according to the star rating), in a variety of combinations, beyond those which they are obligated to provide. Specifically, upgraded amenities and services focus mainly on the Rooms/Apartments, as three amenities and services—such as Internet Access (free of charge), Safe, Dressing Table with Mirror—can be found in all 5* and 4* boutique hotels, and in more than 90% of 3* boutique hotels. Furthermore, early breakfast and dry cleaning—laundry—ironing services are found in all 5* boutique hotels, in more than 80% of 4* boutique hotels, and in 75% of 3* boutique hotels, correspondingly. Regarding the amenities, the Room central light switch at the entrance and by the bed and the autonomous temperature regulation inside the room are found in more than 88% of all boutique hotels categories. Additionally, upgraded room services include changing the towels upon request, breakfast in bed capabilities, and providing bathrobes and bathroom slippers per bed in more than 80% of 3* and 4* boutique hotels (these services are obligatory for 5* hotels by the official ranking system). Also, more than 84% of all boutique hotels categories have satellite television, their website provides direct booking capabilities, and their staff speaks at least one foreign language. A hotel operation manual is dispensed to the staff in more than 92% of 5* boutique hotels, in more than 75% of 4* BH and 65% of 3* boutique hotels. Furthermore, express checkout services are available in more than 84% of 5* and 4* boutique hotels, and more than 60% of 3* boutique hotels. More than 88% of 5* boutique hotels have a laptop or tablet provision service, which is found to a lesser degree in 4* and 3* boutique hotels (in approximately 65%). More than 70% of all boutique hotels categories clean their mattresses annually, and approximately 90% of 5* boutique hotels provide pillow selection (a service found to a lesser extent in 4* and 3* boutique hotels—approximately 55%).

5 Boutique Hotels*. Specifically, 5* boutique hotels have been found to provide the following additional upgraded amenities and services: more than 92% provide food or breakfast to guests in special containers, upon request, they have added new technology televisions in the suite bedrooms, and a pool. More than 80% have increased the obligatory size of the hotel rooms (square metres), providing beds that are larger than the minimum, they offer transfer services to and from the airport, port, railway station, they provide front door and window safe-locking. There is a scale in the room, and they offer recreation and well-being services with rejuvenation centres that include at least two of the following: Sauna, Steam Bath, Jacuzzi, two types of therapy or massage and gym facilities that cover an area of at least 20 m^2, with at least 4 items of modern gym equipment (e.g. treadmill, stationary bicycle, rowing machine, benches, dumbbells). The staff consists of Tourism school graduates. More than 70% have magnetic/electronic keys and a Salon de beaute that provides—at least—facials, hand and foot therapies, make-up, depilation, etc. More than 60% have provisions for special dietary needs (for breakfast and/or meals), a welcome drink/gift in the room, a hydromassage bathtub or shower, a telephone in the bathroom, an espresso machine, a CD—DVD —MP3 PLAYER with speakers, a multi-use hall, a library, valet parking, a doorman, daily newspapers and magazines and an organized guest complaint management system.

4 Boutique Hotels*. Regarding 4* boutique hotels, they were found to provide the following additional upgraded amenities and services: more than 93% offer a recreational shop (cafeteria/bar), as well as a bathtub and shower. More than 80% provide for special dietary needs (for breakfast and/or meals), meals or breakfast for guests in special containers, upon request, a welcome drink/gift in the room, beds that are larger than the bare minimum (minimum requirements for single beds are 0.90 m × 1.90 m and for double beds are 1.60 m × 2.00 m) and their staff consists of graduates of tourism schools. More than 70% have new modern televisions in their suite bedrooms and provide safe door and window locking. They have a concierge and groom/pageboy, and in general, at least 10% of their staff participates in training programmes related to tourism and security of a minimum duration of 20 h annually. Furthermore, they have a pool and an iron and ironing board in the room. More than 60% have unique/special dishes for breakfast and/or signature dishes, magnetic/electronic keys, a hydromassage bathtub or shower, safety grips in the bathtub and a turndown service at night. They also provide services such as babysitting and transfer to and from the airport, port, railway station, they have a library and provide daily newspapers and magazines. The hotel website has web reviews.

3 Boutique Hotels*. Regarding 3* boutique hotels have been found to provide the following additional upgraded amenities and services: more than 91% have a recreational shop (cafeteria/bar) and common use visitor toilets, they provide baggage storage in a separate space and room service. It should also be noted that in comparison to 5* and 4* hotels, they have the highest number of staff members to the number of hotel beds. More than 80% provide for special dietary needs (breakfast and/or meals) and provide safe door and window locking. They have a small table, or a table or desk, or another work surface inside the room, and provide common use computers to guests and printing, faxing, photocopying and scanning services. More than 70% have a 24 h reception service, and they provide a baggage-transfer service and a hotel service guide (printed or electronic) in at least two languages. They have a restaurant, they serve breakfast for at least three hours, and provide food or breakfast to guests in special containers upon request. Bed dimensions in the rooms are larger than the obligatory minimum dimensions, and they have a kettle and coffee and tea making facilities, various types of hangers—at least two—as well as additional cosmetics (shower cap, cotton buds, single-use razors, cotton swabs, body lotion or cream, hair cream, nail file, toilet rubbish bin liners, a toothbrush with single-use toothpaste, paper tissues). More than 60% have increased the obligatory size of the hotel's apartments (square metres) and have magnetic/electronic keys. Inside the room, they offer an iron and ironing board, as well as other personal services (dry cleaning laundry bag, sewing materials, correspondence envelope, writing materials, shoe polish and shoehorn. Furthermore, they have a library, their staff consists of tourism school graduates, and in general, at least 10% of their staff participates in training programmes related to tourism and security of a minimum duration of 20 h annually. It is ascertained that catering is an indispensable aspect of the overall boutique hotels experience. International experience (mainly in the USA and the UK) has indicated that boutique hotels invest in the Food & Beverage sector, as this sector has an increased contribution to the profits of these units, reaching, on average, up to 20% of their overall profits. For that reason, Greek boutique hotels focus on creating atmospheric catering spaces (bars and/or restaurants), creating differentiated menus with signature dishes and cocktails, providing unique experiences, such as meeting the chef, wine tastings, etc. One of the main conclusions is that the presence of catering, recreation and entertainment spaces (such as restaurants, rejuvenation and beauty centres, sports facilities, etc.) are often found in Greek boutique hotels, but are not a necessary condition for a hotel to be characterized as a boutique hotel. It is important that these spaces also operate with a boutique hotel mentality, namely that the spaces have a unique aesthetic, offering singular products and tailored services, following the overall concept of the accommodation.

11.7 Demand Characteristics

Boutique hotels have gained a competitive advantage in the hospitality industry and are aiming to cover the new and different needs of a specific group of consumers. According to the findings of the research, a large portion of guests (~44%) belongs to the 30–50 age group, followed by guests over the age of 50. Greek boutique hotels mainly attract tourists from Europe (~42%), and the rest are either Greek or non-European foreign tourists. Boutique hotels guests mainly travel with their partner (~44%), followed by guests with friends (~25%) and children (~21%), while ~10% travel alone and belong to the corporate travel sector, with an average stay of one to two nights in boutique hotels. Also, according to Aliukeviciute (2012), the great increase in the numbers of the elderly in the western world has significantly extended the boutique hotel market, as interest in culture often increases along with the age of the individual. Furthermore, pensioners are the main part of the guests in luxury accommodations.

11.8 Final Conclusions

The findings of this effort concern the main characteristics of Greek boutique hotels, mainly from a supply perspective. Therefore, Greek boutique hotels are characterized by low capacity (a limited number of rooms) and their location, which is an indispensable "ingredient" of the product, the unique experience it aims to offer its guests. In that framework, the "ideal" location for a boutique hotel is not determined so much by the infrastructure and superstructure at its disposal; instead, it is determined by the unique natural and human-made elements that characterize it. Furthermore, they stand out for their unique design concept, seeing as they are hospitality units with unique architectural design and decoration that characterizes them and sets them apart from their competition. The set up of their frequent use spaces and rooms is approached with a "set-design" mentality, to use the various means of design to intensify the emotional charge of the experiences the hotel offers (privacy and social events, recreational services, entertainment and relaxation, cultural activities, etc.). Greek boutique hotels aim and work towards upgraded amenities and tailored services, as unique design concepts alone are not enough to provide guests with a unique experience. The unique design concept and its application in all aspects of hotel operation, in combination with the tailored, differentiated services provided, may secure a unique, memorable experience, which will remain in the memory of guests long after they leave the accommodation. The conclusions of the 2016 Summit on boutique hotels found that Experience and Consistency are defining characteristics of the product in question. Greek boutique hotels usually invest in providing technology, especially inside the rooms, which is at least on par with what their clientele would have at home. This means that developing technological infrastructure and amenities that are user friendly and practical. Lastly, the findings of the research on the target market of Greek boutique hotels, as analyzed in the previous section, agree with international experience, according to which the target market for boutique hotels over the years has been guests aged just below 20 and up to 55, with an average to high income. The elderly are also a significant target market, especially for 5* boutique hotels. At present, consumers operate with the "self-realization" of their holiday expectations as their main motivation. Necessarily, they are differentiated shoppers who have an understanding and perspective on design, and they expect personal communication with the staff of the accommodation, they are looking for a higher level of service, and desire more than merely observing; they want to be the main participants in a unique experience.

References

Adner, R. (2003). Lifestyles of the rich and almost famous: The boutique hotel phenomenon in the United States. Report of the high-tech entrepreneurship and strategy group project. Insead, New York

Aggett, M. (2007). What has influenced growth in the UK's boutique hotel sector? *International Journal of Contemporary Hospitality Management, 19*(2), 169–177.

Albazzaz, A., Birnbaum, B., Brachfeld, D., Danilov, D., Kets, V., & Moed, O. J. (2003). Lifestyles of the rich and almost famous: The boutique hotel phenomenon in the United States. High Tech Entrepreneurship and Strategy Group Project, Fontainebleau: Insead Business School.

Aliukeviciute, M. (2012). Boutique hotels' evolution. *International Journal of Contemporary Hospitality Management* 1–3.

Anhar, L. (2001) The definition of boutique hotels. https://www.hospitalitynet.org/news/4010409.html.

BLLA. (2012). Boutique and lifestyle hotels and properties—definitions. https://www.blla.org/lifestyle-hotels.html.

Callan, R. J., & Fearon, R. (1997). Townhouse hotels-an emerging sector. *International Journal of Contemporary Hospitality Management, 9*(4), 168–175.

Caterer Search. (2005). Market snapshot: Boutique hotels. In Caterer and Hotelkeeper. https://www.thecaterer.com/news/hotel/market-snapshot-boutique-hotels.

Denzin, NK. (1978). *The research art in sociology.* New York: McGraw-Hill.

Drewer, P. (2005). Key note market report plus 2005: Hotels (20th ed.). Hampton: Key Note. https://www.keynote.co.uk.

Erkutlu, H. V., & Chafra, J. (2006). Relationship between leadership power bases and job stress of subordinates: Examples from boutique hotels. *Management Research News, 29,* 285–297.

Flick, U. (2006). *An introduction of qualitative research* (3rd ed.). London: Sage.

Forsgren, S., & Franchetti, C. (2004). *The marketing role of unique concepts for hotels in Sweden.* Tourism and hospitality management Master Thesis, No 2004: 53, Graduate Business School, School of Economics and Commercial Law, Goteborg University. Elanders Novum.

Freund de Klumbis, D. (2004) Seeking the ultimate hotel experience. http://www.bibemp2.us.es/turismo/turismonet1/economia%20del%20turismo/host/ena/seeking%20the%20ultimate%20hotel%experience.pdf.

Freund de Klumbis, D., & Munsters, W. (2005) Developments in the hotel industry: Design meets historic properties. Retrieved from https://bibemp2.us.es/turismo/turismonet1/economia%20del%20turismo/hosteleria/development%20in%20the%20hotel%20industry.pdf.

Fuentes-Moraleda, L., Lafuente-Ibáñez, C., Muñoz-Mazón, A., & Villacé-Molinero, T. (2019). willingness to pay more to stay at a boutique hotel with an environmental management system a preliminary study in Spain. *Sustainability, 11*(18), 1–15.

Goh, Y. N. (2014). Customers revisit on Penang Boutique Hotel at George Town UNESCO's World Heritage Site: A conceptual paper. In *Proceeding 2nd International Congress of Interdisciplinary Behavior and Social and Sciences 2013*, Jakarta, Indonesia, 4–5 November.

Hartesvelt, M. V. (2006). Building a better boutique hotel. *Lodging Hospitality, 15,* 32–44.

Horner, S., & Swarbrooke, J. (2004). Case study 13—Boutique hotels. International cases. In *Tourism management* (pp. 148–156). Spain: Butterworth-Heinemann, Elsevier.

Horner, S., & Swarbrooke, J. (2005). *Leisure marketing: A global perspective.* Oxford: Butterworth-Heinemann.

Jones, L. D., Day, J., & Quadri-Felitti, D. (2013). Emerging definitions of boutique and lifestyle hotels: A Delphi study. *Journal of Travel & Tourism Marketing, 30,* 715–731.
Judd, D. R. (2006). Commentary: Tracing the commodity chain of global tourism. *Tourism Geographies, 8,* 323–326.
Lea, K. (2002). The boutique hotel: Fad or phenomenon? *Locum Destination Review, 7,* 34–39.
Lim, W. M., & Endean, M. (2009). Elucidating the aesthetic and operational characteristics of UK boutique hotels. *International Journal of Contemporary Hospitality Management, 21*(1), 38–51.
McNeill, D. (2009). The airport hotel as business space. *Geografiska Annaler. Series B. Human Geography, 91,* 219–228.
Mintel Group Ltd. (2011). Boutique hotels in the US. https://academic.mintel.com.proxy2lib.uwo.ca:2048/sinatra/oxygen_academic/display/id=545401.
Olga, A. (2009). The alternative hotel market. In *Proceedings International Conference on Management Science Engineering*, Moscow, Russia.
Prentice, R. C. (1997). Cultural and landscape tourism. Facilitating meaning. In S. Wanab & J. Pigram (Eds.), *Tourism development and growth* (pp. 209–236). London: Routledge.
Prentice, R. C., & Anderson, V. A. (2000). Evoking Ireland: Modeling tourist propensity. *Annals of Tourism Research, 27,* 490–516.
Rogerson, J. M. (2010). The boutique hotel industry in South Africa: Definition, scope, and organization. *Urban Forum, 21,* 425–439.
Sarheim, L. (2010). Design or lifestyle? A review of London's boutique hotel scene. HVS. https://www.hvs.com/Content/3025.pdf.
Schrager, I. (2015). *The boutique hotel concept*. CTBUH Research paper.
Schmitt, B. (1999). Experiential Marketing. *Journal of Marketing Management, 15,* 53–67.
Teo, P. & Chang, T. C. (2009). Singapore's postcolonial landscape: boutique hotels as agents. In T. Winter, P. Teo, & T.C. Chang (Eds.) *Asia on tour: Exploring the rise ofAsian tourism* (pp. 81–96). Routledge: Abingdon.
Teo, C. C. J., Chia, G. H., & Khoo, H. P. M. (1998). Size does matter (when you're small): The critical success factors behind boutique hotels in Singapore. http://www.sbaer.ucaedu/research/1998/ICSB/o))3.html.
Timothy, D., & Teye, V. (2009). *Tourism and the lodging sector*. Oxford: Butterworth-Heinemann.
Victorino, L., Verma, R., Plaschka, G., & Dev, C. (2005). Service innovation and customer choices in the hospitality industry. *Managing Service Quality, 15,* 555–576.

Chapter 12
Airbnb and Overtourism: An Approach to a Social Sustainable Model Using Big Data

María Jesús Such-Devesa, Ana Ramón-Rodríguez, Patricia Aranda-Cuéllar, and Adrián Cabrera

Abstract Tourism has been proved an important driving force for economic growth and development. Despite that, the overtourism phenomenon, hand in hand with the sharing economy, has been proved to affect destinations in multiple ways, tourism rejection or the rising of the housing prices, among others. Over the last decade and mainly since United Nations 2030 Agenda was announced, the sustainability of the cities has become an explicit global objective of development. Tourism has been understood as a tool for improving economic and social aspects in contexts of countries in development, but the phenomenon of tourism saturation and concentration around few neighbourhoods in solid destinations from developed economies could be distancing this achievement from the cities. This chapter presents a comparative analysis between two of the top urban Spanish destinations, Madrid and Barcelona, showing the current reality between tourism, real estate prices and mean household income in a neighbourhood-level approach. Does overtourism help destinations in the goal of reaching regional convergence in terms of urban sustainability or does it worsen the situation?

Keywords Overtourism · Social sustainability · Airbnb · Big data · SDG 11 · Spain

M. J. Such-Devesa (✉) · P. Aranda-Cuéllar · A. Cabrera
Faculty of Economics, Business, and Tourism, Department of Economics, University of Alcalá, Madrid, Spain
e-mail: mjesus.such@uah.es

P. Aranda-Cuéllar
e-mail: p.aranda@uah.es

A. Cabrera
e-mail: adri.cabrerab@gmail.com

A. Ramón-Rodríguez
Department of Applied Economic Analysis, University of Alicante Spain, Alicante, Spain
e-mail: anar@ua.es

D. Balsalobre-Lorente et al. (eds.), *Strategies in Sustainable Tourism, Economic Growth and Clean Energy*, https://doi.org/10.1007/978-3-030-59675-0_12

12.1 Introduction

The tourism sector has experienced accelerated changes in recent years due to its spreading across several social groups and the addition of other countries into the collective worldview. In 2019, worldwide tourists totalled 1.5 billion, according to UNWTO (2020), representing a 4% increase on the previous year, maintaining the trend of the previous years. This rapid growth has led to multiple and complex challenges, including the arising of the sharing economy and its impact on the residents' quality of life and perception of overtourism in the destinations (Huete and Mantecón 2018; Koens et al. 2018; Milano 2018; Oklevik et al. 2019).

According to Milano et al. (2019), overtourism has its emergence in the quick development of unsustainable mass tourism practices that have had their impact on the detrimental exploitation of urban, rural and coastal areas for tourism purposes. Nonetheless, overtourism is also related to the tourist overcrowding experiences when enjoying the destination (Dodds and Butler 2019), showing that it has outweighed its carrying capacity (UNWTO 2018), which can lead to situations of tourismphobia or the rejection of tourists by residents (Milano 2018; Milano et al. 2019). As it can easily be observed, overtourism is listed as the converse scenario of responsible and sustainable tourism (Jørgensen and McKercher 2019; Peeters et al. 2018). The discussion and research on overtourism can be classified, attending to whether the studies use a qualitative or a quantitative approach. Most of the analyses take a qualitative perspective, frequently based on the similarities and differences between several case studies (Peeters et al. 2018; Milano 2018; Alcalde-García et al. 2018; UNWTO 2018; Seraphin et al. 2018; Namberger et al. 2019; Smith et al. 2019). These investigations portray locations suffering from overtourism problems, addressing its causes and effects and suggesting alleviating solutions to policymakers, which, depending on the case, are more or less focused on solving the existing problems between tourists and residents.

Nonetheless, the quantitative approach towards this research topic has recently added substantial contributions (Alcalde-García et al. 2018; Capocchi et al. 2019; McKinsey&Company 2017; Peeters et al. 2018; Perles-Ribes et al. 2020a, b), aiming to find which of the related variables can be considered as key determinants of this phenomenon and, based on them, to be able to presume which locations have more potential to suffer from overtourism situations. Notwithstanding, until now, there is not an explicit variable that can give support to an early warning system on the probability of having overtourism problems on tourist destinations (Peeters et al. 2018). Recent studies from Perles-Ribes et al. (2020a, b) show that it is statistically significant to assume that a situation of overtourism is going to be linked to the existence of a substantial amount of Airbnb accommodations. This allows us to better explore the effects of these non-formal accommodations on the social and economic dynamics of the neighbourhoods, in order to be able to focus more on the social changes that occur in these overcrowded destinations. The same authors' analyses state that overtourism can be understood as an excess of success of a destination when this achievement does not translate into an improvement of the living conditions of the residents.

On that basis, this article explores a particular relationship, the existent between the process of urban gentrification and the presence of Airbnb and hotel accommodation offered in two of the most critical urban destinations in Spain: Madrid and Barcelona. It's important to highlight that this analysis is carried out using Big Data, which allows us to get to a neighbourhood level and determine which areas of the cities are more vulnerable to the overtourism consequences. This perspective enables a profound study of the effects of the emergence of collaborative economy platforms on residents and cities' dynamics. Furthermore, it allows us to understand under what circumstances a high volume of tourists concentrated in a very specific area of the city leads to problems of touristification, expulsion of middle and lower social classes from their neighbourhoods, rising of residential housing prices, overtourism or even tourismphobia.

Gentrification has been known for decades now, understood as a process in which rehabilitation on buildings and also in the urban context, with the activities and jobs generated by it, increase the value of housing (Porter and Shaw 2013). This may lead to the displacement or exclusion of populations with lower incomes. Nonetheless, the called tourism gentrification refers to "*the loss of place experienced by residents as the consumption of space by visitors effectively displaces them from the places they belong to*" (Cócola-Gant 2018). This is now a new reality in many cities with the emergence of the sharing economy phenomenon.

The research questions for this study are: Is the presence of a substantial amount of Airbnb accommodations related to a tourism gentrification process? Are the households' incomes higher in the neighbourhoods with more Airbnb offer? Are people living in the touristified neighbourhoods changing houses more than those living in non-touristic areas? Are housing prices higher in neighbourhoods with more Airbnb offer?

This research questions are strongly related to the sharing economy appearance, that has developed a new way of sharing the urban space between residents and visitors, causing a change in accommodation patterns and the whole tourism value chain (Perles-Ribes et al. 2020a, b). Although it is still in development, this sharing economy could form a key element in managing demand pressure on and inside the destinations, preventing residents from severe consequences, and even rejection, of tourism.

As far as the authors are aware, this is the first time that this type of analysis has been pursued at a neighbourhood level, representing a step forward concerning the literature regarding the topic.

The chapter is structured as follows: after this introduction, Sect. 12.2 reviews the existing literature on overtourism; Sect. 12.3 describes the methodology and data employed for the analysis; Sect. 12.4 shows the data analysis; and eventually, in Sect. 12.5, we discuss the reached conclusions and offer some recommendations.

12.2 Literature Review

Overtourism has been an emerging concept over the last few years. The growth that the activity has experienced worldwide has been followed with the crowding of specific areas, overexploiting the carrying capacity of the destinations (Schneider 1978) and putting their sustainability at risk (Kowalczyk 2010). The earliest references made to this concept were made by the media and academic journals in 2017 (Huete and Mantecón 2018). However, its effects, main causes and consequences have been a regular element in debates on tourism sustainability (Vera-Rebollo and Ivars-Baidal 2003; Alcalde-García et al. 2018; Perkumiene and Pranskuniene 2019). Thus, the term has started to be used in recent literature to address specific patterns of tourism development and some city destinations with severe issues of lack of sustainability (Coca-Stefaniak et al. 2016; Goodwin 2017, 2019; Muler-González et al. 2018; Capocchi et al. 2019).

Nonetheless, there is literature from the 1960s discussing how tourism externalities affected destinations, harming local environment and creating problems between residents and visitors (Forster 1964; Wagar 1964). Some years after that, there are investigations concerning the possibility that the tourism industry could be disrupting residents' wellbeing during the highest seasons (McCool and Martin 1994). As stated in Perles-Ribes et al. (2020a, b), this definition is a good approach to what it has understood as *overtourism* nowadays.

Despite that, the concept of overtourism was not significant until it became a synonym to the Spanish "turismofobia" at the end of 2016 (Koens et al. 2018; Capocchi et al. 2019). This is understandable because the rapid growth of the activity that has induced the current overcrowding of some popular destinations has taken place in recent years. It is also important to note that it is probable that the Spanish neighbourhoods were one of the first to suffer from this phenomenon, as the term was first coined in Spanish.

There is a challenge in current literature to find a valid measure of overtourism that can adapt to all destinations. However, their individual characteristics such as the existence of irregular distribution of profits or the point of the destination on the tourism life cycle it is at, among others, is hindering the situation (Perles-Ribes et al. 2020a, b). This suggests that, when analysing this phenomenon, authors should have important knowledge of economic, social and environmental realities within communities (Harrill 2004).

However, the evidence points to two fundamental facts: there is an unfavourable attitude of the residents towards tourism growth when the carrying capacity of the destinations is surpassed (Muler-González et al. 2018; Swiader 2018) and when the growing demand for accommodation in residential areas has a significant effect on social sustainability (Francis 2019).

Deepening on the current ongoing debate about whether this growth on tourism experienced by some destinations turns into an enhancement on residents' well-being, there is an important factor on understanding this impact on the quality of life: the degree of tourism development. The key aspect of this is the way of understanding tourism development. Attending to literature, the more developed the tourism industry in the destination, the more opposed the residents become (Sharpley 2014; Rasoolimanesh et al. 2017). Nonetheless, for the authors of this chapter, it is important to distinguish between sustainable development, that leads to a successful implementation of the activity in the destination and the opposite of that, the overtourism, that generates several conflicts and diseconomies.

Nevertheless, this is not the only factor in having an impact on the behaviour of residents. Seasonality (Vargas-Sánchez et al. 2014), the type of tourist visiting the destination (Tung et al. 2019) and the development of the sharing economy (Postma and Schmuecker 2017) are also crucial determinants.

The debate of the role played by the sharing economy and the collaborative accommodation platforms in the feeling of loss of wellbeing of the residents is an important topic in literature nowadays (Sarantakou and Terkenli 2019; Yang and Mao 2018). Authors such as Guttentag (2015) or Heo (2016) highlight that these platforms contribute to creating a more authentic experience for the tourists. In addition to it, from a supply perspective, this accommodation product allows a better diversification in terms of price, location and quality (Gutiérrez-Taño et al. 2019; Sovani and Jayawardena 2017; Wang et al. 2016).

Simultaneously, the presence and growth of this type of accommodations can raise costs of living and housing, deteriorate residents' identification with place, the loss of the destination authenticity or the privatization of spaces that were intended to be publicly accessed, leading to exclusion and segregation (Benner 2019). These consequences portray a negative perspective from the tourism activity, influencing and creating rejection towards it from the residents.

There is literature addressing the growing presence of tourists and linking it to an acceleration of the pressure of the called tourism gentrification (Cócola-Gant 2018), that we would address as an overtourism gentrification, understanding that sustainable tourism activity can cause, as previously stated, many benefits to both residents and tourists. Overtourism, exceeding the carrying capacity of destinations, intensifies the use of land, pushing up the value of residential properties (Garau-Vadell et al. 2018; Benner 2019; Oklevik et al. 2019). According to Logan and Molotch (2007), property owners are interested in promoting tourism in their areas since these new spaces of consumption can increase their land values.

Once a destination becomes popular, whether it is because of its tourist attractions or for its easy access (low-cost airlines having an essential role in this aspect), the pressure of the demand creates the need of more tourist accommodation, deriving in residential displacement (Cócola-Gant 2018), due to the conversion of residential homes into tourism accommodations thanks to the platforms of the sharing economy.

The higher value of the properties along with the opportunity of conversion into tourism accommodation brings a central actor into the problematic: investors. These platforms allow individuals and enterprises of storing capital in the housing market of popular destinations while renting them to visitors. Evidence from several studies shows that the main profile offering accommodation on Airbnb are investors and enterprises renting their properties all year long, more than families that rent their spare bedroom for some extra incomes (Arias-Sans and Quaglieri-Domínguez 2016).

There is also commercial gentrification that replaces the services and stores the residents need for their daily living for tourism-oriented ones (Cócola-Gant 2015; Terhorst et al. 2003). Then, the hastening of the gentrification process by tourism happens, mainly, because of two reasons: the rising of the house prices cause that low-income residents cannot afford to stay in their neighbourhoods and only wealthy individuals have enough money to use the services of the area, whose prices have also been risen (Franquesa 2011; Spirou 2011; Vives Miró 2011; Wortman et al. 2016; Cócola-Gant 2019).

In the specific case of the two most important cities in Spain representing urban tourism, such as Barcelona and Madrid, whose neighbourhoods are analysed in this chapter, there is existing literature connecting the quick increase in the supply of accommodation by Airbnb deriving in adverse important effects such as gentrification, the discomfort of neighbours or increased land prices (Adamiak et al. 2019; Gil and Serquera 2018), which give solid evidence for linking these effects to situations of overtourism.

Big data used in this chapter allows us to propose a detailed level of analysis of these negative consequences. Within the city, it is easy to observe the higher profitability of rentals in popular neighbourhoods as opposed to residential ones, a phenomenon that has led to a significant change of use in the first ones, from residential to tourism. All of this may end into price rises and difficulties in accessing housing for the most disadvantaged population in the areas, having to move to a more affordable area (Opillard 2016; Vivés-Miró and Rullan 2017; Alcalde-García et al. 2018).

Some of the most important and recent studies on this topic (Peeters et al. 2018; Perles-Ribes et al. 2020a, b) find the spread of non-regulated accommodation offer as a cause of overtourism and the Airbnb platform as the instigator of it.

According to it, this chapter considers the greater or lesser degree of implantation of the collaborative economy offer in each of the neighbourhoods as the dependent variable of the overtourism situations of the destinations. Therefore, the model proposed they must be able to predict the probability that a neighbourhood from Madrid or Barcelona has to suffer from overtourism, based on the variation of the housing prices for the neighbourhood between 2015 and 2017, the within-city movements of the population of 2017 and the household incomes of each of the neighbourhoods on 2017.

12.3 Methodology and Data

As mentioned above, the present study will measure the problem of overtourism in two of the most important urban destinations in Spain: Madrid and Barcelona. To accomplish this, different techniques and methodologies have been employed.

First, we wanted to test whether there are differences in the comparison of two independent groups, neighbourhoods presenting more than a 5% of their residential properties announced for rent on Airbnb and those presenting lower rates. For the whole analysis, properties have only been taken into account if they were offered as an entire rent, dismissing the announces offering rooms to share. This is because we wanted to test the effects of having these empty houses for rent and its relationship with the above-mentioned gentrification process. For this, the nonparametric Mann–Whitney U test was used (Mann and Whitney 1947) as a complement to the logit analysis.

This test is used to compare two independent samples with quantitative variables. It is the nonparametric version of the parametric t-test, and its great advantage is that it can be used for small samples of subjects. Values from both samples need to be comparable in size and measurable on an ordinary scale. The null hypothesis (H_0) of the Mann–Whitney U test specifies that the two groups come from the same population. That is to say, that the two independent groups are homogeneous and have the same distribution (McCabe et al. 2010).

Second, discrete choice models are commonly used as an instrument capable of measuring the probability of an event occurring or not between a finite set of alternatives. Among the discrete choice models, logit models are the most common, due to their ease of interpretation and the interpretative richness of the obtained results (Bowden 2006).

The present study includes a logit model that estimates the probability of presenting a problem of overtourism, explained by different explanatory variables. The dependent variable, overtourism, takes the value 1 if more than 5% of the houses in the neighbourhood are announced on Airbnb and the value 0 if not. As explanatory variables, the variation rate of the house price between the years 2015 and 2017, the rate of intra-municipal mobility, and the average household income have been included. The expression of the logit model is the following:

$$P_i = P(\text{overtourism}_i = 1) = F(x_i, \beta) = \frac{e^{x_i\beta}}{1 + e^{x_i\beta}};$$

where P_i is the probability of a neighbourhood to present a problem of overtourism, x_i is the vector of explanatory variables and β is a vector of parameters. Besides, the probability that each neighbourhood has a positive outcome has been calculated, that is, $P(Y_i = 1) = F(x_i, \beta)$.

The main database used for our analysis has been Experimental Statistics, from the Spanish National Statistical Office (INE). This database includes information on average income levels, demographic indicators and percentages of the population below certain poverty lines. The level of detail of the database allowed us to obtained information at a neighbourhood and a district level for each Spanish city. Therefore, this database is complete enough to allow the proposed analysis and is perfectly adapted to the objectives of the study. To complete this database, we have also used the census information collected on the official pages of the Madrid and Barcelona city council and the Inside Airbnb data for each year and city.

The national database includes information from the years 2015, 2016 and 2017. However, for our study, we have decided to do a cross-sectional analysis corresponding to the year 2017. This is due to the arising of the collaborative platforms in Spain, such as Airbnb, and because that is the moment when overcrowding problems began to have a higher incidence.

In practice, the database collects information for a total of 2,459 census sections in the city of Madrid, and a total of 1,068 census sections in the city of Barcelona. These census sections are the highest level of detail that can be obtained using this database. For our study, we needed information about the neighbourhoods of each city, which are a larger territorial unit than the census sections. For this reason, the census sections have been grouped around their corresponding neighbourhoods to obtain information on each one, having a total of 129 neighbourhoods in Madrid and a total of 73 neighbourhoods in Barcelona.

12.4 Results

The proposed logit model shows the variation rate of the housing prices between the years 2015 and 2017, and the rate of intra-municipal mobility has a significant influence on levels of overtourism in Madrid and Barcelona, at a 5% level. The fact that both coefficients are significant to mean two important things: (a) neighbourhoods with an important housing price variation are more likely to have more than 5% of their residential properties announced on Airbnb and (b) neighbourhoods with more intra-municipal mobility of their population are more likely to have more than 5% of their residential properties announced on Airbnb (Table 12.1).

First, there is a positive relationship between the variation rate of the housing prices between the years 2015 and 2017 and the likelihood of the neighbourhoods to fall onto our overtourism-affected group. This implies that neighbourhoods presenting higher levels of price variation of the properties could be expected to have a greater likelihood of presenting overtourism problems. On the other hand, the rate of intra-municipal mobility is negatively related to the likelihood of falling within the overtourism-affected group.

This seems a counter-intuitive result since, as stated in the literature review from Sect. 12.2, overtourism seems to have an effect of residential displacement for the conversion of the houses into touristic accommodations. Nonetheless, this does not seem to be an effect for the neighbourhoods in this study for the chosen period. Following what the model is telling us, higher levels of intra-municipal mobility led to a lower likelihood of falling into the overtourism-affected group.

Mean household income is not significant at a 5% level, but it can be accepted at 10%. Although we must address these results carefully, it suggests that the relationship it presents with the dependent variable is negative (Table 12.2).

Results obtained from this analysis are coherent with the previous one, telling us that the mean household income is not significant, with an odds ratio close to 1,

Table 12.1 Logit regression output, coefficients

Number of observations						202
LR chi2(3)						39.46
Prob > chi2						0.000
Pseudo R2						0.193
More than 5% residences offered on Airbnb: overtourism	Coef	Std. err	z	*P* > |z|	[95% conf. interval]	
Housing price variation (2015–2017)	6.670558	1.850638	3.60	0.000	3.043375	10.29774
Within-city movement of population	−0.0751463	0.0239451	−3.14	0.002	−0.1220778	−0.0282147
Mean Household Income	−0.0000268	0.0000163	−1.65	0.100	−0.0000586	5.11e−06
Constant	−2.093016	0.8372507	−2.50	0.012	−3.733997	−0.4520344

Table 12.2 Logit regression output, odds ratio

Number of observations					202	
LR chi2(3)					38.98	
Prob > chi2					0.000	
Pseudo R2					0.1939	
More than 5% residences offered on Airbnb: overtourism	Odds ratio	Std. err	z	P > |z|	[95% conf. interval]	
Housing price variation (2015–2017)	812.3995	1506.237	3.61	0.000	21.4581	30,757.28
Within-city movement of population	0.9277428	0.0222306	−3.13	0.002	0.885179	0.9723532
Mean Household Income	0.9999734	0.0000163	−1.63	0.100	0.9999415	1.000005
Constant	0.1204886	0.1010304	−2.52	0.012	0.0232921	0.6232801

and that the within-city population mobility presents a negative relationship with the likelihood of presenting overtourism.

The most important outcome from this analysis is the presence of such a high value of the odds ratio in the housing price variation variable. The odds ratio tells us how much the odds of the dependent variable change for each unit of change in the independent one. Consequently, for each unit change in the price variation of the residential properties, the odds of the neighbourhood for suffering from an overtourism situation rise by 812. Thus, the relationship between the rise of housing prices and the presence of Airbnb accommodations is positive and robust from our evidence.

Income results needed more detail, so a new logit model is carried out, dividing household income into three categories, as stated in the Table 12.3.

Adding these new categories to the model generated the following results (Table 12.4).

First of all, and as expected, the variables that were previously included and analysed in the first model do not change significantly. Regarding the new income variables, the fact that the neighbourhood has a medium mean household income (between 31.501€ and 41.000€) is now significant at a 5% level for the model. This implies that, compared with the omitted variable (high-income neighbourhoods), the

Table 12.3 Mean household income categories

Low income	20.500€–31.500€
Medium income	31.501€–41.000€
High income	41.001€–89.215€

Table 12.4 Logit regression output, mean household income categories added

Number of observations	202					
Chi2(3)	43.56					
Prob > chi2	0.000					
Pseudo R2	0.2167					
More than 5% residences offered on Airbnb: overtourism	Coef	Std. err	z	*P* > |z|	[95% conf. interval]	
Housing price variation (2015–2017)	6.683519	1.87232	3.57	0.000	3.01384	10.3532
Within-city movement of population	−0.0729551	0.0242923	−3.00	0.003	−0.1220567	−0.025343
Mean household income	−0.0000268	0.0000163	−1.65	0.100	−0.0000586	5.11e−06
Low income	−0.430023	0.5568201	−0.08	0.938	−1.13435	1.048345
Medium income	1.116642	0.4757851	2.35	0.019	0.1842201	2.049163
Constant	−2.093016	0.8372507	−2.50	0.012	−3.733997	−0.4520344

neighbourhoods in the medium range of income present a positive relationship with the likelihood of having more than 5% of the properties announced on Airbnb. We can understand from this analysis that the overtourism and touristification phenomenon has a higher degree of implementation in the medium-income neighbourhoods.

In addition to this, a prediction has been made for the probabilities of having overtourism problems in each neighbourhood of Madrid and Barcelona, available in Annex 12.1. This analysis, carried out at a neighbourhood level, will allow us to determine which areas of both cities are more vulnerable to the overtourism consequences and to check if the proposed model has a good fit. Obtained probabilities have been divided into three groups: the group of neighbourhoods presenting a high incidence ($P \geq 0.7$), the group of neighbourhoods with a medium one ($0.7 \geq P \geq 0.4$) and the group of neighbourhoods reporting low probabilities ($P < 0.4$). The map below shows the results in each city (Fig. 12.1).

Among the group of neighbourhoods with a high probability of having overtourism problems, attending to our study, there are a total of seven: in Barcelona, Sants, Poble Sec–Parc Montjuïc, El Raval, Sant Pere and Santa Caterina i la Ribera; in Madrid, Sol, Universidad and Embajadores. These neighbourhoods are located in central and tourist areas and may, therefore, be prone to tourist overcrowding problems. Real values of the Airbnb offer in January 2018 put these neighbourhoods in the

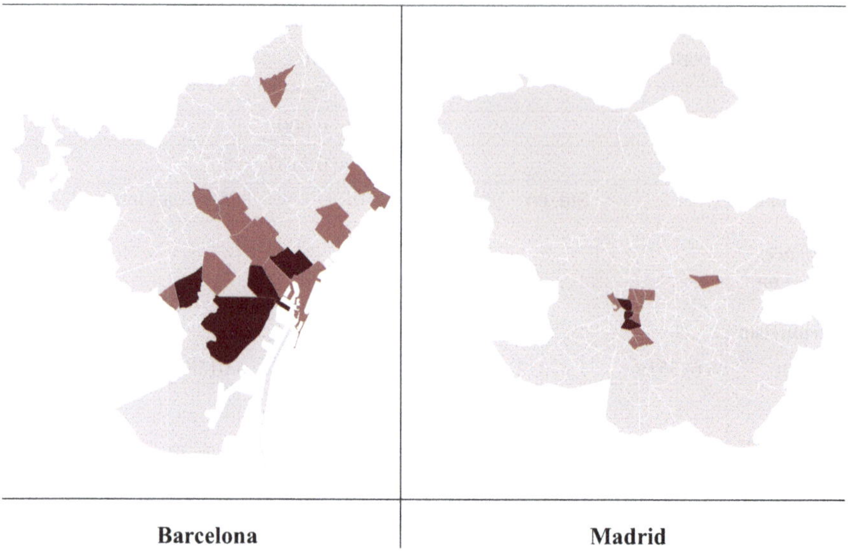

Fig. 12.1 Neighbourhoods classified attending to their prediction of the probability of suffering from overtourism

top positions of both cities. This fact suggests that the model is probably presenting a good fit.

On the other hand, among the group with a medium incidence, we have a total of 21 neighbourhoods. In Barcelona, Les Roquetes, Nova Esquerra de l'Eixample are the neighbourhoods with the highest values within this group, while Chopera is the highest from Madrid. It can be highlighted the presence of neighbourhoods such as Vila de Gràcia, Barceloneta or Barri Gòtic standing out in Barcelona while, in Madrid, Cortes, Castellana or Argüelles are a sign of coherent results from the prediction. Finally, in the group of neighbourhoods with a low incidence of overtourism, we find neighbourhoods that, generally, do not present problems of tourist overcrowding. In this group, we can find the majority of the neighbourhoods in both cities.

In addition to this, we have performed goodness of fit analysis based on the predictions. Assuming the model predicted an overtourism situation (1) if the probability of having more than 5% of resident properties announced on Airbnb is higher than 0.7.

As Table 12.5 evidences, our model was able to correctly predict 83,66% of the results, 162 negative and 7 positive predictions. This, in addition to the previous measures of fit offered by the logit model, endorses a good adjustment of the model to data.

To complement the logit analysis and check the dependent variables of the model behaviour in other tests, we perform nonparametric tests to check whether the distribution of these variables is the same within the neighbourhoods taking a value of 1, presenting more than 5% of their housing announced on Airbnb, and the ones taking

Table 12.5 Comparative table of predicted and real values with the logit model

	More than 5% of residential properties announced on Airbnb		Total
Predicted choice	0	1	
0	162	33	195
1	0	7	7
Total	162	40	202

Table 12.6 Mann–Whitney U for independent samples test results

	Null hypothesis	Significance	Decision
1	The distribution of the price variation rate between 2015 and 2017 is the same among the overtourism categories	0.000*	Reject null hypothesis
2	The distribution of the Average Household Income is the same among the overtourism categories	0.921	Keep null hypothesis
3	The distribution of the Intra Municipal Mobility of the population is the same among the overtourism categories	0.381	Keep null hypothesis

Note Asymptotic significances are shown. Significance level is set at 0.05

a 0 value and presenting lower rates. Results from this analysis are shown in the Table 12.6.

Attending to the Mann–Whitney U for independent samples test, price variation rate between 2015 and 2017 does change significantly between the neighbourhood with more than 5% of their properties announced on Airbnb and those with lower rates. Even though the logit model already told us this result, since it presented a very significant odds ratio—for each unit change in price variation rates, the odds of the neighbourhoods suffering from overtourism boost by 812—authors considered it important to double-check this with another nonparametric measure since data doesn't fit a normal distribution.

This test strengthens the hypothesis that the presence of a significant amount of Airbnbs in a neighbourhood does raise housing prices for our data from Madrid and Barcelona, in Spain. The rest of the variables do not reach a significant level, although this result could be expected due to their odds ratio and overall impact being minor.

12.5 Conclusions

The main result shown in the previous section is the existence of a direct and statistically significant relationship between the increase of housing prices of neighbourhoods and their probability of suffering an overtourism situation. Ostensibly, this result is in line with some of the literature reviewed in Sect. 12.2 (Porter and Shaw

2013; Garau-Vadell 2018; Benner 2019; Oklevik et al. 2019). Nonetheless, some authors talk about a gentrification process going beyond housing prices, expelling population from their neighbourhoods (Opillard 2016; Vivés-Miró and Rullan 2017; Alcalde-García et al. 2018). According to our analysis, this effect is not yet significant in Madrid and Barcelona, although Spain was the first country to acknowledge this situation as a risk to their tourism industry (Koens et al. 2018). It is possible that the population segregation effect has not yet happened or that it is not visible for the 2018 analysis. Nevertheless, the fact that the Spanish sector was the first to call out and identify this situation is an important aspect, showing the constant effort to develop a sustainable tourism model able to offer a win–win situation to both residents and tourists.

Regarding our first research question, we can conclude that there are some characteristics associated with the overtourism gentrification process visible and measurable in these neighbourhoods, but there are others that we cannot observe yet. This means that the overtourism effect in these cities is, up to 2018, purely related to real estate prices. This conclusion is positive, meaning the composition of neighbourhoods has not yet been altered, but it inevitably harms the sustainability of the cities, as we will see further on. Nonetheless, another notable result derived from our analysis is related to which neighbourhoods, in terms of mean household incomes, are more likely to suffer from this phenomenon. To the knowledge of the authors, this is the first time a study attempts to identify the risk of suffering from overtourism in terms of income. Big data analysis allowed us to classify and identify which areas of the cities are more overtourism-prone, concluding this phenomenon is mostly present on medium-income neighbourhoods. This has important implications. As stated before, this analysis was carried out, taking into account only the properties that were announced as entire, dismissing the offer of room rentals that implied sharing the place with the owners. It means that the empty properties announced on Airbnb in these cities that can be taken as an example of urban tourism in Spain are more likely to be on medium-income neighbourhoods than on higher income ones.

Nonetheless, the presence of an Airbnb in a neighbourhood does not imply that the revenues coming from that renting stay in that same neighbourhood since the owners do not necessarily live there. In addition to this, and as stated in the literature review, there is an important amount of companies owning Airbnb announced properties. We can conclude, then, that benefits and adverse effects of the tourism promoted by Airbnb suffer from what we can call an "offshoring process". Residents of these neighbourhoods are the ones suffering the negative impact of having tourists staying overnight in their flat buildings, but the economic benefit of this activity does not necessarily return to the same area.

This effect can be expected in regulated accommodation, since there are no prospects of a better redistribution of the benefits generated by the hotel, to put an example. Nonetheless, when talking about Airbnb and it is alleged belonging to the sharing economy, and income redistribution effect is supposed. To sum up, this chapter contributes to the current research on overtourism in several ways. First of all, the use of big data is key to be able to disaggregate results into more detailed levels of information. In this chapter, it allows us to conclude that neighbourhoods with higher price variation rates between 2015 and 2017 are more likely to suffer from overtourism, and the same applies to mean income neighbourhoods. Certain traits allow us to identify the situation in these cities with a gentrification process, like the increase in housing prices, while other characteristics, like population expelling, are not noticeable yet. Next steps in this line of investigation include extending the sample of neighbourhoods, trying to find different phases of the overtourism phenomenon implantation. Once this is achieved, it will be easier to determine a threshold roughly adjusted to the country, based on the study of its cities and neighbourhoods. Hence, this chapter will undoubtedly be the first step to build a social early warning system to overtourism situations. Authors consider it important to understand the tourism activity not only as an essential asset and driving force of the economy but also as a key aspect to accomplish the Sustainable Development Goals proposed by United Nations within the 2030 Agenda. Such a transversal industry as tourism has an impact on several of the designed goals, but the topic addressed on this chapter focuses mainly on the relationship between tourism and the 11th goal: "Make cities and human settlements inclusive, safe, resilient and sustainable".

More specifically, this analysis addresses targets 1 and 3 of the 11th goal. As to the first one, "ensure access for all to adequate, safe and affordable housing and basic services and upgrade slums", the main finding of our model highlights the significant and direct relationship present between Airbnb and the rise in housing prices, as stated before. Therefore, neighbourhoods presenting more than 5% of their residential properties on Airbnb are moving away from this target since housing has become less affordable in these areas. Regulations and even limitations of the activity may have to be considered to allow access to proper housing to the population, as well as offering some accommodations for tourists. Notwithstanding, and from the authors' point of view, there is the need of finding a comfortable threshold for each destination to enable a fruitful coexistence of tourists and residents.

The third target of the 11th goal is about "enhancing inclusive and sustainable urbanization and capacity for participatory, integrated and sustainable human settlement planning and management in all countries". In a similar approximation as with the previous target, the findings of this chapter evidence the need of better managing and planning of the activity, aiming it to be a real way of sharing experiences and redistributing revenues among the population of the touristified areas. This cannot be achieved other than by regulating, putting some limits to the activity and making these touristic neighbourhoods real ecosystems of cultural richness, exchange and value creation for the cities and its residents. Finally, there are some limitations to the study known by the authors that must be addressed. The measure of 5% of the residential properties announced on Airbnb is a criterion proposed by the authors that can be classified as an objective measure of the overtourism activity. As stated in the literature review section, there is no scientific agreement around a valid and universal measure for this phenomenon. Considering the data available at the moment of the making of this chapter, authors propose a simple measure that, when working in the logit model, throws interesting and, somehow, accurate results. Nonetheless, it is possible that for future investigations and with a better-trained model, we will be able to determine a better-adjusted threshold for the neighbourhoods and destinations, attending to their different characteristics (whether they are urban destinations, sun-and-beach ones, etc.). Albeit, this chapter constitutes one of the first approximations to the effects of Airbnb on the neighbourhoods it is settled on. An attempt to understand the changing process it produces within the cities to help policymakers to make better decisions on the topic, with the ultimate purpose of developing a quality tourism activity both for the tourists and the residents, as a key aspect to achieve a sustainable tourism model attending to the principles of the 2030 Agenda.

Annex 1

Logit detailed predictions of the overtourism risk by neighbourhood in Madrid and Barcelona.

High-Risk/High Airbnb Incidence (>0.7)

Barcelona		Madrid	
Sants	0.9650505	Sol	0.9047003
El Poble Sec-AEI Parc Montjuïc (1)	0.8437184	Universidad	0.7198946
El Raval	0.8053334	Embajadores	0.7179555
Sant Pere, Santa Caterina i la Ribera	0.7252064		

Medium-Risk/Medium Airbnb Incidence (0.7 > x > 0.4)

Barcelona		Madrid	
Les Roquetes	0.6770074	Chopera	0.6015115
La Nova Esquerra de l'Eixample	0.6335196	Cortes	0.568284
La Vila de Gràcia	0.5765759	Castellana	0.5199268
El Putxet i el Farró	0.5376721	Palos de Moguer	0.4651968
El Besòs i el Maresme	0.5016922	Argüelles	0.4520841
Sants-Badal	0.4966001	Justicia	0.4487841
La Barceloneta	0.4831527	Delicias	0.4381615
Verdun	0.4776263	Almagro	0.4232714

(continued)

(continued)

Barcelona		Madrid	
La Dreta de l'Eixample	0.4586901	Trafalgar	0.406309
El Poblenou	0.4156013	Salvador	0.4268486
El Barri Gòtic	0.4012293		

Low-Risk/Low Airbnb Incidence (0.7 > x > 0.4)

Barcelona		Madrid	
Navas	0.3961606	Palacio	0.373479
El Camp de l'Arpa del Clot	0.3775169	Cuatro Caminos	0.3583884
El Fort Pienc	0.35751	Comillas	0.3474085
La Salut	0.3395878	Pueblo Nuevo	0.3419373
La Vall d'Hebron	0.3392858	Apóstol Santiago	0.3321286
La Sagrera	0.292703	Arcos	0.3066629
Sant Martí de Provençals	0.2777808	Berruguete	0.2965476
Hostafrancs	0.2677673	Ibiza	0.2687543
El Camp d'en Grassot i Gràcia Nova	0.2669331	Puerta del Ángel	0.2618302
Moscardó	0.2585256	Ciudad Jardín	0.2585992
Les Corts	0.2405439	Numancia	0.2564068
El Parc i la Llacuna del Poblenou	0.2258069	Arapiles	0.2558724
El Bon Pastor	0.2240449	Goya	0.254212
Montbau	0.2192923	Ríos Rosas	0.2520477
La Maternitat i Sant Ramon	0.2116256	Ventas	0.2505273
La Sagrada Família	0.1945976	Recoletos	0.230523
El Clot	0.1867945	Pacífico	0.2274574
El Coll	0.1858625	Gaztambide	0.2274385
Porta	0.1806007	Horcajo	0.2272723
L'Antiga Esquerra de l'Eixample	0.1718151	Vista Alegre	0.2271481
El Turó de la Peira	0.1676654	Imperial	0.2182999
Sant Gervasi-la Bonanova	0.1468758	Ambroz	0.2106245
El Guinardó	0.1460638	Quintana	0.2104702
Diagonal Mar i el Front Marítim del Poblenou	0.1440515	Acacias	0.2085941
La Verneda i la Pau	0.1338284	Fuente del Berro	0.2081723

(continued)

(continued)

Barcelona		Madrid	
El Carmel	0.1328516	Aluche	0.2054898
El Congrés i els Indians	0.1239728	Lista	0.1907092
La Font d'en Fargues	0.1204125	Canillejas	0.1890756
La Bordeta	0.118691	San Diego	0.1860807
Sant Andreu	0.1177793	Media Legua	0.1851215
Horta	0.1172699	Pilar	0.1744766
Vilapicina i la Torre Llobeta	0.1162698	Pradolongo	0.1741776
Ciutat Meridiana	0.1149507	Campamento	0.1687373
Vallcarca i els Penitents	0.1141023	Portazgo	0.1630411
Vallvidrera, el Tibidabo i les Planes	0.108456	Valdeacederas	0.1619721
La Marina de Port	0.1048182	Lucero	0.16195
Sant Antoni	0.1047706	Guindalera	0.1615356
La Trinitat Vella	0.0969887	San Juan Bautista	0.159678
La Prosperitat	0.0958681	Almendrales	0.1475912
Can Peguera	0.0943234	Casco Histórico de Vallecas	0.1465741
Sarrià	0.0916327	La Paz	0.1460169
El Baix Guinardó	0.0873048	Atalaya	0.1452564
La Vila Olímpica del Poblenou	0.076847	Vallehermoso	0.1406388
La Guineueta	0.0756544	Adelfas	0.1404797
Sant Gervasi-Galvany	0.0746399	Atocha	0.1403796
La Font de la Guatlla	0.0739311	Legazpi	0.1401717
La Trinitat Nova	0.0664196	Fontarrón	0.138699
Vallbona	0.0571642	Opañel	0.1383292
Les Tres Torres	0.0537542	Almenara	0.1366364
La Teixonera	0.0480068	Vinateros	0.1357149
Can Baró	0.047766	Bellas Vistas	0.1314478
Pedralbes	0.035658	Castillejos	0.1309924
Sant Genís dels Agudells	0.0280545	Zofío	0.1299021
Torre Baró	0.0270397	Pavones	0.1236867
Provençals del Poblenou	0.0197183	Casco Histórico de Vallecas- La Gavia	0.123452
La Marina del Prat Vermell-AEI Zona Franca (2)	0.0192455	Ángeles	0.1228487
Canyelles	0.0148974	Los Jerónimos	0.1188799
La Clota	0.0000166	Hellín	0.1182556
		Águilas	0.1175716
		Palomeras Bajas	0.1151587

(continued)

(continued)

Barcelona		Madrid	
		Entrevías	0.1148736
		San Cristóbal	0.1130546
		Amposta	0.1112683
		Hispanoamérica	0.1108404
		Pinar del Rey	0.1064952
		Abrantes	0.1063834
		Rejas	0.1048103
		El Viso	0.1039993
		Marroquina	0.1039565
		San Pascual	0.1035322
		Casa de Campo	0.103273
		Ciudad Universitaria	0.1026617
		San Isidro	0.101342
		Rosas	0.1011157
		Prosperidad	0.1007023
		Concepción	0.0981398
		Orcasitas	0.0908892
		Fuentelarreina	0.0854804
		Puerta Bonita	0.0821026
		El Plantío	0.0797935
		Simancas	0.0795764
		Estrella	0.0742826
		Buenavista	0.0741288
		Peñagrande	0.0725124
		Los Rosales	0.0697517
		Palomeras Sureste	0.0690511
		Vicálvaro	0.065988
		Alameda de Osuna	0.0590114
		Casco Histórico de Barajas	0.0586634
		Aeropuerto	0.0565656
		Los Cármenes	0.054578
		Orcasur	0.0535975
		Costillares	0.0498005
		Valdezarza	0.0486893
		Castilla	0.0467094
		Canillas	0.0466428
		Valverde	0.0462628

(continued)

(continued)

Barcelona		Madrid	
		Colina	0.0462592
		Niño Jesús	0.0461661
		Valdemarín	0.0453971
		Nueva España	0.0453068
		Palomas	0.0436773
		San Fermín	0.0423674
		San Andrés	0.0375365
		El Pardo	0.0358251
		Valdefuentes	0.0264855
		Cuatro Vientos	0.0263317
		El Goloso	0.0200991
		Mirasierra	0.0133658
		Piovera	0.0131758
		Aravaca	0.0116863
		Corralejos	0.0114454
		Butarque	0.0057837
		Santa Eugenia	0.0049095
		Timón	0.0026068

References

Adamiak, C., Szyda, B., Dubownik, A., & García-Álvarez, D. (2019). Airbnb offer in Spain-spatial analysis of the pattern and determinants of its distribution. *International Journal of Geo-Information, 8,* 155. https://doi.org/10.3390/ijgi8030155

Alcalde-García, J., Guitart-Casalderrey, N., Pitarch-March, A., & Vallvé-Fernández, O. (2018). De la turismofobia a la convivencia turística: el caso de Barcelona. Análisis comparativo con Ámsterdam y Berlín. *ITEA*, *8*(2), 25–34.

Arias-Sans, A., & Quaglieri-Domínguez, A. (2016). Unravelling Airbnb. Urban perspectives from Barcelona. In: P. Russo & G. Richards (Eds.), *Reinventing the local in tourism. Travel communities and peer-produced place experiences* (pp. 209–228). London: Channel View.

Benner, M. (2019). *From overtourism to sustainability: A research agenda for qualitative tourism development in the Adriatic*, Heidelberg University. MPRA Paper No 92213. https://mpra.ub.uni-muenchen.de/92213/.

Bowden, J. (2006). A logistic regression analysis of the cross-cultural differences of the main destination choices of international tourists in China's main gateway cities. *Tourism Geographies, 8*(4), 403–428.

Capocchi, A., Vallone, C., Pierotti, M., & Amaduzzi, A. (2019). Overtourism: A literature review to assess implications and future perspectives. *Sustainability, 11*(12).

Coca-Stefaniak, A., Morrison, A. M., Liu, C., Pearce, D. G., Stepchenkova, S., Richards, G. W., et al. (2016). Editorial. *International Journal of Tourism Cities*, *2*, 273–280.

Cócola-Gant, A., & Palou i Rubio, S. (2015) Tourist promotion and urban space in Barcelona. Historic perspective and critical review, 1900–1936. *Documents d'Anàlisi Geogràfica, 61*(3), 461–482.

Cócola-Gant, A. (2018). Tourism gentrification. In L. Lees & M. Phillips (Eds.), *Handbook of gentrification studies*. Cheltenham and Northampton: Edward Elgar Publishing

Cócola-Gant, A. (2019). Gentrificación turística. In E. Cañada & I. Murray (Eds.), *Turistificación global. Estudios críticos en turismo*. Barcelona: Icaria, 291–308.

Dodds, R., & Butler, R. W. (2019). *Overtourism*. Berlin: De Gruyter

Forster, J. (1964). The sociological consequences of tourism. *International Journal of Comparative Sociology, 5,* 217–227.

Francis, J. (2019). Overtourism, what is It and How Can We Avoid It? 2018. Retrieved July 8, 2020, from www.responsibletravel.com/copy/what-is-overtourism.

Franquesa, J. (2011). We've lost our bearings": Place, tourism, and the limits of the "Mobility Turn" . *Antipode, 43*(4), 1012–1033.

Garau-Vadell, J. B., Gutierrez-Taño, D., & Diaz-Armas, R. (2018). Economic crisis and residents' perception of the impacts of tourism in mass tourism destinations. *Journal of Destination Marketing & Management, 7*, 68–75.

Gil, J., & Serquera, J. (2018). Expansión de la ciudad turística y nuevas resistencias. El caso de Airbnb en Madrid. *Revista De Metodología De Ciencias Sociales, 41,* 15–32.

Goodwin, H. (2017). The Challenge of Overtourism, Responsible Tourism Partnership Working Paper 4, October 2017. Retrieved April 10, 2020, from https://haroldgoodwin.info/pubs/RTP/T1/textquoterightWP4Overtourism01/T1/textquoteright2017.pdf.

Goodwin, H. (2019). Barcelona: Crowding out the locals. In R. Dodds & R. Butler (Eds.), *Overtourism: Issues, realities and solutions*. de Gruyter Studies in Tourism.

Gutiérrez-Taño, D., Garau-Vadell, J. B., & Díaz-Armas, R. J. (2019). The influence of knowledge on residents' perceptions of the impacts of overtourism in P2P accommodation rental. *Sustainability, 11,* 1043.

Guttentag, D. (2015). Airbnb: Disruptive innovation and the rise of an inform tourism accommodation sector. *Current Issues in Tourism, 18*(12), 1192–1217.

Harrill, R. (2004). Residents' attitudes toward tourism development: A literature review with implications for tourism planning. *Journal of Planning Literature, 18*(3), 251–266.

Heo, C. Y. (2016). Sharing economy and prospects in tourism research. *Annals of Tourism Research, 58*(C), 166–170.

Huete, R., & Mantecón, A. (2018). El auge de la turismofobia ¿hipótesis de investigación o ruido ideológico? *Pasos. Revista De Turismo Y Patrimonio Cultural, 16*(1), 9–19.

Jørgensen, M., & Mckercher, B. (2019). Sustainability and integration–the principal challenges to tourism and tourism research. *Journal of Travel & Tourism Marketing, 36*.

Koens, K., Postma, A., & Papp, B. (2018). Is overtourism overused? Understanding the impact of tourism in a city context. *Sustainability, 10*(4348), 1–15.

Kowalczyk, A. (2010). *Turystyka Zrównowazóna*. Warszawa, Poland: Wydawnictwo Naukowe PWN.

Logan. J. R., & Molotch, H. L. (2007). *Urban fortunes: The political economy of place* (2nd ed.). Berkeley and Los Angeles: University of California Press.

Mann, H. B., & Whitney, D. R. (1947). On a test of whether one of 2 random variables is stochastically larger than the other. *Annals of Mathematical Statistics, 18,* 50–60.

McCabe, S., Joldersma, T., & Li, C. (2010). Understanding the benefits of social tourism: Linking participation to subjective wellbeing and quality of life. *International Journal of Tourism Research, 12*(6), 761–773.

McCool, S. F., & Martin, S. R. (1994). Community attachment and attitudes toward tourism development. *Journal of Travel Research, 32*(3), 29–34.

McKinsey&Company. (2017). Coping with success. Management overcrowding in tourism destinations. (Travel, Transport & Logistics). World Travel & Tourism Council.

Milano, C. (2018). Overtourism, malestar social y turismofobia. Un debate controvertido. *Pasos. Revista de Turismo y Patrimonio Cultural, 16*(3), 551–564.
Milano, C., Cheer, J. M., & Novelli, M. (Eds.). (2019). *Overtourism: Excesses, discontents and measures in travel and tourism.* CABI.
Muler-González, V., Coromina, L. L., & Galí, N. (2018). Overoturism: Residents' perceptions of tourism impact as an indicator of resident social carrying capacity–case study of a Spanish heritage townTourism Reviewhttps://doi.org/10.1108/TR-08-2017-0138.
Namberger, P., Jackish, S., Schmude, J., & Karl, M. (2019). Overcrowding, overtourism and local level disturbance: How much can munich handle?. *Tourism Planing & Development*, 1–21.
Oklevik, O., Gössling, S., Hall, M. C., Steen Jacobsen, J. K., Grøtte, I. P., & McCabe, S. (2019). Overtourism, optimization, and destination performance indicators: A case study of activities in Fjord Norway. *Journal of Sustainable Tourism, 27*(12), 1804–1824.
Opillard, F. (2016). From San Francisco's 'Tech Boom 2.0' to Valparaíso's UNESCO World Heritage Site: Resistance to tourism gentrification in a comparative political perspective. In C. Colomb & J. Novy (Eds.), *Protest and Resistance in the Tourist City* (pp. 129–151). London: Routledge.
Peeters, P., Gössling, S., Klijs, J., Milano, C., Novelli, M., Dijkmans, C., et al. (2018). *Research for TRAN committee-overtourism: Impact and possible policy responses.* European Parliament, Policy Department for Structural and Cohesion Policies, Brussels.
Perkumiene, D., & Pranskuniene, R. (2019). Overtourism: Between the right to travel and residents' rights. *Sustainability, 11*(2138).
Perles-Ribes, J., Ramón-Rodriguez, A., Moreno-Izquierdo, L., & Such-Devesa, M. (2020a). Tourism competitiveness and the wellbeing of residents: A debate on registered and non-registered accommodation establishments. *European Journal of Tourism Research, 24.*
Perles-Ribes, J., Ramón-Rodriguez, A., Moreno-Izquierdo, L., & Such-Devesa, M. (2020b). Machine learning techniques as a tool for predicting overtourism: The case of Spain. *International Journal of Tourism Research.* https://doi.org/10.1002/jtr.2383.
Porter, L., & Shaw, K. (2013). *Whose urban renaissance? An international comparison of urban regeneration strategies.* Routledge.
Postma, A., & Schmuecker, D. (2017). Understanding and overcoming negative impacts of tourism in city destinations: Conceptual model and strategic framework. *Journal of Tourism Futures, 3,* 144–156.
Rasoolimanesh, S. M., Roldán, J. L., Jaafar, M., & Ramayah, T. (2017). Factors influencing residents' perception toward tourism development: Differences across rural and urban world heritage sites. *Journal of Travel Research, 56*(6), 760–775.
Sarantakou, E., & Terkenli, T. (2019). Non-institutionalized forms of tourism accommodation and overtourism impacts on the landscape: The case of santorini, Greece. *Tourism Planning & Development*, 1–23.
Schneider, D. (1978). The carrying capacity concept as a planning tool. Chicago, IL, USA: American Planning Association.
Seraphin, H., Sheeran, P., & Pilato, M. (2018). Over-tourism and the fall of Venice as a destination. *Journal of Destination Marketing & Management, 9,* 374–376.
Sharpley, R. (2014). Host perceptions of tourism: A review of the research. *Tourism Management, 42,* 37–49.
Smith, M., Szive, I., & Olt, G. (2019). Overtourism and resident resistance in Budapest. *Tourism Planning & Development, 16*(4), 376–392.
Sovani, A., & Jayawardena, Ch. (2017). How should Canadian tourism embrace the disruption caused by the sharing economy? *Worldwide Hospitality and Tourism Themes, 9*(4), 464–470.
Spirou, C. (2011). *Urban tourism and urban change: Cities in a global economy.* New York and London: Routledge.
Swiader, M. (2018). The implementation of the concept of environmental carrying capacity into spatial management of cities: A review. *Management of Environmental Quality, 29,* 1059–1074.

Terhorst, P., van de Ven, J., & Deben, L. (2003). Amsterdam: It's all in the mix. In L. Hoffman, S. Fainstein, & D. R. Judd (Eds.), *Cities and visitors: Regulating people, markets, and city space* (pp. 75–90). Oxford: Blackwell.
Tung V. W. S., King, B. E. M., & Tse, S. (2019). The tourist stereotype model: Positive and negative dimensions. *Journal of Travel Research*, 1–15.
UNWTO. (2018). Centre of Expertise Leisure, Tourism & Hospitality; NHTV Breda University of Applied Sciences & NHL Stenden University of Applied Sciences (2018). *'Overtourism'?- Understanding and Managing Urban Tourism Growth beyond Perceptions*. Executive Summary. UNWTO, Madrid.
UNWTO. (2020). World Tourism Barometer (English version), 18:1, 1–48. UNWTO, Madrid.
Vargas-Sánchez, A., Porras-Bueno, N., & Plaza-Mejía, M. A. (2014). Resident's attitude to tourism and seasonality. *Journal of Travel Research, 53*(5), 581–596.
Vera-Rebollo, J. F. V., & Ivars-Baidal, J. A. I. (2003). Measuring sustainability in a mass tourist destination: Pressures, perceptions and policy responses in Torrevieja, Spain. *Journal of Sustainable Tourism, 11*(2–3), 181–203.
Vives Miró, S. (2011). Producing a 'Successful City': Neoliberal urbanism and gentrification in the tourist city—the case of Palma (Majorca). *Urban Studies Research*, 1–13.
Vivés-Miró, S., & Rullan, O. (2017). Desposesión de vivienda por turistización? Revalorización y desplazamientos en el Centro Histórico de Palma (Mallorca). *Revista De Geografía Norte Grande, 67,* 53–71.
Wagar, J. A. (1964). The carrying capacity of wild-lands for recreation. *Forest Science, 10,* a0001.
Wang, D., Li, M., Guo, P., & Xu, W. (2016). The impact of sharing economy on the diversification of tourism products: Implications for tourist experience. In A. Inversini, & R. Schegg (Eds.), *Information and communication technologies in tourism.* Switzerland: Springer International Publishing.
Wortman, T., Donaldson, R., & Westen, G. (2016). They are stealing my island': Residents' opinions on foreign investment in the residential tourism industry in Tamarin Mauritius. *Singapore Journal of Tropical Geography, 37*(2), 139–157.
Yang, Y., & Mao, Z. (2018). Welcome to may home! An empirical analysis of Airbnb supply in US cities. *Journal of Travel Research*, 1–14.

Chapter 13
Determination of Standard of Living for People Involved with Tourism in Digha by Ordinal Regression Analysis

Subhankar Parbat, Payel Chatterjee, Sourav Sen, and Adwitiraj Banerjee

Abstract Tourism in India has evolved considerably during the last decade, and Bengal as a state has performed considerably well for the growth of tourism. Tourism in Bengal has not only contributed to the growth of the economy in the state of Bengal but has developed the state's remote area which has now turned into a key tourist spot. One such area is the region of Digha in the Purba Medinipur district of West Bengal, and it has turned out to be the most popular sea resort in West Bengal. The Digha Sankarpur Development Authority has taken key measures to develop tourism in Digha. The basic objective of this paper will be to determine the standard of living among the people of Digha who are associated with tourism after the various schemes to promote tourism by the state government. We have used ordinal regression analysis as most of our data were qualitative in nature and interpretation was done accordingly for the results obtained. Through our findings, we have tried to bring out the various problems associated with tourism in Digha along with the suggestions to improve it.

Keywords Tourism · Development · Project · Income · Schemes

JEL Code Z32 · Q57 · Q26 · L83

S. Parbat
Indian Institute of Management Calcutta, St. Xavier's College Kolkata, Kolkata, India
e-mail: subhankar02.parbat@gmail.com

P. Chatterjee
Indian Institute of Management Calcutta, The University of Burdwan, Bardhaman, India
e-mail: payel.chatterjee11@gmail.com

S. Sen (✉)
Indian Institute of Management Calcutta, Midnapore Treasury, Midnapore, India
e-mail: sensourav9012@gmail.com

A. Banerjee
Indian Institute of Management Calcutta, Kolkata, India
e-mail: adwitiraj@gmail.com

D. Balsalobre-Lorente et al. (eds.), *Strategies in Sustainable Tourism, Economic Growth and Clean Energy*, https://doi.org/10.1007/978-3-030-59675-0_13

13.1 Introduction

The growing influence of tourism on the economy and its share in improving the quality of people's lives by creating diverse large-scale employment both directly or indirectly are irrefutable. It promotes diverse cultural heritage with unity in diversity, strengthens peace of living in harmony and nurtures values of nature and environment. To assess and implement the profile of international and domestic tourism, periodical surveys and researches are conducted. In the year 2018, 17.42 million[1] international tourists visited India. During the period 2019, Rs. 2,10,981 crores (Provisional estimates) and US$29.962 billion[1] (Provisional estimates) with a growth of 8.3[1] and 4.8%[1] of foreign exchange earnings were recorded. According to the Tourism Satellite Account for India (TSAI), the estimates of the contribution of tourism to GDP during the year 2016–17 are 5.06%[1] (direct 2.63% and indirect 2.43%). The estimate of share in employment generated in the year 2018–19 is 12.75%[1] (direct 5.56% indirect 7.19%). Therefore, 87.5 million[1] employments generated through tourism. Domestic tourism continues to be a major contributor to the sector. There were 1854.93 million[1] tourists' visits all over the country during the year 2018. This study focuses on the tourism sector of West Bengal. The diversified flora and fauna make West Bengal an attractive tourist destination. About 74.5 million[2] domestic tourists visited West Bengal during the year 2016, i.e. 4.51%[2] of the overall domestic tourists. Foreign visitors are about 1.53 million[2] i.e. 6.19% of overall foreign tourists visited India. The government is identifying potential tourist spots to develop tourism infrastructure on the PPP model. As per the state budget 2018–19, the Government of West Bengal has allocated US$ 57.18 million for the development of the tourism sector. As per our convenience, we have chosen Digha, Purba Medinipur, West Bengal. It is the most popular tourist destination with beaches located south of Kolkata. It is described as 'Brighton of the East' best for a holiday. The main purpose of selecting Digha for the study is due to its popularity and tourism being the primary source of livelihood of the rural population. Our chapter which is based on a primary survey of 50 respondents which includes vendors, storekeepers, hotel workers and those involved in auto or toto driven mode of the passenger vehicle was carried on to understand the impact of tourism in their daily life. The main objective of our survey was to find out the change in the standard of living (SOL) after the growing scale of development in Digha by the West Bengal Government. The chapter is divided into five sections and following the introduction, we have the literature review of works done on this field. Next, we have a methodology which focuses on the survey and the area of study along with the specific technique adopted. The fourth section includes the model and its analysis. The fifth represents the final interpretation of the results, along with the conclusion obtained from this analysis.

---

[1]From Annual Report 2018–19, Ministry of Tourism, Government of India.

[2]From https://www.ibef.org/download/West-Bengal-March-20181.pdf.

13.2 Literature Review

Ghatage and Kumbhar (2005) in their paper "Growth and Performance of Tourism Industry in India" takes into account the performance of the tourism sector in the Indian Economy and the level of growth experienced by the tourism industry. The growth attained has been examined by observing the Foreign Tourist Arrivals (FTA) rate as well as Foreign Exchange Earnings (FEE) earned by India. With the help of secondary data, this paper showed the growth in the tourism sector in India from 1997 to 2013. From this paper, it was found that India has developed in the tourism sector as we can observe an improved global ranking of India in terms of the tourism industry. India's tourism sector has been ranked eighth among Asia and Pacific countries. This can be achieved due to the periodical review in tourism policy by the central as well as state governments. However, major issues are also there for which a consistent growth in FTA and FEE has not been observed. Dash et al. (2018) found out the effect of tourism on the economic growth of India. India has been presented as a case study to establish the linkage between economic growth and tourism. Many studies across the world have been referred to in this paper, and these studies show a multidirectional relationship between tourism receipts and economic growth. While conducting the research, the data of Foreign Tourist Arrivals in India and the contribution, both direct and indirect, of FEE to GDP has been taken into consideration. The time-series data has been taken from 1999 to 2015. The Augmented Dickey–Fuller (ADF) test and the Phillips–Perron test has been undertaken to prove that the time-series data are of stationery series in nature. Through the regression test on the Autoregressive Distributed Lag (ARDL) model, it has been found that the equilibrium relationship exists between different variables used in this study. Moreover, investments in human, as well as physical capital, are significant in tourism led to economic growth. On analysing various factors, it can be found that factors like exchange rate depreciation can cause only short-term economic benefits, while investments in physical capital have a positive impact on economic growth. Sharma (2018) explored how tourism leads to economic as well as the financial development of a country. The objective of this study was to show the economic growth caused by tourism development in a country. As far as Foreign exchange Earnings (FEE) was concerned, this study shows that, except in the year 2009, the FEE has increased considerably over the past concerning year between the period of 2007 and 2017. The time-series data of FEE and GDP has been used as the data to prove the fact that tourism leads to the economic growth of a country. An ADF test has been performed to prove that the GDP of India was stationary data which means it can indicate a possible future behaviour in the tourism sector of India. The cointegration between the two main data series, i.e. GDP & FEE, also does not exist. With the help of the Granger causality test, it has been established that tourism receipts in India are not affected by growth in GDP. However, tourism receipts do contribute to GDP growth. From this study, it can be concluded that like many other countries, India's tourism sector was also attributed to economic development. It has been established that tourism leads to economic growth, and hence, the government should provide the required assistance to the tourism industry by providing various infrastructural facilities. It was necessary to grow domestic tourism so that a wave of economic development can be achieved in various places within a country. Wang et al. (2014) estimated the satisfaction performance of Mainland China tourists who are travelling to Taiwan. The authors have used grey relational analysis in grey mathematics to analyse the satisfaction question items along with the predictive accuracy of the least mean regression model and each quantile regression model. For this, they have taken the first eight satisfaction items as independent items and the overall satisfaction level as the dependent variables. The empirical results from the paper concluded that gender, tourist attractions, hotel facilities, night fair culture and street cleanliness affected the overall satisfaction performance of the mainland tourists and had significant differences in both the high and low quantiles. Battour et al. (2017) aimed to test the relationship between tourist motivations and tourist satisfaction by using partial least square. The result shows that religion significantly moderates the relationship by using push and pull motivations which influence the overall tourist satisfaction. The study successfully develops a theoretical framework linking tourism motivation and religion (Islam) for a better understanding of Muslim tourist behaviour. Eygu and Gulluce (2017) tried to determine the factors that were relevant to the satisfaction of customers in conservative hotels, which were expected to have their incomes rise above 200 million dollars in the upcoming year. A survey was conducted on the guests staying at conservative hotels, and the data so obtained as a result of the study have been modelled using logistical regression method to process the results related to hotel satisfaction. It investigates whether the satisfaction of hotel enterprises is different according to the individual variables (gender, age, marital status, education status and monthly income) and sociocultural variables (nationality) for domestic and foreign customers staying in Islamic hotels in Turkey. The study shows that the customers who stay in conservative hotels are satisfied with the services provided as per the expectations of the customers. Although various studies have been conducted related to the tourism aspect in Digha, we were not able to find any relevant study based on the prospect of financial and change in the SOL for those associated with the tourism sector. As a result of which, we have decided to base our research on this front, to understand how certain factors can justify in influencing the SOL among those associated with the tourism industry in Digha. The main objective of our study is to determine the SOL among the people of Digha who are associated with tourism.

We had to face the following limitations during our course of the study:

- Unavailability of large-scale data due to Covid-19, which restricted our study to only 50 respondents as the whole nation went into lockdown after a week of our first phase of data collection.
- The dubious response of the respondent as they were biased on specific issues.
- Too much reliance on qualitative data as government reports for development was unavailable.
- The various assumptions associated with ordinal regression analysis.

13.3 Methodology

The study is conducted in India at the beaches of Digha which is situated in Purba Midnapore District of West Bengal on the coast close to Orissa border at a distance of 180 km from Kolkata. Digha is located at 21°38′18″N 87°30′35″E. Our respondents for the first study being the local people of Digha who have set up stall provided by the government to them, and also the various businesses and store owners that cater to tourism. The purpose of the study is to find the work done by the state government in the last 5 years in promoting tourism in Digha. Our objective is to see how their SOL changed in the last 5 years, and the effect in their daily and monthly income due to the policies adopted by the government and how effective these policies were in consideration to their SOL. We asked them individually venturing stalls and enquired whether the tourist arriving here increased due to the steps taken by government to make Digha a tourist hotspot. We asked them questions regarding the change in the footfall of tourist, how their income changed over the last 5 years, how developed Digha has been due to the steps taken by the government and about their livelihood as well, like whether they were getting the basic amenities or whether they were getting better SOL from the past. From the data-driven survey of 50 respondents across several sectors, numerous implications are evident. If we study the data thoroughly, we can see different important angles which make our study more interesting. We have based our study into two broad spheres, namely the present scenario and previous (5 years before). As observed, the mean income of 50 respondents previously is Rs. 11,728 and now present it is around Rs. 13,492. So, the respondents were better off as their income has increased over time by 15.04%. This can be neglected as we are not considering the general inflation rate over the years in our study. The average mean change in income is 55.11 and the standard deviation is 111.0748. Secondly, if we see the sector-wise classification, then presently, the shares of employment in sectors like the hotel, stores and transportation has increased significantly than before. The percentage share of employment in stores has got doubled than before from 32 to 64%. So, it can be seen that a large number of people are engaging themselves in business by setting up stores related to tourism than before. Interestingly the sectoral share of employment in agriculture before is 26%, but now presently, no such share in agriculture can be witnessed. So as per the Clerk–Fisher hypothesis of development, a structural shift in the pattern of employment of the local economy from primary (Agriculture) to tertiary (Stores, Transportation and Hotel related to tourism) sector is evident from the data. Now if we discuss the sector-wise classification of employment nature, then it can be seen that presently the percentage share of permanent employment has increased to 32%, which was just 10% before. The seasonal and temporary shares of employment have decreased from 22 and 20% to 12% and 8%, respectively. Moreover, a slight increase in the self-employed nature of employment can be seen from 42 to 48%, which shows us that a few portions of the mass tend to become independent and self-reliant. Indeed 6% of people among the respondents before 5 years or so were unemployed, but we simply cannot conclude from here that unemployment has decreased as we exactly don't have any data about their past work activities.

Variables Specification and Scale of Analysis

In this study, an ordinal regression is constructed to examine the relationship between the SOL of the respondent from the time of the infrastructural changes (IC) taking place which we have considered as the last 5 years since the new government has taken steps to rebuilt Digha as a significant tourist spot in West Bengal, from the period 2015 to 2020. The explanatory variable to explain the relationship being majorly the government role over these 5 years and the IC taking place in Digha within the study period. We have also considered certain factors like the use of luxury items (LI) which we felt would be a key to determine the SOL along with the use of loans to develop their business as a tool to enhance the living standards of the respondent.

Detail Analysis of Each Category

Standard of Living (SOL)—The dependent variable—This is our dependent variable, and it states the SOL of the respondent over the period of our study. The SOL of the respondent was taken based on ranks ranging from 1 to 5 (1: strongly disagree, 2: disagree, 3: undecided, 4: agree, 5: strongly agree). This estimate is taken to find out the impact of tourism on the lives of the vendors/store owners or the persons associated to cater the needs of the tourist. The change that tourism caused in their SOL is a key factor to determine the strength of association of being involved in this tourism sector.

Role of Government (ROG)—This is a relevant category that helped us estimate the view of the respondent in terms of the benefit they received from the state government. The support is mainly biased against the government as we found out that the respondents were expecting something even better in terms of the effort or compensation given to them by the government. The buildup of new stores in a location further away from the beach is the main reason for this dissatisfaction for the government added to that is the condition of the stores. This is followed in the considerable growth of competition as most of the vendors were selling similar items which included seashells to local items. The data obtained were ranked similar to SOL ranking and it has been considered an important factor that could determine the SOL of the respondent.

Infrastructural Changes (IC)—Digha has seen rapid growth in its infrastructural pattern over the years, once considered just a local tour spot, Digha now attracts tourists from all over India. Moreover, this has been possible mainly due to the IC, the newly developed Digha Railway Station, ropeways and water sports for tourists and the growth of new chains of hotels in the region have contributed to this rapid transformation in Digha. It now attracts tourists in thousands. Also, there is no specific season as people are piling up throughout the year, all this is possible due to the newly developed Biswa Bangla park which connects old and new Digha and gives an enormous view for the tourist. This is considered a great opportunity for the respondent to enhance their income due to an unprecedented rise in tourists in Digha and thus we have considered it as a factor influencing the SOL.

Loan Taken (LT)—This category highlights whether the respondent has taken any loan either from Banks, Microfinance Institutions and Self-Help Groups. During the survey, it is found that most loans were taken from Self-Help Group, and people who took loans have mostly cleared their loans. This is also taken as a measure for the SOL as we wanted to observe the dependency of SOL on loan is taken and how much it influences a respondent's business.

Use of Luxury Items (LI)—Living in a place like Digha where the vendors were mainly in mid-income range the SOL can have been estimated from the LI being used by the respondent. During our survey, we asked whether the respondent had a fridge, computer, TV, bike, proper electricity facility, water facility and proper houses. It is found that most respondents did not pose those basic amenities, so we classified certain items like Fridge, computer and bike for personal use as LI. Whether they had (1: YES) or they did not have (0: NO) was considered as an important factor to influence the SOL.

Modelling Standard of Living of those involved with business associated with Tourism

Various types of regression analysis are commonly used to model relationships between random variables. The use of a specific technique depends heavily on the level of data availability, spatial analysis and format and the specific questions to be answered (Norusis 2004, 2005). This paper focuses on the ordinal regression modelling technique that can be applied to the model SOL of those involved with business associated with tourism. Out of all the multiple regression techniques available, we have chosen this due to its advantages and after considering the various literature available on this kind of research. It does not assume that the response variable and the error terms are distributed normally (Norusis 2004). Secondly, it can take into consideration and introduce into the calculations some of that extra information in the ordinal scale of the response variable compared to logistic regression models. Finally, and most importantly, it allows investigating the influence and significance of all individual categories of categorical independent variables (Polyzos and Dionysis 2011).

The logit link takes the form link $\gamma_{ij} = ln(\frac{\gamma}{1-\gamma})$. The general model for ordinal regression is.

$$\ln(\frac{\gamma}{1-\gamma}) = \left\{\begin{matrix} \beta_{ROG-1} & ROG-1 \\ \beta_{ROG-2} & ROG-2 \\ \beta_{ROG-3} * & ROG-3 \\ \beta_{ROG-4} & ROG-4 \\ \beta_{ROG-5} & ROG-5 \end{matrix}\right\} + \left\{\begin{matrix} \beta_{IC-1} & IC-1 \\ \beta_{IC-2} & IC-2 \\ \beta_{IC-3} * & IC-3 \\ \beta_{IC-4} & IC-4 \\ \beta_{IC-5} & IC-5 \end{matrix}\right\}$$

$$+ \left\{\begin{matrix} \beta_{LT-0} \\ \beta_{LT-1} \end{matrix} * \begin{matrix} LT-0 \\ LT-1 \end{matrix}\right\} + \left\{\begin{matrix} \beta_{LI-0} \\ \beta_{LI-1} \end{matrix} * \begin{matrix} LI-0 \\ LI-1 \end{matrix}\right\}$$

13.4 Empirical Analysis

We have conducted ordinal logistic regression to show how the SOL (dependent variable) has improved among the respondents based on the variables like ROG, IC by the government, LT and the use of LI. For this study, we have used a Likert scale with five variables (1: strongly disagree, 2: disagree, 3: undecided, 4: agree, 5: strongly agree) for the ordinal data which were SOL, ROG in improving tourism, IC by the government in Digha. For the nominal data like a loan, we asked the respondent whether they had taken a loan or not and ordered it accordingly (0: No, 1: Yes) and similarly LI based on average income are considered to be fridge, air conditioner, bikes, cars and personal computers. Any respondent having any of the items were ranked 1, and those who did not have were numbered 0. We ran our test on SPSS package 25 and based on the output, we framed our analysis and working research hypotheses were determined as follows:

H.1. There is a significant relationship between a respondent's standard of living and the role of government in spreading tourism in Digha.

H.2. There is a significant relationship between a respondent's standard of living and the infrastructural change in Digha.

H.3. There is a significant relationship between a respondent's standard of living and the use of luxurious items.

H.4. There is a meaningful relationship between a respondent's standard of living and whether they have taken a loan or not.

13.5 Findings

The model with maximum likelihood in which the independent variables given in Table 13.1 that influence the SOL can be obtained is the logit model. The analysis results of the predicted model are summarized in the table. Out of the 50 respondents who participated in the survey, it is found that 13 of them strongly disagreed, 12 disagreed, 5 were neutral, 12 agreed and 7 strongly agreed that their SOL changed in the last 5 years. The goodness of fit test of the model is given using Pearson chi-square and deviation statistics. The goodness of fit test of the model is given using Pearson chi-square and deviation statistics. The model's suitability is determined using the difference between the observed and expected values of the model. Therefore, it is assumed that the model agrees with the assumption that $p > 0.05$ as statistically significant. The R square values of the model are calculated, showing how many per cents of the dependent variable is explained by the independent variables. For R square, we take the Nagelkerke R square, which is at 45.8% for our test. This explains that the independent variables express 45.8% of the variability in the dependent variable. In the model, there are four independent variables (ROG in improving

Table 13.1 Overall Model-fitting information, the strength of association, goodness-of-fit statistics[a]

Model	−2 log-likelihood	Chi-square	d.f	Sig
Intercept Only	127.804	–	–	–
Final	99.091	28.712	10	0.001

Pseudo R-square

Statistics	Value
Cox and Snell	0.437
Nagelkerke	0.458
McFadden	0.185

Goodness-of-fit

	Chi-square	d.f	Sig
Pearson	89.199	102	0.813
Deviance	77.59	102	0.966

[a]Link function: logit

tourism, IC, LT and use of LI) that are found and the probability of these variables is examined. These probability values are the values of the Wald test to determine whether the parameters are meaningful. When the analysis results are examined in Table 13.2, the significance level is found to be statistically significant when p values of some variables were less than 0.05.

The reference category in the study is made according to the interpretations determined as the last category. According to the ordinal logistic regression analysis in Table 13.2, the reference category is determined as the last category for each independent variable, and the interpretations were made accordingly. Three categories of the threshold values calculated in the model are significant. It is found that, when the independent variables explaining the SOL are examined, it is found that a meaningful relationship exists for one category in the case of infrastructure, two categories of government support and one category of LI being statistically significant at 5%. When the value of each of these significant variables increases by one unit, it is observed that the predicted rate of the dependent variable will also increase. In this category, it is observed that due to them disagreeing with the IC, there has been a reflection in their SOL. Similarly, disagreeing with governmental support which is significant for our test, we can conclude that their SOL has also been affected, as almost 25 of the respondents responded they disagreed with the change in the SOL over the 5 years. Also, the use of LI seems to affect the respondent as well ($p < 0.05$ for non-use of LI), which signifies that those who doesn't have LI were not satisfied with the SOL, giving a reason to conclude that SOL and why it has turned out falling for this respondent is because of the lack of government support, lack of use of LI and the lack of IC as felt by some of the respondent. Thus, we can accept the hypotheses 1, 2 and 3, and we have to reject hypothesis 4 based on the test of significance.

Table 13.2 Parameter estimates

Parameter Estimates								
		Estimate	Std. error	Wald	d.f	Sig	95% confidence interval	
							Lower bound	Upper bound
Threshold								
d1	(SOL) Standard of Living = [1] Strongly Disagree	−5.059	1.146	19.507	1	0.000***	−7.305	−2.814
d2	(SOL) Standard of Living = [2] Disagree	−3.615	1.06	11.622	1	0.001***	−5.694	−1.537
d3	(SOL) Standard of Living = [3] Neutral	−3.067	1.031	8.855	1	0.003***	−5.087	−1.047
Δ4	(SOL) Standard of Living = [4] Agree	−0.958	0.922	1.081	1	(0.299)	−2.766	0.849
Location								
IC								
X1	IC = [1] Strongly Disagree	−0.609	1.444	0.178	1	(0.673)	−3.44	2.222
–	IC = [2] Disagree	−3.037	1.439	4.452	1	0.035**	−5.857	−0.216
–	IC = [3] Neutral	−1.501	1.185	1.604	1	(0.205)	−3.825	0.822
–	IC = [4] Agree	−1.1	0.675	2.661	1	(0.103)	−2.422	0.222
–	IC = [5] Strongly Agree	0(a)			0			

(continued)

Table 13.2 (continued)

Parameter Estimates

		Estimate	Std. error	Wald	d.f	Sig	95% confidence interval	
							Lower bound	Upper bound
ROG								
X2	ROG = [1] Strongly Disagree	−2.693	1.073	6.301	1	0.012**	−4.796	−0.59
–	ROG = [2] Disagree	−4.887	1.609	9.222	1	0.002***	−8.04	−1.733
–	ROG = [3] Neutral	−1.699	1.291	1.732	1	(0.188)	−4.23	0.831
–	ROG = [4] Agree	−1.12	0.682	2.696	1	(0.101)	−2.456	0.217
–	ROG = [5] Strongly Agree	0(a)			0			
LT								
X3	Loan Taken = [0] No	−0.297	0.616	0.233	1	(0.629)	−1.504	0.91
–	Loan Taken = [1] Yes	0(a)			0			
LI								
X4	Use of Luxury Items = [0] No	−1.749	0.742	5.56	1	0.018**	−3.202	−0.295
–	Use of Luxury Items = [1] Yes	0(a)			0			

Link function: logit. (a) This parameter is set to 0 because it is redundant. *Note* * Significant at 10%, ** Significant at 5%, * Significant at 1%, () Not Significant.

13.6 Analysis of Findings

Model Fit Summary and Goodness of Fit

The final ordinal regression model includes the constant, all the tested variables and the statistically significant two-way interaction effect. The model uses the logit link function. Results about the strength of associations, the predicted ability of the model as well as goodness-of-fit statistics, are presented in Table 13.1. The location coefficients for all of the predictor variables in the model are zero; thus, the results helped us in the test of null hypothesis yielding a significance level of 0.001.

R Square—The pseudo R^2 statistics measure the success of the model in explaining the variations in the data, which is an indication of the strength of associations between the dependent and the independent variables. The pseudo R^2 for Cox and Snell (0.437) and Nagelkerke (0.458) can be considered slightly below satisfactory as it is below 0.50. The pseudo R^2 for McFadden (0.185), which is least satisfactory is a measure of entropy reduction between the intercept-only and the final model. The goodness-of-fit measures of Pearson and Deviance are reliable since it has a significance value of p (0.813) and (0.996) which is higher than (0.05).

Parameter Estimates—The results are displayed in Table 13.2 and considered whether they are statistically significant or not. The estimate indicates that the variables have a significant influence on the SOL of the respondent. The variable IC has a negative coefficient for all the categories. Furthermore, the statistical significance is satisfactory only for the second category. The negative sign of all the categories indicates that the IC were less likely to attract increased income sources for those associated in tourism as a result than where the state predicated that growth in tourism due to infrastructure would lead to a higher level in tourism. Hence, we see that SOL has not been better due to the added infrastructure in Digha. Also, the second factor being significant shows that disagreeing IC has contributed to the fall in the quality of living among the respondent, whereas the higher ranks being insignificant suggest that respondent are not satisfied. Although the provision of incentives to tourism due to massive changes in infrastructure may be of great significance to the tourist its result is not reflected in the growth of livelihood of the people who generate income from tourism-related activities leading to serious disadvantages found in some locations. This is logical since an area to be suitable for tourism development of people is prior important along with the tourist spot and local government must look into this matter to improve the condition of the place. Coming to the ROG, we find a similar negative coefficient for all the categories. In this case, we have a statistically significant result for the first and second categories only, rest all being insignificant. Thus, we can say that the respondents were strongly dissatisfied and disagreed with the support given to them by the local government. While the survey is going on, they expressed this quite vocally and it is reflected in the test result as well. The major shift in their business from the beachside to a concentrated zone along with the lack of diversity in the items for sale by this vendor was a major reason why they were not able to earn much from their business. And for this, they have held the local government responsible along with signifying the government's inability to attract tourists. Next, the two binary variables representing the LT and use of LI out of which the former is statistically insignificant, while the latter is statistically significant at 5%. The variable concerning LI facilities has a negative sign for code [0] (absence of LI) meaning that the prefectures without LI have led to the fall in the quality of living than those with such facilities. The significance of LI can be critical as in rural Bengal possession of LI is often considered as a symbol of status, and people observe their standard based on the number of LI one has. The one who has such LI behaved to be in a better position than the one who did not have during our survey. In regards to LT, we see it too has a negative coefficient for the loan not taken [0] but had an insignificant result which states that respondents were not that certain whether LT could have significantly influenced their SOL. This loan is a source of funding for them which they take at a higher interest rate from the Self-Help Group who urges them for this loan so they consider it more as a liability than a significant tool for investment generation. This could be improved by proper training by the Microfinance Institutions who are working at those locations who can improve the financial literacy levels among the vendors and thus help in improving their SOL.

13.7 Conclusion

This empirical research helped us in understanding the views of the respondent and to determine the factors influencing their SOL, the grievances and the expectations of the respondent were also highlighted during the research and proper steps must be taken to address this issue. According to the primary response, the government did everything to renovate Digha and make it attractable to the tourist by making artificial beaches and cemented pavements, but it had led to the demise of their hard-earned business which they had set up all these years and had put on the hard work, they are still waiting for the government to provide the "real" development which they had hoped for which were reflected in the result. Consequently, this should be taken into consideration by the relevant authority and there is a need to better certain factors which would benefit both the tourist and the people associated to this sector, especially in hotspots like Digha.

References

Battour, M., Ismail, M. N., Battor, M., & Awais, M. (2017). Islamic tourism: An empirical examination of travel motivation and satisfaction in Malaysia. *Current Issues in Tourism, 20*(1), 50–67.

Dash, A. K., Tiwari, A. K., & Singh, P. K. (2018). Tourism and economic growth in India: An empirical analysis. *Indian Journal of Economics, 392,* 29–49.

Eygu, H., & Gulluce, A. C. (2017). Determination of customer satisfaction in conservative concept hotels by ordinal logistic regression analysis. *Journal of Financial Risk Management, 6*(03), 269.

Ghatage, L. N., & Kumbhar, V. M. (2005). Growth and performance of tourism industry in India. In *International Conference on Recent Trends in Commerce, Economics and Management*, Satara Maharashtra. ResearchGate.

Norusis, M. (2004). *SPSS 13.0 advanced statistical procedures companion*. New Jersey: Prentice-Hall.

Norusis, M. (2005). *SPSS 14.0 statistical procedures companion*. New Jersey: Prentice-Hall.

Polyzos, S., & Minetos, D. (2011). An ordinal regression analysis of tourism enterprises' location decisions in Greece. *Anatolia, 22*(1), 102–119.

Sharma, N. (2018). Tourism led growth hypothesis: Empirical evidence from India. *African Journal of Hospitality, Tourism and Leisure, 7*(2), 1–11.

Wang, W., Cho, W., & Chen, Y. (2014). Analysis of the influence of quantile regression model on mainland tourists' service satisfaction performance. *The Scientific World Journal*.

Chapter 14
The Validation of the Tourism-Led Growth Hypothesis in the Next Leading Economies: Accounting for the Relevant Role of Education on Carbon Emissions Reduction?

Festus Victor Bekun, Festus Fatai Adedoyin, Daniel Balsalobre-Lorente, and Oana M. Driha

Abstract Over the last few decades, a significant volume of research has been documented on the tourism-led growth hypothesis (TLGH). However, the role of education over environmental degradation is yet to be given the desired attention. This study explores the impact of air transport over economic growth between 1994 and 2014 in China, India and the US, the three economies predicted to be the largest in forthcoming years. This way, TLGH is tested while also introducing the connection between education and pollutant emissions (CO_2) for these economies. Thus, suggesting how development in air transport contributes positively to enhance economic growth in the long run. In contrast, ascending CO_2 emissions are negatively connected to economic growth contributing to its reduction in selected countries. Further empirical results also confirm the positive effects of energy use and education on economic growth. Based on these results, education is seen to mitigate the pernicious effects of environmental degradation over economic growth's dampening effects.

F. V. Bekun
Faculty of Economics and Administrative Sciences, Istanbul Gelisim University, Istanbul, Turkey
e-mail: fbekun@gelisim.edu.tr

Department of Accounting, Analysis and Audit, School of Economics and Management, South Ural State University, 76, Lenin Aven, Chelyabinsk, Russia 454080

F. F. Adedoyin
Department of Accounting, Finance and Economics, Bournemouth University, Poole, UK
e-mail: fadedoyin@bournemouth.ac.uk

D. Balsalobre-Lorente (✉)
Department of Political Economy
and Public Finance, Economics and Business Statistics and Economic Policy, University of Castilla-La Mancha, Ciudad Real, Spain
e-mail: daniel.balsalobre@uclm.es

O. M. Driha
Department of Applied Economics, International Economy Institute, Institute of Tourism Research, University of Alicante, Alicante, Spain
e-mail: oana.driha@ua.es

D. Balsalobre-Lorente et al. (eds.), *Strategies in Sustainable Tourism, Economic Growth and Clean Energy*, https://doi.org/10.1007/978-3-030-59675-0_14

Keywords Tourism-led growth hypothesis · Education · Environmental degradation · CO_2 emissions · GDP per capita · Air transport

14.1 Introduction

Globally, tourism-led growth hypothesis (TLGH) has been an epitome of debate owing to its direct and indirect significances in the all-round policy formation for developed and emerging economies worldwide (Brida and Risso 2008; Lee and Chang 2008; Holzner 2011; Brida et al. 2016; Balsalobre-Lorente et al. 2020a, b). The World Travel and Tourism Council (2018) revealed that tourism is one of the leading growth sectors next to the manufacturing sector with over 3% and generating more than 10% contribution of economic activity to the global economy. The concerns about its impact and how it can achieve sustained environment and growth is still rising, as the validity of the hypothesis is still questionable. TLGH, which was coined from the Export-Led Growth Hypothesis (ELGH), hypothesizes that strong long-run growth of economies is not only a result of productive factor inputs within a country, but a result of tourism channelled activities in the country (Brida et al. 2016). In other words, it proposes that a rise in tourism-led activities in an economy translates to increased growth for the economy. To this end, several researchers have confirmed this by indicating that tourism is an essential catalyst that spurs economic growth through channels including job creation, foreign exchange earnings, the inflow of investment and innovation, transfer of technology, development of infrastructure, energy systems expansion, transportation, human capital, research and development (see Schubert et al. 2011; Shahzad et al. 2017a, b; Fahimi et al. 2018). On the contrary, others have shown little or no evidence of this growth discovering instead that it stimulates pressure on energy sources, induces spikes in carbon emission and raises concerns on income growth and environmental sustainability (Katircioglu 2009; Khan et al. 2018).

These studies, however, note that tourism-led growth is dependent on some factors including specific or regional countries, developed or emerging economies, level of tourism dependence or independence, tourist infrastructures or carbon emissions, among others. Unlike previous studies focused on analysing the TLGH, this research includes additional explanatory variables exploring the linkage between trends of passenger transportation by air (ATP) and gross domestic product per capita (GDPPC), the expenditure on education (EDU), energy use (EU), per capita carbon emission (CO_2PC) in China, India and the United States of America (hereafter the US) between the period 1974 and 2014. During this period, many differences are observed in the evolution of the variables considered. A different pattern can be appreciated between the US, on the one hand, and China and India on the other hand. This might be not just due to the economic development and its potential in the medium and long run but also due to both trade and financial globalization. The increase of CO_2 emissions is not always accompanied by a similar level of energy use, economic or air transportation passenger growth. Moreover, this period also covers several economic drawbacks of different complexity levels as well as geopolitical, social and environmental episodes that might explain shifts in the patterns expected, especially in the tourism sector. Also, in the case of education, during the whole period, China and India increase the expenditure in education at a higher rhythm than the US except during the second half of the 1980s and the beginning of the 1990s (Fig. 14.1).

The trends of China, India and the US pose questions on whether these economies growth may or may not have also been influenced by the tourism-led activities (transportation, education and carbon emission).

Following the classification of the International Monetary Fund (IMF 2017), this study dwells on the next largest economies China, India and the US (Fig. 14.2). It is expected that the largest economies will also contribute more to CO_2 emissions (OECD 2012) on their deliberate growth trajectory. Even though these countries are not highly tourism-dependent, in 2017, the sector alone contributed more than 11%, 7% and 9%, respectively to their GDP (WTTC 2018). Second, these economies account for nearly 70% of the global energy consumption with a consistent rise in the CO_2 emitted by the countries (IEA 2019). Additionally, statistics reveal that more recently progressive investments in transport and tourist infrastructure in these economies account for human and transport induced emissions, which increased global emission by approximately 3.5% in these regions (Solaymani 2019).

Furthermore, many studies that assessed the TLGH considered tourism (inbound and outbound), GDP growth, transportation infrastructure (air, road, rail), CO_2 emissions and/or energy consumption, as essential and control variables (Tang and Tan 2013; Shahzad et al. 2017a, b; Dogru and Bulut 2018; Hu et al. 2019). However, education was not among them even though education (human capital investment) flowing through tourism expectedly achieves economic growth and reduces degradation of the environment in the long run (Fahimi et al. 2018). Thus, this study is also testing the effect of education, including it as an explanatory variable. In the light of these highlighted remarks, the motivation for this current study arises first from the preceding and suggestions of Shahzad et al. (2017a, b) on the importance of using a group of economies to understand the TLG hypothesis to discover divergence among economies. Second, this study contributes to the growing debate by incorporating air transport in confirming the growth led by tourism in these economies. Third, this study adopts education as a channel of tourism; considering its role over the degradation of the environment and therefore adds to the limited studies in this regard. Thus, the present study complements the plethora of literature on the TLGH in terms of scope and the incorporation of timely interaction of education to see its impact on environmental degradation/mitigation. This is timely and worthwhile for the selected countries and serves as a policy blueprint for other countries alike. Following the introduction, the study goes further to first provide a review of related literature on the subject matter. The second section explains the data and methodology and, lastly, summarize the findings, conclusions and appropriate policy direction following study-specific results.

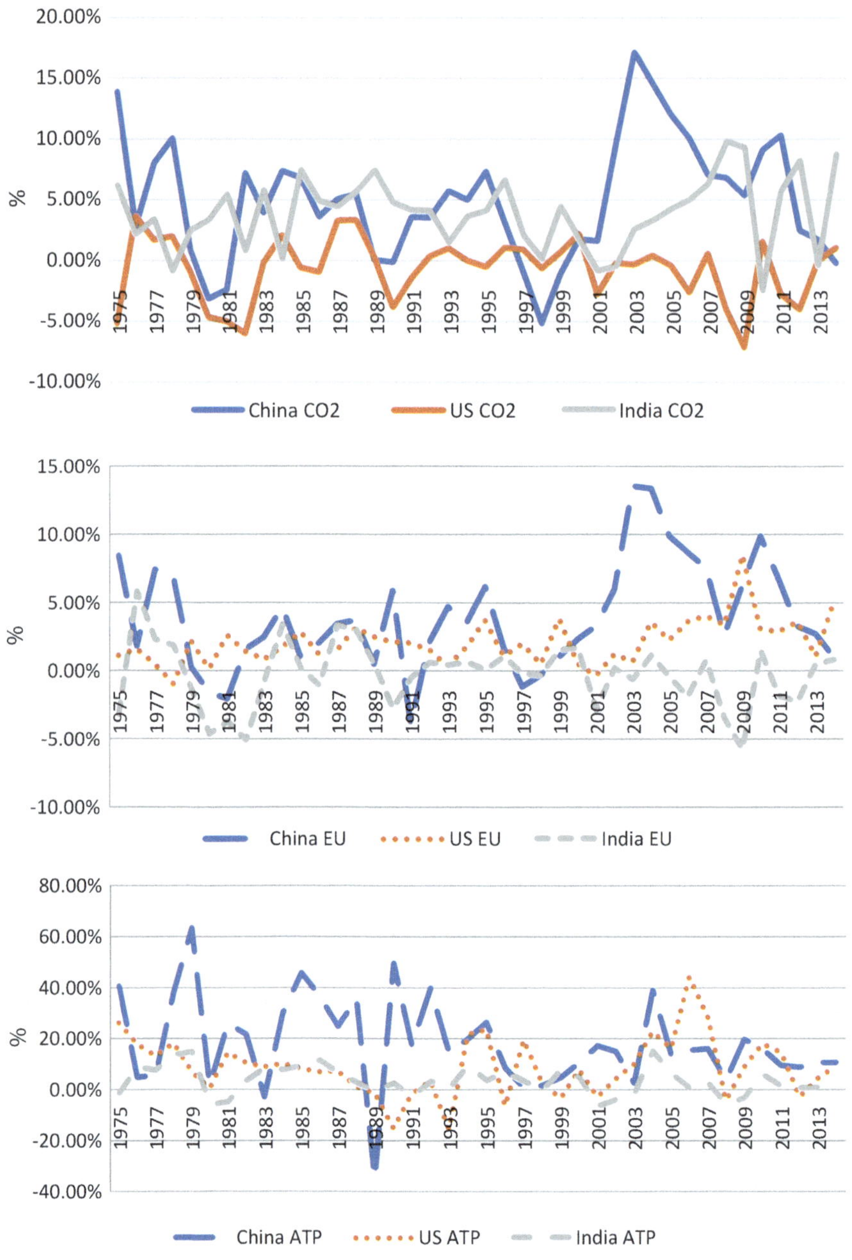

Fig. 14.1 Internal growth rate of CO_2 emissions, energy use, passenger transportation by air and expenditure on education. *Source* World Bank (2020)

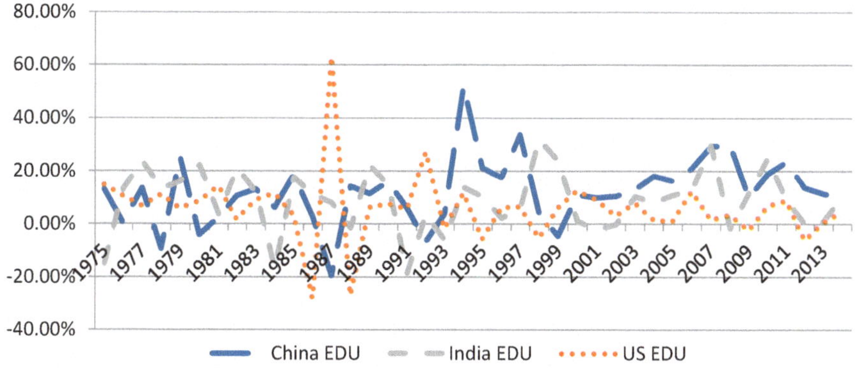

Fig. 14.1 (continued)

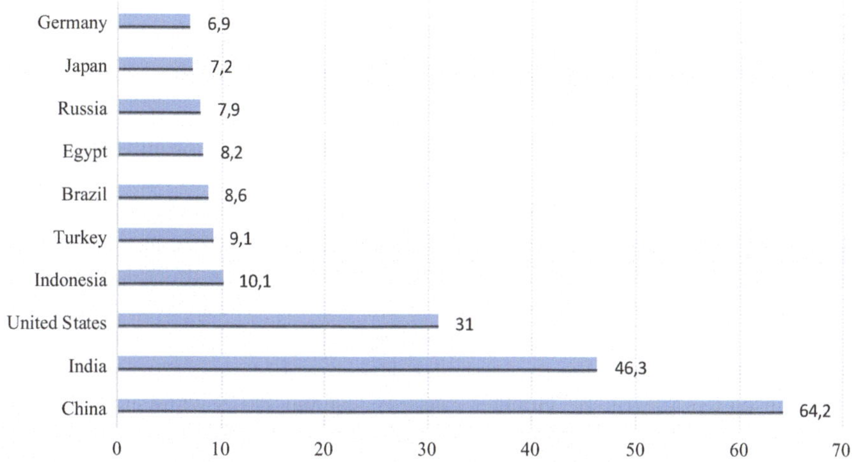

Fig. 14.2 Projections of World's ten biggest economies in 2030 (GDP PPP international dollars, trillion). *Source* IMF (2017). World Economic Outlook, April 2017

14.2 Literature Review

Previous studies while investigating the TLG hypothesis employed different methodologies and tourism variables including tourist arrivals (Katircioglu 2009; Tang and Tan 2013; Fahimi et al. 2018; Ahmad et al. 2019), tourist receipts (Perles-Ribes et al. 2017; Dogru and Bulut 2018), tourist income/expenditure (Ma et al. 2015), transport sector (Saidi and Hamamami 2017; Saidi et al. 2018; Hu et al. 2019; Solaymani 2019) and investment in human capital (Fahimi et al. 2018; Xu and Lin 2018; Shahbaz et al. 2019) for varying periods and in country-specific and panel country studies. These studies emphasized their impacts on GDP and CO_2 emissions. Some findings revealed mixed effect across countries, some others discovered positive effect, bidirectional and unidirectional causality effects, while only a few have stated insignificant effects between the variables.

In fact, despite energy efficiency improvements, energy-related CO_2 emissions are predicted to continue increasing, power generation, industry and transport being the leading sectors in CO_2 emissions (OECD 2012). The transport sector with specificity on air transport s considered in this study as aviation industry contributes to the growth in tourism, economic growth and energy use, and subsequently increases in carbon emissions (Solaymani 2019). Based on the preceding, this section, as shown in Table 14.1, concisely summarizes the central studies considered for the literature review regarding the air transport and GDP connection. Secondly, the link between air transport and CO_2 owing from the fact that rise in growth implies increases in energy use and invariably more levels of pollution. Lastly, the section captures the connection between education and CO_2 that is, the impact of education as tourism led activity on environmental degradation.

14.2.1 Impact of Air Transport on Economic Growth

Using cointegration and Granger causality test to explore the relationship between air transport and GDP,Marazzo et al. (2010) and Schubert et al. (2011) discovered that a direct connection between the two variables in Brazil and Aigua (and Barbuda), respectively. Studies by Baltaci et al. (2015) and Eric et al. (2020) revealed similar findings. The former employed 2SLS method while examining this relationship in 26 sub-regions in Turkey during the period 2004–2011.[1] The latter applied CGE, SAM and GMA techniques in Kenya for the period 2002–2014. In explaining the causality effects between the two variables, Hakim and Merkert (2016) employed the Johansen Co-integration, Granger causality test and TSCS while analysing the linkage between air transport and GDP in South Asia during the period 1973–2014. Their results indicated a unidirectional causality in the long-run flowing from GDP to air transport activity.

Marazzo et al. (2010) found unidirectional causality in the case of Brazil. Kucukonal and Sedefoglu (2017) results also indicated a unidirectional causality flowing from GDP to air transport but in the short run while employing Pesaran CDLM, Friedman test and Granger causality test in 28 OECD countries. On the other hand, Schubert et al. (2012) observed a unidirectional causality flowing from Air transport to GDP in Aigua and Barbuda due to the economy's reliance on tourism for growth. Baker et al. (2015) employed the granger causality while investigating airport activity and total airport passenger movement on growth in 88 regional airports in Australia. They stress a bidirectional relationship between the variables that are

[1]For brevity, here fixed and random effect Model are panel estimation techniques where fixed effect approach constant across individuals, but random effects vary. For easy of understanding 2SLS means two-stage least squares also a panel estimation technique that ameliorate for endogeneity issues with the use of instrumental variable (IV) approach. In summary, all highlighted test are panel estimators.

Table 14.1 Literature reviewed

Paper	Scope	Country/Territory (s)	Tourism variables	Estimation technique	Results on economic growth and CO_2
Abate (2016)	2000–2005	20 African city routes	Passenger fare, demand and departure frequency	2SLS random effect	Not statistically significant effect
Alonso et al. (2014)	Not stipulated	EU countries	Air transport (number of flights, passengers or freight cargo tones)	Projections (scenarios)	Positive effect
Arvin et al. (2015)	1961–2012	G-20 countries	Transportation intensity	Panel VAR	Positive effect
Baker et al. (2015)	1985–1986; 2010–2011	88 regional airports in Australia	Real Aggregate taxable income, level of airport activity, total airport passenger movement	Granger causality	Bidirectional effect (A-GDP)
Baltaci et al. (2015)	2004–2011	26 sub regions in Turkey	Airport, education, airline traffic	2SLS	Positive effect
Du et al. (2019)	2002, 2007 and 2012	China	Air transport	HEM (hypothetical extraction method)	Positive effect (export driven internal tourism)
Eric et al. (2020)	2002–2014	Kenya	Distance and trip fare	CGE, SAM, Gravity model analysis	Positive effect
Fahimi et al. (2018)	1995–2015	11 microstates	Tourist arrival, human capital	Granger causality	Positive effect
Hakim and Merkert (2016)	1973–2014	South Asia	Air transport activity (passenger number and freight volumes)	Johasen co-integration, Granger and Wald causality test, TSCS,	Unidirectional causality in the long run (GDP-A)
Hu et al. (2019)	2011	Sweden	Transport services	Input–output	Positive effect (net importer)
Jalil et al. (2013)	1972–2011	Pakistan	International tourism	ARDL	Positive effect in the long run

(continued)

Table 14.1 (continued)

Paper	Scope	Country/Territory (s)	Tourism variables	Estimation technique	Results on economic growth and CO_2
Khan (2018)	1990–2015	40 heterogenous countries	Air transport freight and passengers, rail transport goods and passengers	Panel econometric techniques, IRFVDA	Mixed results across countries
Kucukonal and Sedefoglu (2017)	2000 and 2013	28 OECD countries	Air transport, number of tourist arrivals	Pesaran CDLM and Friedman test, granger causality	Unidirectional (GDP-A) short run
Lo et al. (2018)	1997–2011	Italy	Air transport (route distance, aircraft size and seat, price of fuel, flight cap)	Log linear model, random effect model	Positive effect (increases total emissions and reduces emission per seat)
Loo and Banister (2016)	1990–2012	15 countries	Transport externality	Decoupling framework	Positive result
Ma et al. (2015)	2002–2011	China	Annual tourist aggregates and tourist income	β convergence model	Significant effect
Morazzo et al. (2010)	1966–2006	Brazil	Air transport demand	Co-integration and Granger test, IRFVD	Positive effect (A and Gdp); causality (GDP-A)
Qureshi et al. (2017)	1995–2015	80 international tourist destination cities in 37 countries	Inbound tourism	GMM	Positive effect on growth. reduce GHG
Saidi and Hammami (2017)	2000–2014	75 heterogeneous countries	Freight transport	Simultaneous equation and GMM estimator	Bidirectional (t-GDP); unidirectional (T-CO_2)
Schubert et al. (2011)	1970–2008	Aigua and Barbuda	GDP due to its dependence	Co-integration and Granger causality tests	Positive relationship and causality effect (T-GDP)
Shahbaz et al. (2019)	1975–2016	US	Education	Bootstrapping ARDL bounds test, VECM	Positive effect
Solaymani (2019)	2000–2015	7 top carbon emitter countries	Transport	LMDI	Mixed results

(continued)

Table 14.1 (continued)

Paper	Scope	Country/Territory (s)	Tourism variables	Estimation technique	Results on economic growth and CO_2
Song et al. (2019)	2005–2015	China	Economic openness and human capital investment	Fixed effect model	Positive effect and negative effect
Xu and Lin (2018)	2000–2015	30 provinces in China	Freight and passenger turnover	Quantile regression	Positive effect
Zaman et al. (2016)	2005–2013	(i) East Asia and Pacific, (ii) European Union (iii) High income OECD and Non-OECD regions	Tourists index, energy consumption income per capita (GDP) gross capital formation, health expenditure and pollutant emission (CO_2)	Two-stage least square and Dumitrescu–Hurlin causality test	Energy-induced emissions Tourism-induced carbon emissions

airport activities cause changes in income, which results in air transportation growth as well as economic growth. Saidi and Hammami (2017) observed similar results while analysing the relationship between freight transport and growth in 75 heterogeneous countries using a simultaneous equation and GMM estimator for the period 2000–2014. Khan et al. (2018) employed econometric panel techniques and IRFVD while investigating this relationship in 40 heterogeneous countries. Finding revealed mixed results across countries. Contrarily, Abate (2016) found this relationship to be non-significant in 20 African city routes while employing the 2SLS random effect. This review suggests that there is no consensus on the causality direction and effect between the two variables. These studies also neglected the effect of air transport over environmental pollution (CO_2).

14.2.2 Air Transport and Carbon Emission

Alonso et al. (2014) assessed the relationship between the two variables in EU countries and observed that air transport positively influenced higher levels of carbon emissions. Arvin et al. (2015) investigated the relationship between the two variables using the panel VAR in G-20 countries during the period 1961–2012. They discovered that air passenger and airfreight transport facilities flowing through increased urbanization positively influenced carbon emission. Loo and Banister (2016) noted that there is a significant relationship between the two, which resulted in increased fatal incidents across 15 countries. Lo et al. (2018) also confirmed similar findings while using the log-linear and random effect model to investigate this subject matter in Italy during the period 1997–2011. In the same vein, a study by Xu and Lin (2018) indicated the same result when quantile regression technique was employed on freight turnover, passenger turnover and CO_2 from 30 provinces in China during the period 2000–2015. Furthermore, Du et al. (2019) used the Hypothetical Extraction Method (HEM) to ascertain the determinants of CO_2 emissions in the transport sector of China for the period 2002, 2007 and 2012. Their findings stated that the demand from internal sectors of the economy influenced the contribution of air transport to the degradation of the environment. On the other hand, Hu et al. (2019) noted that the external sources drove the contribution of air transport to environmental pollution in Sweden. Also, Ahmad et al. (2019) while using FMOLS on data from three lower income South Asian economies discovered that there were mixed results across these economies. From the review, it appeared that there are very few studies in this aspect. The literature also noted that recent studies reported a positive effect of air transport over CO_2 emission in the different countries, recognizing that there should be a solution before the effect of CO_2 engenders growth and the environment. This leads to the next line of the review.

14.2.3 Connection Between Education and CO_2

Jalil et al. (2013) analysed the tourism growth nexus in Turkey using ARDL for the period 1972–2011. They recognized that capital stock in the form of human capital is a positive channel of tourism, sustained environment and growth increase. Mirza and Uddin (2014) discovered a long-run relationship between education and carbon emission in Bangladesh during 1974–2010 while using the VECM estimation technique. The findings showed that a unidirectional causality exists flowing from education to environmental degradation that is education reduces emission and leads to income growth for the economy. Baltaci et al. (2015) also observed that education is a channel through which tourism flow enhances progressive growth and reduced emissions in 26 sub-regions in Turkey during the period 2004–2011. Also, Alsabbagh (2017) empirical findings revealed the same results, noting that social learning contributes to reductions in transport emissions in Bahrain. Likewise, Xu and Lin (2018) or Balsalobre-Lorente et al. (2019) noted that investment and funding in R&D induce energy use improvements and adoption of emission reduction technologies. Fahimi et al. (2018) incorporated human capital as a pathway to reduce the degradation of the environment and enhance the growth of 11 microstates. Their results indicated that adequate human capital investment maximizes and sustains an economy's growth and its environs. Also, Song et al. (2019) while employing the fixed effect model in the case of China showed that increase in human capital investment (research and development scale) reduced GHG emissions in the environment and therefore promote the growth of a green economy in the long run. Similarly, Shahbaz et al. (2019) employed Bootstrapping, ARDL, VECM estimation technique to the relationship between these variables and discovered that a long-run linkage exists such that education positively influences demand for efficient energy equipment's, which contribute to less CO_2 emission in the US. The review shows that there is a negative relationship between education and CO_2 emissions such that as the levels of expenditure or investment in education increases, the fewer levels of environmental pollution and rising levels of economic growth.

Based on the reviewed literature, it is worth mentioning at least two main conclusions. First, there is no evidence of consensus on the connection between air transport and economic growth. Secondly, the existing literature confirms the positive relationship between air transport and CO_2. More noticeable is that there have been recent studies on the link between tourism, CO_2 and GDP Notwithstanding, specificity on air transport is still limited. Similarly, there is a dearth on the incorporation of education into these studies having known the contributing role in mitigating degradation of the environment and promoting growth. Interestingly, the literature lacks evidence of those above on the combination of the next three largest economies, namely China, India and the US seeing that their air transport sector and expenditure on education expectedly contribute immensely to the growth in transport and tourism, GDP and carbon emission globally. As a result, this study fills this gap in knowledge by using these economies, incorporating both air transport and education ascertaining the TLG hypothesis across these economies.

14.3 Data and Empirical Methodology

This study explores some of the variables that might impact economic growth, mainly focused on the impact of international tourism and education (Eqs. 14.1–14.4).

$$\mathrm{LGDP}_{it} = \alpha_0 + \alpha_1 \mathrm{LAT_{it}} + \alpha_3 \mathrm{LCO2_{it}} + \alpha_4 \mathrm{LEU_{it}} + \alpha_5 \mathrm{LEDU_{it}} + \varepsilon_{\mathrm{it}} \tag{14.1}$$

$$\mathrm{LGDP_{it}} = \alpha_0 + \alpha_1 \mathrm{LAT_{it}} + \alpha_3 \mathrm{LCO2_{it}} + \alpha_4 \mathrm{LEU_{it}} + \alpha_5 \mathrm{LEDU_{it}} + \alpha_6 \mathrm{LAT} * \mathrm{CO}_{2_{\mathrm{it}}} + \varepsilon_{\mathrm{it}} \tag{14.2}$$

$$\mathrm{LGDP_{it}} = \alpha_0 + \alpha_1 \mathrm{LAT_{it}} + \alpha_3 \mathrm{LCO2_{it}} + \alpha_4 \mathrm{LEU_{it}} + \alpha_5 \mathrm{LEDU_{it}} + \alpha_7 \mathrm{LEDU} * \mathrm{CO}_{2_{\mathrm{it}}} + \varepsilon_{\mathrm{it}} \tag{14.3}$$

$$\mathrm{LGDP_{it}} = \alpha_0 + \alpha_1 \mathrm{LAT_{it}} + \alpha_2 \mathrm{LAT^2}_{\mathrm{it}} + \alpha_3 \mathrm{LCO}_{2_{\mathrm{it}}} + \alpha_4 \mathrm{LEU_{it}} + \alpha_5 \mathrm{LEDU_{it}} + \varepsilon_{\mathrm{it}} \tag{14.4}$$

Using the annual information facilitated by the World Bank database (WDI 2020), the sample includes the period 1974–2014. The selection of these three countries (China, India and US) is based in the fact that these countries will be the next largest economic systems in forthcoming years (World Economic Forum 2019; Farhani and Balsalobre-Lorente et al. 2020). Where $\mathrm{LGDP_{it}}$ stands for the logarithm of GDP per capita (current US$) and $\mathrm{LAT_{it}}$ is the logarithm of air transport, passengers carried. The logarithm of adjusted savings: education expenditure (current US$) ($\mathrm{LEDU_{it}}$)

tries to evaluate the impact of education on $LGDP_{it.}$ Per capita, carbon emission ($LCO_{2_{it}}$) and energy use (logarithm of koe per capita) (LEU_{it}) are considered for testing its impact of environmental and energy inputs on economic growth and carbon emissions. Equation 14.4 also considers the existence of a U-shaped connection between economic growth and tourism, to confirm a non-linear behaviour between air transport and economic growth (Balsalobre-Lorente et al. 2020a, b). Equation 14.4 presents a non-linear model (LAT_{it} and LAT_{it}^2), which expects a U-shaped connection between air transport and economic growth (Rasekhi et al. 2016; Balsalobre-Lorente et al. 2020a, b). Equation 14.4 also includes as economic growth's driving factors, $LEDU_{it}$, $LCO_{2_{it}}$ and LEU_{it} (Table 14.2).

To investigate the effect of proposed independent variables on economic growth, we first explore the cross-sectional dependence (Table 14.3).

Table 14.3 illustrates how the null hypothesis of cross-sectional independence is rejected by most of the test statistics. Testing first and the second generations of unit root test (Tables 14.4 and 14.5), we evaluate the stationarity properties of selected variables. The confirmation of non-stationary variables would alter the accuracy of the empirical results. In a first step, Im, Pesaran and Shin (IPS) (Im et al. 2003) and Fisher-type Augmented Dickey–Fuller test (ADF) (Choi 2001) the first generation of panel unit root tests were applied. Secondly, we run CADF and CIPS (Pesaran 2003) the second generation of panel unit root tests, which admits cross-sectional

Table 14.2 Main statistics and correlation matrix

	LGDP	LATP	LCO_2	LEDU	LEU
Mean	7.572798	17.86257	1.234787	24.37181	7.236949
Median	6.935696	17.83824	0.942346	24.08143	6.704614
Maximum	10.91569	20.45239	3.089815	27.40525	9.040548
Minimum	5.052394	13.47302	−0.964788	21.49334	5.608422
Std. dev	2.021643	1.934048	1.387975	1.855118	1.297900
Skewness	0.385305	−0.254328	0.110226	0.139779	0.361424
Kurtosis	1.559858	1.888328	1.513297	1.635612	1.443673
Jarque–Bera	13.67273	7.659543	11.57679	9.941000	15.09139
Probability	0.001074	0.021715	0.003063	0.006940	0.000528
Sum	931.4541	2197.097	151.8788	2997.732	890.1447
Sum sq. dev	498.6191	456.3462	235.0300	419.8586	205.5145

Correlation matrix

	LGDP	LATP	LCO_2	LEDU	LEU
LGDP	1.000000	0.953094	0.933331	0.985294	0.950653
LATP	–	1.000000	0.884796	0.967649	0.879162
LCO_2	–	–	1.000000	0.891921	0.990798
LEDU	–	–	–	1.000000	0.900947
LEU	–	–	–	–	1.000000

Table 14.3 Cross-sectional dependence test for panel data. Null hypothesis: cross-sectional independence

Test	Statistic	d.f	Prob.
Breusch–Pagan Chi-square	15.97597*	3	(0.0011)
Pearson LM normal	4.072671*		(0.0000)
Pearson CD normal	−3.346514*		(0.0008)
Friedman Chi-square	29.47271	40	(0.8894)
Frees normal	0.279582*		(0.0098)

Note *, **, ***, significance at 1%, 5%, 10% respectively. Null hypothesis states cross-sectional independence ~$N(0, 1)$

Table 14.4 First generation of panel unit root test

	Null: unit root (assumes individual unit root process)					
	Im, Pesaran and Shin W-stat		ADF–Fisher Chi-square		PP–Fisher Chi-square	
	t-Statistic	Prob.	t-Statistic	Prob.	t-Statistic	Prob.
	Level					
LGDP	0.68750	(0.7541)	4.34027	(0.6307)	5.70671	(0.4568)
LATP	0.64099	(0.7392)	3.53247	(0.7396)	1.54083	(0.9567)
LCO_2	0.19628	(0.5778)	4.04854	(0.6701)	1.91928	(0.9270)
LEU	1.64766	(0.9503)	2.40356	(0.8791)	1.03295	(0.9843)
LEDU	−0.58591	(0.2790)	7.61923	(0.2673)	16.2682*	(0.0124)
	At first difference					
ΔLGDP	−3.29136*	(0.0005)	21.1138*	(0.0018)	48.6206*	(0.0000)
ΔLATP	−5.14587*	(0.0000)	34.6862*	(0.0000)	93.5453*	(0.0000)
ΔLCO_2	−3.16952*	(0.0008)	20.7784*	(0.0020)	39.3602*	(0.0000)
ΔLEU	−2.78734*	(0.0027)	18.4428*	(0.0052)	39.3168*	(0.0000)
ΔLEDU	−7.25931*	(0.0000)	51.8209*	(0.0000)	194.924*	(0.0000)

Note *, ** and *** significance at 1%, 5% and 10%, respectively

Table 14.5 Second Generation of Panel unit root test

Variables	CADF		CIPS	
	Level	First difference	Level	First difference
LGDP	−2.281	−4.558*	−2.258	−5.695*
LATP	−1.053	−5.440*	−1.004	−5.650*
LEU	−1.778	−4.861*	−1.934	−5.431*
LCO_2	−1.860	−4.793*	−1.333	−4.873*
LEDU	−1.829	−6.190*	-2.927^{b}	−6.420*

Note *, ** and *** significance at 1%, 5% and 10% respectively

dependence (Table 14.2) to circumvent for spurious analysis. Table 14.4 presents the Pedroni (1999) cointegration test, which presents, the panel cointegrating—*within-based*—and the group mean panel—*between-based*—statistics, with four components; a non-parametric *panel-v* statistic, a *panel-rho*, a *panel-PP* and a *panel-ADF* statistic. This test also includes three statistics based on pooling the residuals along the between-dimension of the panel. Kao (1999) tests cross-sectional intercepts and homogeneous coefficients on the first-stage regressors (see Table 14.4). To do so, Dickey–Fuller (*DF*) and ADF tests were employed. Lastly, Johansen's (1991) cointegration test is used combining individual tests and connecting tests from individual cross section. Subsequently, the proposed CADF and CIPS second generation of unit root test (Pesaran 2003) are based in the individual ADF statistics (CADF), to avoid the needless influence of results that could be found in small samples. Moreover, CIPS is a revised version of IPS *t-bar* test built on the individual CADF statistics, for cross-sectional augmented IP. Both generations of unit root tests will evaluate the unit root and their alternative of the non-existence of a unit root. Tables 14.4 and 14.5 present the stationary properties showing if the variables are cointegrated in order 1 I(1). If we confirm I(1), we have to validate the long-run relationship of the panel. Our study proposes Pedroni (1999), Kao (1999) and Johansen (1991) cointegration tests.

Previously to test FMOLS and DOLS econometric techniques, once we have validated confirm the long-run relationship between variables, we will employ the heterogenous Dumitrescu and Hurlin (2012) causality test to determine the linkage and the directions of the causalities. Dumitrescu and Hurlin (2012) offer reliable result for small and unbalanced samples and with cross-sectional dependence Dumitrescu and Hurlin (2012) that is considered a heterogeneous causality test. At last, once we have determined, though the proposed cointegration tests, the validation of a long-run connection between the variables, we employ the Fully Modified Ordinary Least Square (FMOLS), established by Phillips and Hansen (1990) and Dynamic Ordinary least square (DOLS), proposed by Saikkonen (1991) and Stock and Watson (1993). These econometric techniques are suitable when exits serial correlation and endogeneity, generating unbiased and normally distributed coefficient outcomes. In advance, the DOLS estimation process considers the orthogonality in the cointegrating equation error term, being a technique asymptotically efficient estimator that eradicates feedback in the cointegrating system.

14.4 Empirical Results and Interpretations

Tables 14.4 and 14.5 present the stochastic properties of the selected series. Further, both the first and the second generations of unit root tests are included after checking the existence of cross-sectional dependence as reported in Table 14.3.

Table 14.4 presents the traditional unit root tests aiming at confirming that all the selected variables are non-stationary at level form and stationary after at the first difference. This result indicates indicating that all of the variables are integrated to order one I(1). Same results are obtained from the second generation of CAP and CIPS unit root tests (Table 14.4), showing that variables are integrated to order one I(1) in nature.

Confirmed that the selected variables are stationary at the first difference (Tables 14.4 and 14.5), Table 14.6 presents the results of the proposed Pedroni, Kao and Fisher-Johansen cointegration tests.

The cointegration tests confirm the long-run connection between proposed variables. Table 14.7 shows the Dumitrescu–Hurlin causality results (Fig. 14.3).

Table 14.6 Pedroni, Kao and Johansen Fisher Panel cointegration tests

(a) Pedroni cointegration test				
Alternative hypothesis: common AR coeffs. (within-dimension)				
			Weighted statistic	
	t-Statistic	Prob.	t-Statistic	Prob.
Panel v-statistic	1.646589*	(0.0498)	1.506937*	(0.0659)
Panel rho-statistic	−0.320907	(0.3741)	−0.234963	(0.4071)
Panel PP-statistic	−2.694629*	(0.0035)	−2.703803*	(0.0034)
Panel ADF-statistic	−3.270941*	(0.0005)	−3.580379*	(0.0002)
Alternative hypothesis: individual AR coeffs. (between-dimension)				
	t-Statistic	Prob.		
Group rho-statistic	0.318199	(0.6248)		
Group PP-statistic	−2.713392*	(0.0033)		
Group ADF-statistic	−3.309480*	(0.0005)		
(b) Kao cointegration test				
	t-Statistic	Prob.		
ADF	−4.690007*	(0.0000)		
Residual variance	0.003535			
HAC variance	0.002461			
(c) Johansen Fisher panel cointegration test. Unrestricted cointegration rank test (trace and maximum eigenvalue)				
Hypothesized no. of CE(s)	Fisher Stat.# (from trace test)		Fisher stat.# (from max-eigen test)	
	t-Statistic	Prob.	t-Statistic	Prob.
$r \leq 0$	48.78*	(0.0000)	27.46*	(0.0001)
$r \leq 1$	27.09*	(0.0001)	18.60*	(0.0049)
$r \leq 2$	13.32*	(0.0383)	9.836	(0.1318)
$r \leq 3$	7.678	(0.2627)	5.917	(0.4326)
$r \leq 4$	6.858	(0.3341)	6.858	(0.3341)

Note *, **, *** Indicates statistical significance at 1%, 5% and 10%, respectively; # Probabilities are computed using asymptotic Chi-square distribution

Table 14.7 Dumitrescu–Hurlin analysis

Null hypothesis	Causality	F-statistic	Prob.
LATP does not Granger Cause LGDP	LATP → LGDP	2.36019*	(0.0178)
LGDP does not Granger Cause LATP		1.37574	(0.2090)
LCO_2 does not Granger Cause LGDP	LGDP → LCO_2	1.58913	(0.1273)
LGDP does not Granger Cause LCO_2		1.97930*	(0.0482)
LEDU does not Granger Cause LGDP	LGDP → LEDU	0.75559	(0.6702)
LGDP does not Granger Cause LEDU		3.16229*	(0.0021)
LEU does not Granger Cause LGDP	LGDP ↔ LEU	2.75943*	(0.0061)
LGDP does not Granger Cause LEU		2.13112*	(0.0325)
LCO_2 does not Granger Cause LATP	LATP → LCO_2	1.35989	(0.2165)
LATP does not Granger Cause LCO_2		1.79135*	(0.0776)
LEDU does not Granger Cause LATP	LATP → LEDU	0.57480	(0.8292)
LATP does not Granger Cause LEDU		2.83399*	(0.0050)
LEU does not Granger Cause LATP	LATP → LEU	0.87934	(0.5563)
LATP does not Granger Cause LEU		2.85990	(0.0047)
LEDU does not Granger Cause LCO_2	LEDU $\neq$ LCO_2	0.90055	(0.5372)
LCO_2 does not Granger Cause LEDU		1.51006	(0.1535)
LEU does not Granger Cause LCO_2	LCO_2 ↔ LEU	1.96541*	(0.0499)
LCO_2 does not Granger Cause LEU		4.23025*	(0.0001)
LEU does not Granger Cause LEDU	LEU ↔ LEDU	2.75956*	(0.0061)
LEDU does not Granger Cause LEU		0.87342	(0.5616)

Note *, **, *** Indicates statistical significance at 1%, 5% and 10%, respectively. Here does not Granger Cause denotes null hypothesis of "does not Granger cause". Rejection of null hypothesis suggests causal interaction between the considered pair of variables

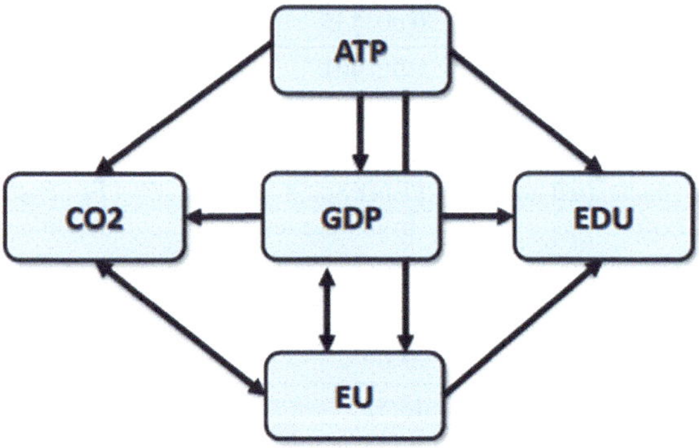

Fig. 14.3 Dumitrescu-Hurlin causality test scheme

The causality test shows a unidirectional causality running from LAT_{it} to $LGDP_{it}$. This test shows that air transport drives economic growth in the next largest economies and not vice versa. Furthermore, there is a one-way causality between $LEDU_{it}$ $LGDP_{it}$ and $LCO_{2_{it}}$. The unidirectional causality between economic growth and emissions conforms with previous studies (Esso and Keho 2016; Adedoyin et al. 2020a, b). We confirm the feedback hypothesis between $LGDP_{it}$ and LEU_{it} which is also in line with Shahbaz et al. (2013). Overall, bidirectional causality is also observed between $LCO_{2_{it}}$ and LEU_{it} and $LEDU_{it}$ and $LCO_{2_{it}}$. The causality in the former is not alarming since environmental degradation has been proven in the literature to be a consequence of the consumption of energy. On the other hand, the latter shows an interesting causal linkage expenditure on education causes environmental degradation and vice versa.

Table 14.8 presents the FMOLS and DOLS econometric outputs. The empirical results confirm the TLGH in all proposed models.

The econometric results of Model 1, Model 2 and Model 3 reveal a direct connection ($\beta_1 > 0$) between LAT_{it} and $LGDP_{it}$ validating the TLGH in selected countries during the period 1974–2014. Besides, Model reveals a non-linear connection ($\beta_1 < 0$, $\beta_2 > 0$) between LAT_{it} and $LGDP_{it}$, revealing how development in air transport contributes positively to enhance economic growth in the long run. In addition, the econometric results reveal a negative connection ($\beta_3 < 0$) between CO_2 emissions ($LCO_{2_{it}}$) and economic growth, confirming that ascending emissions will reduce economic growth in selected countries. Our empirical results also validate the positive effects of energy use ($\beta_4 > 0$) and education ($\beta_5 > 0$) on economic growth. Model 2 also includes the interaction between ait transport and carbon emissions ($LATP_{it}$ * $LCO_{2_{it}}$) as an explanatory variable, showing that this interaction dumps the effect of emissions over economic growth ($\beta_6 > 0$). Moreover, the air transport accompanied by an increase in the fuel consumption boosted CO_2 emissions growth in the last decade and seemed to contribute also to the acceleration of the economic growth, despite the advances in the air transport industry efficiency. In consequence, the contribution of air transport to climate change is around 3% of global fossil fuel consumption and 12% of transport-related CO_2 emissions (Amizadeh et al. 2016), it has grown quicker than other dirty sources (Simone et al. 2013; Mayor and Tol 2010) and is expected to continue doing so in the next decades (OECD 2012). Hence, the present study concludes that the linkage between air transport and emissions supports the process of economic growth. Thus, it implies the necessity of regulating efficiency in air transport to reach a more sustainable economic growth, related with tourism industry rather than dirty economic growth by the considerable demand for conventional energy fuel by the aviation industry (Air transport). The International Air Transport Association underlines that in 2012, the global total energy cost for airlines was over 160 billion dollars (IATA 2015). Besides, over 0.676 billion tons of CO_2 emission was registered at that time. The aviation industry is one of the few sectors where energy consumption has increased at a rate of more than 6% since the middle of last decade, where, energy/fuel efficiency of airlines was verified only very recently (Cui and Li 2016, 2017a, b; Chen et al. 2017; Cui et al. 2016; Ko

Table 14.8 FMOLS and DOLS econometric results

Dependent variable: **LGDP** (1974–2014)								
	Equation 14.1		Equation 14.2		Equation 14.3		Equation 14.4	
	FMOLS	DOLS	FMOLS	DOLS	FMOLS	DOLS	FMOLS	DOLS
LATP	0.094522*	0.112122*	0.060032*	0.137377*	0.104844*	0.136171*	−2.003358*	−1.489799*
	[3.118237]	[3.774230]	[2.007088]	[4.574048]	[3.849623]	[4.269069]	[−8.796535]	[−4.241595]
	(0.0023)	(0.0003)	(0.0471)	(0.0000)	(0.0002)	(0.0001)	(0.0000)	(0.0001)
LATP2	–	–	–	–	–	–	0.068799	0.053704
	–	–	–	–	–	–	[9.305688]	[4.662288]
	–	–	–	–	–	–	(0.0000)	(0.0000)
LCO_2	−1.505982*	−1.462623*	−2.106829*	−3.111735*	−2.450742*	−2.767994*	−0.422759*	−0.756964*
	[−10.29071]	[−9.534549]	[−9.076152]	[−8.683982]	[−7.386359]	[−5.829748]	[−2.984805]	[−3.968600]
	(0.0000)	(0.0000)	(0.0000)	(0.0000)	(0.0000)	(0.0000)	(0.0035)	(0.0002)
LEU	1.732598*	1.838461*	0.617662*	1.064165*	0.837964*	1.344904*	0.457330*	1.031044*
	[8.331753]	[8.652960]	[1.665376]	[3.464116]	[2.565269]	[3.595800]	[2.494858]	[4.080812]
	(0.0000)	(0.0000)	(0.0986)	(0.0010)	(0.0116)	(0.0006)	(0.0141)	(0.0001)
LEDU	0.836773*	0.771404*	0.758978*	0.577457*	0.688480*	0.601500*	0.485010*	0.472881*
	[22.82476]	[20.60567]	[20.35138]	[12.42781]	[12.70399]	[9.210000]	[11.90942]	[7.077920]
	(0.0000)	(0.0000)	(0.0000)	(0.0000)	(0.0000)	(0.0000)	(0.0000)	(0.0000)
LATP * LCO_2	–	–	0.087612	0.133404	–	–	–	–
	–	–	[3.357138]	[4.856925]	–	–	–	–
	–	–	(0.0011)	(0.0000)	–	–	–	–
LEDU * LCO_2	–	–	–	–	0.075114*	0.080185	–	–

(continued)

Table 14.8 (continued)

Dependent variable: **LGDP** (1974–2014)

	Equation 14.1		Equation 14.2		Equation 14.3		Equation 14.4	
	FMOLS	DOLS	FMOLS	DOLS	FMOLS	DOLS	FMOLS	DOLS
	–	–	–	–	[3.199446]	[3.260716]	–	–
	–	–	–	–	(0.0018)	(0.0018)	–	–
R-squared	0.998057	0.999286	0.998497	0.999630	0.998417	0.999501	0.999196	0.999645
Adjusted R-squared	0.997953	0.998863	0.998403	0.999315	0.998318	0.999075	0.999145	0.999343
S.E. of regression	0.091274	0.067829	0.080634	0.052644	0.082749	0.061184	0.058984	0.051588
Long-run variance	0.016124	0.005389	0.012647	0.002071	0.012807	0.003074	0.005596	0.002201
Mean dependent var	7.603300	7.596209	7.603300	7.596209	7.603300	7.596209	7.603300	7.596209
S.D. dep var	2.017611	2.011988	2.017611	2.011988	2.017611	2.011988	2.017611	2.011988
Sum squared resid	0.941400	0.326651	0.728204	0.169057	0.766913	0.228351	0.389656	0.162339

Note *, **, *** Indicates statistical significance at 1%, 5% and 10%, respectively

et al. 2017). Previous studies found that the Great Recession of 2008–2009 generated adverse effects on the airlines' energy efficiency (Cui and Li 2017a) despite the relevance of sustainable growth. The tourism industry could enhance sustainable growth, as it is not all about economic growth but also environment and society.

Model 3 offers evidence of a positive connection ($\beta_7 > 0$) between the interaction of education and carbon emissions ($LEDU_{it} * LCO_{2_{it}}$) and economic growth. This evidence implies that the interaction between education and economic growth would reduce the global pernicious effect of emission on economic growth. Empirical results are in line with Ledley et al. (2017), who proved that increasing public efforts and understanding of climate change contributes positively to sustainable growth. In the same line, Chang and Pascua (2017) concluded that climate change education is the transfer and use of knowledge to concern society to engage in climate change to advance positively in the reduction of emission levels.

14.5 Discussion of Results

The baseline equation for this study present advances in comparison with previous empirical literature, considering a battery of variables that include not only the effect of tourism on economic growth, it also considers the impact of education, energy and emissions on economic growth. Models 1, 2 and 3 confirm a direct connection between air transport and economic growth, validating the TLGH (Jiao et al. 2019; Mitra 2019; Etokakpan et al. 2019; Balsalobre-Lorente et al. 2020a, b). These results contain several induced effects on economic systems for the investigated countries, for instance, the positive impacts of tourism on economic growth implies the existence of positive externalities like the expanding labour and the expansion of tourism-related industries (Brida et al. 2016; Risso 2018; Etokakpan et al. 2019). Consequently, the TLGH represents one of the main topics not only in tourism but also in economic growth and social development, being key research in tourism literature. This is due to the complementary nature of the several services provided by the tourism industry alongside the external social and economic benefits to the economy (Aratuo and Etienne 2019). These outcomes are insightful for the different countries studied herein. The connection between tourism and the aviation sector (air transportation) is revealing against previous literature on just tourism index such as tourism arrival, receipt and expenditure (Akadiri et al. 2019). This study extends the frontier of discussing on the TLGH to show the role of air transportation as a key growth driver in China India and the US, that promotion of the aviation sector as there seems to be complementary role between airline passengers and tourism industry has the far-reaching economic benefit which is worthwhile government investment (Balsalobre-Lorente et al. 2020a, b). However, the connection is plagued with a trade-off for the quality of the environment with the usage of conventional energy/fuel that induce pollutant emissions. This is a call for more policy-mix and strategies to ameliorate pollutant-induced growth by aviation industry (air transport) decision that ranges from the gradual drift to more renewable and new technologies and innovations that a known to be cleaner and environmentally friendly and much more sustainable in the long run (Alvarez-Herranz et al. 2017; Balsalobre-Lorente et al. 2018) (Fig. 14.4).

The econometric results favour the TLGH (Models 1, 2 and 3). First, we observe the direct linkage between air transport and economic growth, even in Model 4, we also consider a non-linear relationship (Balsalobre-Lorente et al. 2020a, b) a between economic growth and air transport, that verifies the long-run positive effects of tourism on economic growth. Concerning the output variables, the interaction between education and environmental degradation, suggests that education contributes positively to reduce environmental pressure, even it reduces in a first stage the positive connection between dirty inputs and economic growth. This situation relates to the scale effects that imply the direct connection between environmental degradation and economic growth. This reflects the direct inverse liner nexus between pollutant emission and the selected countries (China, India and the US) growth path. Thus, suggesting the trade-off between the outlined variables, indicating the emphases on economic growth in jeopardy of quality of the environment at the early stage of growth till a threshold before a decline in pollution emission and improved quality of the environment see (Shahbaz and Sinha 2019). When we consider education as a key to social development, we can assume the existence of a transition from a developing to a developed stage via education. This process implies that societies need to generate advances to reduce the pernicious effects of dirty inputs, even in the first moment, it can decelerate economic growth (Balsalobre and Álvarez 2017).

Model 4 (Fig. 14.5) provides evidence of a U-shaped linkage between air transport and economic growth, confirming the existence of an induced transition from a poor specialized tourism sector to the high-specialized tourism industry.

Figure 14.5, based in econometric results (Eq. 14.4), illustrates a non-linear connection between $LATP_{it}$ and $LGDP_{it}$ ($\beta_1 < 0$; $\beta_2 > 0$). This result is in line with previous empirical literature that suggests that the linkage between tourism and economic growth is not conclusive (Song et al. 2012). While numerous empirical studies have shown a positive connection between tourism and economic growth, some evidence concludes that tourism would impact negatively on economic growth as a consequence of an emerging or deficient formed tourism industry (Javier et al. 2007; Zuo and Huang 2018; Balsalobre-Lorente et al. 2020a, b). These results are based on the symptoms of Dutch disease (Corden and Neary 1982), which illustrated the existence of adverse economic effects of tourism expansion. So, the negative relationship between tourism and economic growth is connected with resources scarcity and labour from other industries to tourism-oriented sectors, ascending local land and residence prices, or reductions in local social welfare (Copeland 1991; Chao et al. 2006; Nowak et al. 2007; Sheng and Tsui 2009; Holzner 2011).

In consequence, the econometric results assume the heterogeneity impact of tourism on economic growth. In Eq. 14.4 (Model 4), we reflect how counties with different circumstances in tourism specialization, industrial and labour structure would experiment with different effects on the tourism–economic growth linkage (Zhang and Cheng 2019). So, we validate both perspectives, illustrating that in a

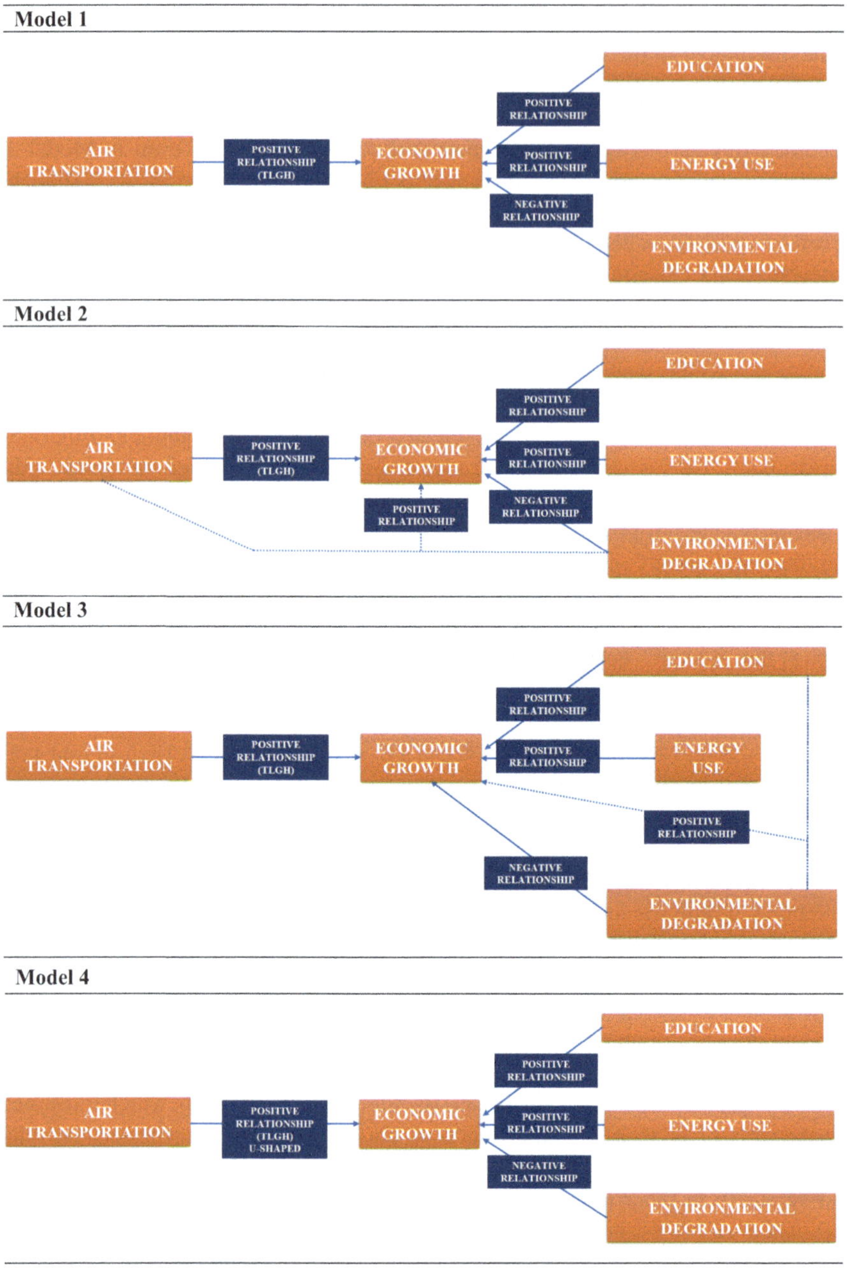

Fig. 14.4 Graphical abstracts of proposed models

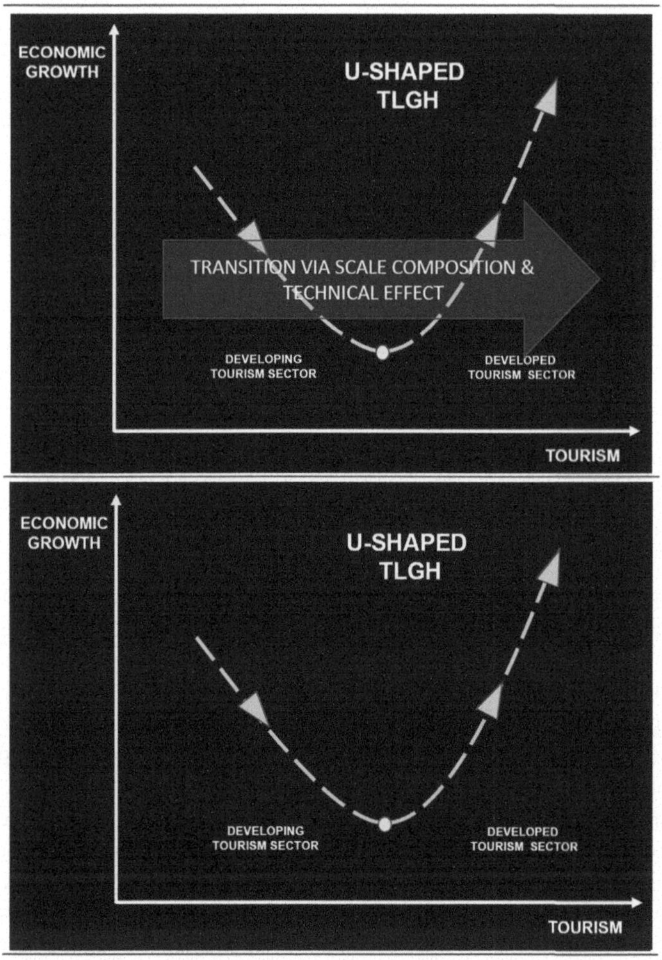

Fig. 14.5 U-shaped linkage between economic growth–tourism (Model 4)

first stage, the tourism industry reduces economic growth till a certain level of development of tourism industry, where increases in the tourism industry will enhance economic growth. In consequence, these results reveal that tourism industry significantly contributes the economic growth, in long term, validating the presence of a transition from a developing to a developed stage (Fig. 14.5) in air transport, which generates positive effects on economic growth (Balsalobre-Lorente et al. 2020a, b).

The results of this study contain policy implications. Empirical results of the present study support the existence of TLGH for the largest economies, China, Indian and the US validating the direct connection between economic growth and tourism industry. Also, we support the existence of a dampening effect between education and environmental degradation over economic growth. This evidence suggests that education transcends beyond enhances economic development but has a ripple effect with induced implications over the reduction of environmental degradation as citizenry awareness is awakening on the need for a cleaner ecosystem devoid of pollution emission in line with the UN sustainable development goals 8 and 13. That decouples pollutant emission (Ramadhan et al. 2019).

14.6 Final Conclusions and Policy Implications

The pivotal role of the tourism industry stretches beyond economic growth as created by employment opportunities, tax revenue generation and gain for foreign exchange earnings. It is on this premise that the present study empirically explores the nexus between tourism (air transportation) and economic growth while accounting for the contribution of education and pollutant emission to the literature. Our empirical finding traces cointegration relations among the outlined variables over the sampled period for the case of India, China and the US The TLGH affirmed for the countries under review.

This paper examines the connection between economic growth and air transport, confirming the TLGH for a panel of the largest economies (China, India and the US), including as additional driving forces energy use, education and environmental degradation during 1974–2014. We have considered air transport as a proxy of tourism, for the increased availability of data. This availability allows us to explore all more extensive range of years and in consequence, a more robust analysis. The impact of air transport on economic growth was analysed performing FMOLS and DOLS econometric techniques. Consequently, this study offers an advanced approach, testing the existence of a dampening effect between education and environmental degradation over economic growth. The empirical results confirm that in the early stages of development, the reduction of dirty environmental inputs with reducing economic growth. The adoption of additional efforts in the budget in education could be increased on the profits that it generates in carbon emissions control (Model 3). These additional revenues could then provide social welfare schemes. In addition, Model 4 presents a non-linear approach that examines a long-run linkage between the tourism industry and economic growth. The econometric results confirm a U-shaped connection between air transport and economic growth. Hence, Model 4 indicates a non-linear behaviour, which implies that the tourism industry needs improvements to transit to a highly specialized and developed stage, where the linkage between tourism and economic growth is positive. This affirms the transition from the scale stage where the emphasis is on economic growth relative to environmental damage. This is prominent in developing economies where economies thrive on the primary sector like agriculture, mining, among others. Subsequently, the economic growth trajectory moves to the composite stage evidence among developed economies, which have embraced clean technologies where the emphasis is on environmental conscious for clean growth. This behaviour implies that in the first stage of development, the impact of air transport over economic growth is not favourable, mainly as a consequence of

creates new structures or excessive public budget expenditure. Nevertheless, in long run, the results confirm the positive effects of the tourism industry over economic systems. These results reinforce our idea about the necessity of making budgetary efforts related to more energy-efficient infrastructures and a cleaner tourism sector. Further, these results confirm that ascending air transport development will impact on subsidiary tourism-related activities, increasing income levels. This stage will be possible if policymakers adopt suitable measures that promote the highly specialized and sustainable tourism industry. Given this argument, these countries should promote effective policies, related to energy efficiency processes, the reduction of dirty tourism-related activities and promoting education levels. From a policy standpoint, policymakers or additional stakeholders, our results suggest the existence of new opportunities to improve economic systems through the promotion of cleaner and more efficient inputs, under a more advanced tourism industry. Assumed that tourism promotes not only economic though but also generates new job opportunities, tax revenues, foreign inflows, countries in the panel of this research should boost trade opportunities with countries in which they obtain the majority of their tourists (Drogu et al. 2018). Also, the empirical results suggest that these countries should generate new trade opportunities to stimulate tourism development (Massida and Mattana 2012). These countries can also generate new tourism niches, creating new tourism segments like ecotourism, medical or food tourism. Based on the findings of the present study, these countries need to advance in the consolidation of a cleaner tourism industry that helps to attract tourists. Inconsequence, policymakers should develop strategies developed in the promotion of clean tourism industry, supported by tourism-related activities and more efficient energy uses. Otherwise, education and social awareness are also crucial to generate positive externalities and advances in the high-quality tourism industry. By contrast, our study also reflects some limitations related to the role of dirty energy sources in the current stage. These countries reflect high dependence of fossil sources with pernicious effect for the environment, where administrations need to assume environmental agreements to reach a sustainable tourism industry. This situation does not look plausible in the short run, especially in the case of the US, which is not in agreement with the 2015 COP21 United Nations Climate Change Conference.

References

Abate, M. (2016). Economic effects of air transport market liberalization in Africa. *Transportation Research Part A, 92,* 326–337.

Adedoyin, F. F., Alola, A. A., & Bekun, F. V. (2020a). An assessment of environmental sustainability corridor: The role of economic expansion and research and development in EU countries. *Science of the Total Environment, 713,* 136726.

Adedoyin, F. F., Gumede, I. M., Bekun, V. F., Etokakpan, U. M., & Balsalobre-Lorente, D. (2020b). Modelling coal rent, economic growth, and CO_2 emissions: Does regulatory quality matter in BRICS economies? *Science of the Total Environment, 710,* 136284.

Ahmad, F., Draz, M. U., Su, L., & Rauf, A. (2019). Taking the bad with the good: The nexus between tourism and environmental degradation in the lower idle-income Southeast Asian economies. *Journal of Cleaner Production, 233,* 1240–1249.

Akadiri, S. S., Akadiri, A. C., & Alola, U. V. (2019). Is there growth impact of tourism? Evidence from selected small island states. *Current Issues in Tourism, 22*(12), 1480–1498.

Alonso, G., Benito, A., Lonza, L., & Kousoulidou, M. (2014). Investigations on the distribution of air transport traffic and CO_2 emissions within the European Union. *Journal of Air Transport Management, 36,* 85–93.

Alsabbagh, M. (2017). Social learning and the mitigation of transport CO_2 emissions. *Climate, 5,* 6.

Alvarez-Herranz, A., Balsalobre-Lorente, D., Shahbaz, M., & Cantos, J. M. (2017). Energy innovation and renewable energy consumption in the correction of air pollution levels. *Energy Policy, 105,* 386–397.

Amizadeh, F., Alonso, G., Benito, A., & Morales-Alonso, G. (2016). Analysis of the recent evolution of commercial air traffic CO_2 emissions and fleet utilization in the six largest national markets of the European Union. *Journal of Air Transport Management, 55,* 9–19.

Aratuo, D. N., & Etienne, X. L. (2019). Industry-level analysis of tourism-economic growth in the United States. *Tourism Management, 70,* 333–340.

Baker, D., Merkert, R., & Kamruzzaman, . (2015). Regional aviation and economic growth: Cointegration and causality analysis in Australia. *Journal of Transport Geography, 43,* 140–150.

Balsalobre-Lorente, D., Driha, O. M., Bekun, F. V., & Adedoyin, F. F. (2020a). The asymmetric impact of air transport on economic growth in Spain: Fresh evidence from the tourism-led growth hypothesis. *Current Issues in Tourism*, 1–17.

Balsalobre-Lorente, D., Driha, O. M., & Sinha, A. (2020b). The dynamic effects of globalization process in analyzing N-shaped tourism led growth hypothesis. *Journal of Hospitality and Tourism Management*, *43,* 42–52.

Balsalobre-Lorente, D., Shahbaz, M., Jabbour, C. J. C., & Driha, O. M. (2019). The role of energy innovation and corruption in carbon emissions: Evidence based on the EKC hypothesis. In *Energy and environmental strategies in the era of globalization* (pp. 271–304). Cham: Springer.

Balsalobre-Lorente, D., Shahbaz, M., Roubaud, D., & Farhani, S. (2018). How economic growth, renewable electricity and natural resources contribute to CO_2 emissions? *Energy Policy, 113,* 356–367.

Baltaci, N., Sekmen, O., & Akbulut, G. (2015). The relationship between air transport and economic growth in Turkey: Cross-regional panel data analysis approach. *Journal of Economics and Behavioural Studies, 7*(1), 89–100.

Brida, J. G., & Risso, W. A. (2008). Multidimensional minimal spanning tree: The Dow Jones case. *Physica A: Statistical Mechanics and its Applications, 387*(21), 5205–5210.

Brida, J., Cortes-Jimenez, I., & Pulina, M. (2016). Has the tourism-led growth hypothesis been validated? A literature review. *Current Issues in Tourism, 19*(5), 394–430.

Chang, C. H., & Pascua, L. (2017). The state of climate change education–reflections from a selection of studies around the world. *International Research in Geographical and Environmental Education, 26*(3), 177–179.

Chao, C. C., Hazari, B. R., Laffargue, J. P., Sgro, P. M., & Yu, E. S. H. (2006). Tourism, Dutch disease and welfare in an open dynamic economy. *The Japanese Economic Review, 57,* 501–515.

Chen, Z., Wanke, P., Antunes, J. J. M., & Zhang, N. (2017). Chinese airline efficiency under CO_2 emissions and flight delays: A stochastic network DEA model. *Energy Economics, 68,* 89–108.

Choi, I. (2001). Unit root tests for panel data. *Journal of International Money and Finance, 20*(2), 249–272.

Copeland, B. (1991). Tourism, welfare, and de-industrialization in a small open economy. *Economica, 58,* 515–529.

Corden, W., & Neary, J. (1982). Booming sector and deindustrialization in a small open economy. *The Economic Journal, 92,* 825–848.

Cui, Q., & Li, Y. (2016). Airline energy efficiency measures considering carbon abatement: A new strategic framework. *Transportation Research Part D: Transport and Environment, 49,* 246–258.
Cui, Q., & Li, Y. (2017a). Airline efficiency measures using a Dynamic Epsilon-Based Measure model. *Transportation Research Part A: Policy and Practice, 100,* 121–134.
Cui, Q., & Li, Y., (2017b). Will airline efficiency be affected by "Carbon Neutral Growth from 2020" strategy? Evidences from 29 international airlines. *Journal Cleaner Production, 164,* 1289–1300.
Cui, Q., Li, Y., Yu, C. L., & Wei, Y. M. (2016). Evaluating energy efficiency for airlines: An application of virtual frontier dynamic slacks based measure. *Energy, 113,* 1231–1240.
Dogru, T., & Bulut, U. (2018). Is tourism an engine for economic recovery? Theory and empirical evidence. *Tourism Management, 67,* 425–434.
Du, H., Chen, Z., Peng, B., Southworth, F., Ma, S., & Wang, Y. (2019). What drives CO_2 emissions from the transport sector? A linkage analysis. *Energy, 175,* 195–204.
Dumitrescu, E., & Hurlin, C. (2012). Testing for Granger non-causality in heterogeneous panels. *Economic Modelling, 29*(4), 1450–1460.
Eric, T. N., Semeyutin, A., & Hubbard, N. (2020). Effects of enhanced air connectivity on the Kenyan tourism industry and their likely welfare implications. *Tourism Management, 78,* 104033.
Esso, L. J., & Keho, Y. (2016). Energy consumption, economic growth and carbon emissions: Cointegration and causality evidence from selected African countries. *Energy, 114,* 492–497.
Etokakpan, M. U., Bekun, F. V., & Abubakar, A. M. (2019). Examining the tourism-led growth hypothesis, agricultural-led growth hypothesis and economic growth in top agricultural producing economies. *European Journal of Tourism Research, 21,* 132–137.
Fahimi, A., Akadiri, S., Seraj, M., & Akadiri, A. (2018). Testing the role of tourism and human capital development in economic growth. A panel causality study of microstates. *Tourism Management Perspectives, 28,* 62–70.
Farhani, S., & Balsalobre-Lorente, D. (2020). Comparing the role of coal to other energy resources in the environmental Kuznets curve of three large economies. *The Chinese Economy, 53*(1), 82–120.
Hakim, M., & Merkert, R. (2016). The causal relationship between air transport and economic growth: Empirical evidence from South Asia. *Journal of Transport Geography, 56,* 120–127.
Holzner, M. (2011). Tourism and economic development: The beach disease? *Tourism Management, 32,* 922–933.
Hu, J., Wood, R., Tukker, A., Boonman, H., & de Boer, B. (2019). Global transport emissions in the Swedish carbon footprint. *Journal of Cleaner Production, 226,* 210–220.
IATA (2015). See https://www.iata.org/publications/Pages/annual-review.aspx (for Annual Review).
Im, K. S., Pesaran, M. H., & Shin, Y. (2003). Testing for unit roots in heterogeneous panels. *Journal of Econometrics, 115*(1), 53–74.
International Energy Agency (IEA). (2019). https://www.iea.org/global%20energy%20demand.
International Monetary Fund. (2017). World Economic Outlook (April 2017). World Economic Outlook.
International Monetary Fund (IMF) 2017. International Financial Statistics of IMF. http://data.imf.org/?sk=4C514D48-B6BA-49ED-8AB9-52B0C1A0179B. (Accessed June 2020).
Jalil, A., Mahmood, T., & Idrees, M. (2013). Tourism-growth nexus in Pakistan: Evidence from ARDL bounds tests. *Economic Modelling, 35,* 185–191.
Javier, C. P., Antoni, R. F., & Rossello, J. (2007). Tourism and long-term growth. A Spanish perspective. *Annals of Tourism Research, 34,* 709–726.
Jiao, S., Gong, W., Zheng, Y., Zhang, X., & Hu, S. (2019). Spatial spillover effects and tourism-led growth: An analysis of prefecture-level cities in China. *Asia Pacific Journal of Tourism Research, 24*(7), 725–734.
Johansen, S. (1991). Estimation and Hypothesis-Testing of Cointegration Vectors in Gaussian Vector Autoregressive Models. *Econometrica, 59*(6), 1551–1580.
Kao, C. (1999). Spurious regression and residual-based tests for cointegration in panel data. *Journal of Econometrics, 90*(1), 1–44.

Katircioglu, S. (2009). Tourism, trade and growth: the case of Cyprus. *Applied Economics, 41*(21), 2741–2750.

Khan, M. J., Khan, H. S., Yousaf, A., Khurshid, K., & Abbas, A. (2018). Modern trends in hyperspectral image analysis: a review. *IEEE Access, 6*, 14118–14129.

Khan, H., Siddique, M., Zaman, K., Yousf, S., Shoukry, A., Gani, S., et al. (2018). The impact of air transportation, railways transportation and port container traffic on energy demand, customs duty, and economic growth: Evidence from a panel of low, middle- and high-income countries. *Journal of Air Transport Management, 70,* 18–35.

Ko, Y. D., Jang, Y. J., & Kim, D. Y. (2017). Strategic airline operation considering the carbon-constrained air transport industry. *Journal of Air Transport Management, 62,* 1–9.

Kucukonal, H., & Sedefoglu, G. (2017). The causality analysis of air transport and socio-economics factors: The case of OECD countries. *Transportation Research Procedia, 28,* 16–26.

Ledley, T. S., Rooney-Varga, J., & Niepold, F. (2017). Addressing climate change through education. In *Oxford research encyclopedia of environmental science.*

Lee, C. C., & Chang, C. P. (2008). Tourism development and economic growth: A closer look at panels. *Tourism Management, 29*(1), 180–192.

Lo, P., Martini, G., Porta, F., & Scotti, D. (2018). The determinants of CO_2 emissions of air transport passenger traffic: An analysis of Lombardy (Italy) *Transport Policy.*

Loo, B., & Banister, D. (2016). Decoupling transport from economic growth: Extending the debate to include environmental and social externalities. *Journal of Transport Geography, 57,* 134–144.

Ma, X., Yu, H., Wang, Y., & Wang, Y. (2015). Large-scale transportation network congestion evolution prediction using deep learning theory. *PloS one, 10*(3), e0119044.

Marazzo, M., Scherre, R., & Fernandes, E. (2010). Air transport demand and economic growth in Brazil: A time series analysis. *Transport Research Part E, 46,* 261–269.

Massida, C., & Mattana, P. (2013). A SVECM analysis of the relationship between international tourism arrivals, GDP and trade in Italy. *Journal of Travel Research, 52*(1), 93–105.

Mayor, K., & Tol, R. S. (2010). Scenarios of carbon dioxide emissions from aviation. *Global Environmental Change, 20*(1), 65–73.

Mirza, M., & Uddin, M. (2014). Causal relationship between education, carbon dioxide, emission, and economic growth in Bangladesh. *Journal of Humanities and Social Science, 19,* 60–67.

Mitra, S. K. (2019). Is tourism-led growth hypothesis still valid? *International Journal of Tourism Research, 21*(5), 615–624.

Nowak, J. J., Sahli, M., & Cortes-Jimenez, I. (2007). Tourism, capital goods imports and economic growth: Theory and evidence for Spain. *Tourism Economics, 13,* 515–536.

OECD (2012). OECD Tourism Trends and policies 2012. OECD Publishing. https://www.oecd-ilibrary.org/industry-and-services/oecd-tourism-trends-and-policies-2012_tour-2012-en. (Accessed May 2020).

Pedroni, P. (1999). Critical values for cointegration tests in heterogeneous panels with multiple regressors. *Oxford Bulletin of Economics Statistics, 61,* 653–670.

Perles-Ribes, J., Ramon-Rodriguez, A., Rubia, A., & Moreno-Izquierdo, L. (2017). Is the tourism-led growth hypothesis valid after the global economic and financial crisis? The case of Spain 1957–2014. *Tourism Management, 61,* 96–109.

Pesaran, H. M. (2003). *A simple panel unit root test in the presence of cross section dependence.* Mimeo: University of Southern California.

Phillips, P. C., & Hansen, B. E. (1990). Statistical inference in instrumental variables regression with I-1 processes. *Review of Economic Studies, 57*(1), 99–125.

Ramadhan, S., Sukma, E., & Indriyani, V. (2019). Environmental education and disaster mitigation through language learning. In *IOP conference series: Earth and environmental science* (Vol. 314, No. 1, p. 012054). IOP Publishing.

Rasekhi, S., Karimi, S., & Mohammadi, S. (2016). Environmental Impacts of Tourism A Case Study of Selected Developing and Developed Countries. 5(16), 21.

Risso, W. A. (2018). Tourism and economic growth: A worldwide study. *Tourism Analysis, 23*(1), 123–135.

Saidi, S., & Hammami, S. (2017). Modelling the causal linkages between transport, economic growth and environmental degradation for 75 countries. *Transport Research Part D, 53,* 415–427.

Saidi, S., Shahbaz, M., & Akhtar, P. (2018). The long-run relationships between transport energy consumption, transport infrastructure, and economic growth in MENA countries. *Transportation Research Part A: Policy and Practice, 111,* 78–95.

Saikkonen, P. (1991). Asymptotically efficient estimation of cointegration regressions. *Econometric Theory, 7*(1), 1–21.

Schubert, S. F., Brida, J. G., & Risso, W. A. (2011). The impacts of international tourism demand on economic growth of small economies dependent on tourism. *Tourism Management, 32,* 377–385.

Shahbaz, M., & Sinha, A. (2019). Environmental Kuznets curve for CO_2 emissions: A literature survey. *Journal of Economic Studies.*

Shahbaz, M., Gozgor, G., & Hammoudeh, S. (2019). Human capital and export diversification as new determinants of energy demand in the United States. *Energy Economics, 78,* 305–349.

Shahbaz, M., Khan, S., & Tahir, M. I. (2013). The dynamic links between energy consumption, economic growth, financial development and trade in China: Fresh evidence from multivariate framework analysis. *Energy Economics, 40,* 8–21.

Shahzad, S., Shahbaz, M., Ferrer, R. & Kumar, R. (2017a). Tourism-led growth hypothesis in the top ten tourist destinations: New evidence using the quantile-on-quantile approach. *Tourism Management, 60,* 223–232.

Shahzad, S. J. H., Shahbaz, M., Ferrer, R., & Kumar, R. R. (2017b). Tourism-led growth hypothesis in the top ten tourist destinations: New evidence using the quantile-on-quantile approach. *Tourism Management, 60,* 223–232.

Sheng, L., & Tsui, Y. A. (2009). General equilibrium approach to tourism and welfare: The case of Macao. *Habitat International, 33,* 419–424.

Simone, N. W., Stettler, M. E., & Barrett, S. R. (2013). Rapid estimation of global civil aviation emissions with uncertainty quantification. *Transportation Research Part D: Transport and Environment, 25,* 33–41.

Solaymani, S. (2019). CO_2 emissions pattern in 7 top carbon emitter economies: The case of transport sector. *Energy, 168,* 989–1001.

Song, H., Dwyer, L., Li, G., & Cao, Z. (2012). Tourism economics research: A review and assessment. *Annals of Tourism Research, 39,* 1653–1682.

Stock, J. H., & Watson, M. W. (1993). Simple estimator of cointegrating vectors in higher order integrated systems. *Econometrica, 61*(4), 783–820.

Tang, C. F., & Tan, E. C. (2013). How stable is the tourism-led growth hypothesis in Malaysia? Evidence from disaggregated tourism markets. *Tourism Management, 37,* 52–57.

World Bank. (2020). World Development Indicators Data. https://data.worldbank.org/datacatalog/world-development-indicators.

World Economic Forum. (2019). India will overtake the US economy by 2030. Retrieved April 22, 2020, from https://www.weforum.org/agenda/2019/01/india-will-overtake-the-us-economy-by-2030.

World Travel & Tourism Council. WTTC. (2018). Travel and Tourism Economic Impact 2018. London: WTTC. Retrieved from https://www.wttc.org/-/media/files/reports/economic-impactresearch/regions-2018/world2018.pdf (Accessed June 2020)

Xu, B., & Lin, B. (2018). Investigating the differences in CO_2 emissions in the transport sector across Chinese provinces: Evidence from a quantile regression model. *Journal of Cleaner Production, 175,* 109–122.

Zaman, K., Shahbaz, M., Loganathan, N., & Raza, S. A. (2016). Tourism development, energy consumption and Environmental Kuznets Curve: Trivariate analysis in the panel of developed and developing countries. *Tourism Management, 54,* 275–283.

Zhang, J., & Cheng, L. (2019). Threshold effect of tourism development on economic growth following a disaster shock: Evidence from the Wenchuan earthquake PR China. *Sustainability, 11*(2), 371.

Zuo, B., & Huang, S. S. (2018). Revisiting the tourism-led economic growth hypothesis: The case of China. *Journal of Travel Research, 57,* 151–163.

Correction to: Developments and Challenges in the Greek Hospitality Sector for Economic Tourism Growth: The Case of Boutique Hotels

Aimilia Vlami

Correction to:
Chapter 11 in: D. Balsalobre-Lorente et al. (eds.), *Strategies in Sustainable Tourism, Economic Growth and Clean Energy*, https://doi.org/10.1007/978-3-030-59675-0_11

The original version of the book was inadvertently published with an incorrect name of the author in Chapter 11, which has now been corrected. The book and the chapter have been updated with the change.

The updated version of this chapter can be found at
https://doi.org/10.1007/978-3-030-59675-0_11

D. Balsalobre-Lorente et al. (eds.), *Strategies in Sustainable Tourism, Economic Growth and Clean Energy*, https://doi.org/10.1007/978-3-030-59675-0_15

The manufacturer's authorised representative in the EU is Springer Nature Customer Service Centre GmbH, Europaplatz 3, 69115 Heidelberg, Germany. If you have any concerns regarding our products, please contact ProductSafety@springernature.com

Printed and bound by CPI Group (UK) Ltd, Croydon, CR0 4YY
30/09/2026
02225695-0006